History, Philosophy and Theory of the Life Sciences

Volume 34

Frederik Nijhout, Duke University, Durham, NC, USA
Samir Okasha, University of Bristol, Bristol, UK
Susan Oyama, The City University of New York, New York, USA
Kevin Padian, University of California, Berkeley, CA, USA
David Queller, Washington University in St. Louis, St. Louis, MO, USA
Stephane Schmitt, Archives Poincaré, Nancy, France
Phillip Sloan, University of Notre Dame, Notre Dame, IN, USA
Jacqueline Sullivan, Western University, London, ON, Canada
Giuseppe Testa, University of Milan, Milan, Italy
J. Scott Turner, SUNY College of Environmental Science and Forestry, Syracuse, NY, USA
Denis Walsh, University of Toronto, Toronto, ON, Canada
Marcel Weber, University of Geneva, Geneva, Switzerland

History, Philosophy and Theory of the Life Sciences is a space for dialogue between life scientists, philosophers and historians – welcoming both essays about the principles and domains of cutting-edge research in the life sciences, novel ways of tackling philosophical issues raised by the life sciences, as well as original research about the history of methods, ideas and tools, which constitute the genealogy of our current ways of understanding living phenomena.

The series is interested in receiving book proposals that • are aimed at academic audience of graduate level and up • combine historical and/or philosophical and/or theoretical studies with work from disciplines within the life sciences broadly conceived, including (but not limited to) the following areas: • Anatomy & Physiology • Behavioral Biology • Biochemistry • Bioscience and Society • Cell Biology • Conservation Biology • Developmental Biology • Ecology • Evolution & Diversity of Life • Genetics, Genomics & Disease • Genetics & Molecular Biology • Immunology & Medicine • Microbiology • Neuroscience • Plant Science • Psychiatry & Psychology • Structural Biology • Systems Biology • Systematic Biology, Phylogeny Reconstruction & Classification • Virology The series editors aim to make a first decision within 1 month of submission. In case of a positive first decision the work will be provisionally contracted: the final decision about publication will depend upon the result of the anonymous peer review of the complete manuscript. The series editors aim to have the work peer-reviewed within 3 months after submission of the complete manuscript. The series editors discourage the submission of manuscripts that contain reprints of previously published material and of manuscripts that are below 150 printed pages (75,000 words). For inquiries and submission of proposals prospective authors can contact one of the editors: Charles T. Wolfe: ctwolfe1@gmail.com Philippe Huneman: huneman@wanadoo.fr Thomas A.C. Reydon: reydon@ww.uni-hannover.de

Maria Elice Brzezinski Prestes
Editor

Understanding Evolution in Darwin's "Origin"

The Emerging Context of Evolutionary Thinking

Springer

Editor
Maria Elice Brzezinski Prestes
Institute of Biosciences
University of São Paulo
São Paulo, Brazil

ISSN 2211-1948 ISSN 2211-1956 (electronic)
History, Philosophy and Theory of the Life Sciences
ISBN 978-3-031-40164-0 ISBN 978-3-031-40165-7 (eBook)
https://doi.org/10.1007/978-3-031-40165-7

This Springer imprint is published by the registered company Springer Nature Switzerland AG
The registered company address is: Gewerbestrasse 11, 6330 Cham, Switzerland

Paper in this product is recyclable.

For Caetano, Bruno, and Fernão

Foreword: From Biology to Darwin

The theory of evolution unifies biology. It is at its very core. The concept of common ancestry guides comparative anatomy and comparative embryology. It organizes taxonomy through phylogeny. It makes sense of varying degrees of similarity and reproductive isolation between species, all aligned with biogeographical patterns. The concept of natural selection is fundamental to morphology and to interpreting the relationship between structure and function. When combined with history—the concept of descent with modification—it makes sense of both homologies and convergent structures. It helps us understand vestigial structures, exaptations, and "contrivances," just as much as apparent design—all without an unscientific appeal to a transcendental teleology. Most remarkably, perhaps, it achieves this at multiple levels simultaneously—molecular, cellular, organismal, behavioral, hereditary, and ecological. There is no area of biology where evolution does not provide deeper insight.

How do students come to understand and appreciate the central concepts of evolution? We may regard today's learners, perhaps, as similar to those who preceded Charles Darwin. Unaware of his insights. What better way to learn, then, than over Darwin's shoulder and along with those who discovered or struggled with the concepts historically? The history of science can inform almost all our teaching. It provides a context that can motivate inquiry, along with a narrative map for how concepts might fruitfully develop step by step.

Modern students have typically been exposed to the rhetoric of evolution. Yet they frequently hold intuitive, often misleading preconceptions—akin to their historical counterparts. Engagement with the original historical arguments and evidence can help students "unlearn" their misconceptions and replace them with more informed understanding. Working through such information in a historical scenario can dispel such modern caricatures as the monkey-to-man "March of Progress" or help students unravel the deceptive allure of the argument from design. The bane of many biology teachers today—Intelligent Design—is really no more than nineteenth-century natural theology, surreptitiously resurrected and relabeled. Using history, we can see how that worldview established a context so familiar to

Darwin's contemporaries (and, indeed, at first, Darwin himself). What led scientists to discard it? How do ideas change? What justifies the current scientific consensus?

Darwin's fabled voyage on the *Beagle* was a landmark step in his intellectual journey. It is not surprising, therefore, that modern biology textbooks typically celebrate its significance. They often include a brief narrative of the adventure, highlighting many of Darwin's noteworthy observations—fossil giant glyptodonts, the two species of rhea on the South American pampas, the mockingbirds and tortoises of the Galápagos Islands, coral atolls, and more. But one cannot assume that students are able to assemble and interpret all this evidence effectively on their own. Their own learning needs to be scaffolded. Sooner or later, they need to engage with the structured arguments that Darwin presented to his colleagues in his now-renowned *Origin of Species*. For that purpose, this book is a much-needed and welcome guide.

Darwin's work was prescient. Of course, he had two decades to develop and explore his ideas before publicly presenting them. Several chapters in this volume nicely map that period of Darwin's thinking. Equally important, Darwin was keenly aware of the reach of his unorthodox interpretation. Thus, in the 1859 *Origin*, he was able to map out the territory quite exhaustively—laying out the full scope of his theory, as described above and now widely acknowledged. The many scholars writing in this volume nicely articulate how each of Darwin's chapters contributed to and fit into that whole.

The special virtue of this volume, however—compared to others that also help elucidate Darwin's work—is its steadfast focus on historical context. To appreciate Darwin's achievement and the revolutionary nature of his ideas, we need to fully situate them in their own time—just how his contemporaries in 1859 saw them. One group of chapters in this volume helps vividly set the scene. They recreate the intellectual perspective leading up to the *Origin*. The historical focus may seem to make Darwin more remote: an artifact of arcane history. But in fact (and far more importantly), it brings the *process* of science much closer.

Contrary to the widespread popular image of science as abstract and decontextualized, scientists always work in a particular culture and amidst a distinct constellation of accepted ideas. Engaging with Darwin's original text in that context opens a much deeper understanding of how science works—not through an idealized (or romanticized) reconstruction but as a result of real people working in real time. Students thereby learn not just about evolution or its emergence as a central idea but also about the very nature of science. And there can hardly be a more important lesson in science education and the philosophy of science.

University of Minnesota
Minneapolis, MN, USA

Douglas Allchin

Acknowledgments

Welcome to this volume that delves into the fascinating world of Charles Darwin's groundbreaking work, *On the Origin of Species*. Amidst the variety of publications on Darwin's *Origin*, this book stands out as a comprehensive exploration of the development of Darwin's ideas in their historical context. It extends previous publications by offering a broad structure that covers the pre-Darwinian concepts of species changes, some key elements of his pursuit of the causes of evolution during almost two decades, his presentation of the theory in the Origin, and the subsequent developments in evolutionary thought. The intention was to encourage the reading of *On the Origin of Species* and to include it in the teaching of evolution.

Meticulously sifting through a vast array of sources, the invited historians and philosophers of science have unearthed exciting insights that enrich our understanding of Darwin's theory. This volume builds upon and sheds fresh light on it.

Moreover, this volume goes beyond traditional conceptual boundaries of Darwin's thoughts by integrating the observational and socio-cultural dimensions of his practice of science and providing a well-rounded examination of Darwin's work. Biology instructors will find this book particularly valuable as it offers didactically fruitful insights to motivate the learning of evolution.

Among the distinguishing features of this volume is the commitment to geographic diversity of contributors. Although gender and race balance was a challenge, the volume strives to include diverse voices and perspectives with rarely grouped scholars from both the Global South and North. While most of the chapters were written by long term scholars on Darwin, some space were openned for younger researchers.

Furthermore, the authors of this volume come from different fields, gathering researchers in the history and philosophy of biology, science education, and evolution, as well as the hybrid community linked in the International Group on History, Philosophy, Sociology and Science Teaching. The transdisciplinary approach creates a rich tapestry of perspectives and encourages a dialogue that transcends traditional disciplinary boundaries, by a collaborative effort that can deepen the appreciation of Darwin's original works and their far-reaching implications. While most of the chapters were written by long-time Darwin scholars, some space were made for young researchers.

The genesis of this volume can be traced back to Bob Richards's influential course on "Darwin's Origin of Species and Descent of Man" at the University of Chicago in 2018. Inspired by the experience, I carried out my old desire to promote a similar course to the students of the University of São Paulo. The support and generosity of Bob Richards were instrumental in bringing this vision to fruition, and culminated in this volume.

The Covid pandemic disrupted our lives and forced us into a period of online classes, oppening the unexpected opportunity to bring together renowned experts in Darwin studies from around the world. As a result, students benefited from the collective expertise of professors from different parts of Brazil and different countries, creating a space of global dialogue on Darwin's *Origin of Species*. From the first moment, it was clear that the course had generated a rich material that deserved to be published.

The completion of this volume would not have been possible without the invaluable contributions of numerous collaborators. Each colleague that embarqued in both projects and dedicated their time giving their lecture and writing their chapter played a crucial role. Nothing would have progressed if it were not for the enthusiastic collaboration of postgraduate students helping with the "new" media for virtual teaching. Marcelo Pavani, Bruno Lima and Maísa Jensen were more than partners in preparing the originals.

Special thanks are extended to Kostas Kampourakis for his encouragement to publish this work for a broader English language audience. The editors of the Springer Series on History and Philosophy of Science embraced this proposal wholeheartedly. The guidance of Phillip Sloan significantly enhanced the quality of the project, as well as the insightful comments from the final anonymous reviewers further enriched the material.

Above all, this volume owes a vibrancy to the enthusiastic response and active engagement of the University of São Paulo students. Their curiosity, lively discussions and participation infused the entire endeavor with fun and a sense of accomplishment.

Contents

Chapter 1
Introduction: Why Read the "Origin of Species"?

Maria Elice Brzezinski Prestes

Abstract *Understanding Evolution in Darwin's "Origin": The Emerging Context of Evolutionary Thinking* aims to encourage the reading of *On the Origin of Species* and to include it in the teaching of evolution. With a comprehensive overview of the development of Darwin's theory, the volume provides relevant aspects of Darwin's life and work in connection with the broader context of his time. The historical and philosophical analysis, mirrored in the sociocultural scope, enables the diachronic reading of the text. It is built on various sources of historians and philosophers of science and sheds fresh light on them. Its uniqueness is the broad structure that covers four parts: the pre-Darwinian concepts of species changes; some key elements of Darwin's pursuit of the causes of evolution, from his voyage on Beagle to the publication of his groundbreaking work; chapter-by-chapter analysis of the "Origin"; and subsequent developments in evolutionary thought.

1.1 Darwin's Own Theory

Science students learn that the theory of biological evolution, encompassing the concepts of common origin and natural selection, was independently formulated in the nineteenth century by Charles Robert Darwin (1809–1882) and Alfred Russel Wallace (1823–1913). They also learn that these concepts became linked with Gregor Mendel's (1822–1884) ideas on heredity and its further developments in the early twentieth century within the emerging field of genetics. In the 1930s and 1940s, prominent scientists from various biological disciplines developed a unified explanatory framework known as the Modern Synthesis, Synthetic theory, or Neo-Darwinism. These terms are now used interchangeably to refer to the current

M. Elice Brzezinski Prestes (✉)
Laboratory of the History of Biology and Teaching, Department of Genetics and Evolutionary Biology, Institute of Biosciences, University of São Paulo, São Paulo, Brazil
e-mail: eprestes@ib.usp.br

M. Elice Brzezinski Prestes (ed.), *Understanding Evolution in Darwin's "Origin"*, History, Philosophy and Theory of the Life Sciences 34,
https://doi.org/10.1007/978-3-031-40165-7_1

evolutionary theory of living organisms. However, science and biology curricula often overlook the significant differences between the Modern Synthesis and the original ideas proposed by Wallace and Darwin, leading to the omission of distinct sets of challenges and complexities.

Darwin introduced his theory in the book titled *On the Origin of Species by Means of Natural Selection, or the Preservation of Favoured Races in the Struggle for Life*, published in London in 1859. Due to various factors, Darwin received sole credit for the theory, and even today, both the general public and biologists themselves commonly refer to it as the "Darwinian theory" of evolution. Apart from the unjust omission of Wallace's name, the issue lie in the unfortunate fate of an author's legacy. The original ideas become overshadowed by the modifications and interpretations introduced by subsequent followers over time. This distortion, known as an "anachronism" in the field of history of science, is condemned as a methodological error that causes historical misinterpretation. Add to this unfair tradiction the fact that even biologists today do not read Darwin's seminal work despite being translated and edited in various countries.

Another consequence of neglecting the original text is its impact beyond the field of biology. The *Origin* "is one of those books whose controversial reputation ensures that many who have never read it nevertheless have an opinion about it" (Endersby, quoted by Kampourakis, 2010, p. 827).

Understanding Evolution in Darwin's Origin: The Emerging Context of Evolutionary Thinking aims to promote the reading of *On the Origin of Species* and its inclusion in the teaching of evolution. Since the mid-twentieth century, the history and philosophy of science have actively demystified the prominent figures of science, humanizing them and placing them within the broader context of human society and culture. In recent decades, the incorporation of postmodern principles has further deconstructed these figures, shifting away from their central role and decolonizing their legacies, as highlighted by Janet Browne (2022, p. 2), who also emphasized that "this does not mean people have lost interest in his life and work." This volume serves as an example of recognizing the value and lessons to be learned from Darwin's original work.

1.2 Primary Sources of the History of Science in Science Education

A common assumption in education is that science books from the past remain in the past, and science teaching relies on regularly updated manuals containing accepted and validated knowledge. However, original texts of the history of science do possess educational value, especially with the constructivist turn toward the contextual teaching of science that emerged about 50 years ago. Primary texts, either in their entirety or in the form of selected excerpts, with or without accompanying comments, have become part of a range of instructional resources for incorporating the history of science in science teaching (McComas, 2020, p. 533).

Various reasons support the utilization of primary texts in science education. From a pedagogical perspective, these texts have the potential to facilitate the acquisition of scientific knowledge by tapping into earlier investigations that were guided by more general questions and fueled by intuitive, common-sense curiosity about natural phenomena.[1] This approach can also motivate students to explore scientific issues essential for everyday life. Additionally, by delving into the historical, philosophical, and social aspects of the time, the diachronic analysis of primary sources from the history of science enables a comprehensive understanding of how science operates—an essential element of scientific literacy.

In the pre-Internet era, so-called "Great books" published excerpts from the most notable scientific works of the past, sometimes thematically, not rarely grouping different fields of knowledge (Moulton & Schifferes, 1957; Hall, 1970; Ruse, 2009). Given the way to online publications, it came the time for the complete works of scholars from all areas, including the variety of preserved documents such as the author's correspondence and manuscripts and analytical essays about them. While until the end of the twentieth century, those materials were exclusive in the special collections of diverse separate libraries, mainly in Europe and North America, their online access considerably impacted the historiography of science. That was the case with the Darwin-online project, which substantially increased the first-hand interaction with the original work of the English naturalist, notably from 1985, when the first volume of his *Correspondence* was issued (Browne, 2022, p. 2).

Among celebrated original works in science, *On the Origin of Species* holds a distinct position. As radically envisioned by David Resnick, the book remains relevant to science today, just as it was in 1859 (2009, p. xii). This premise served as the central motivation behind the elaboration of the present volume, serving as an essential guide for the development of its 25 chapters.

1.3 Significance of the First Edition

Rather than focusing on the more commonly read and translated sixth edition of 1876, often regarded as the "canonical final" edition, this volume prioritizes the examination of the 1859 first edition of *On the Origin of Species*. This choice focuses on Darwin's development of his explaining ideas developed over two decades of investigations, engagement with extensive literature, and the continuous process of constructing a solid theoretical foundation. In a general sense, such a project started

[1]The idea comes from the parallelism thesis of Jean Piaget and Rolando García presented in *Psicogénesis e Historia de la ciencia* (1983/2008). Despite the criticism that the development of science does not follow a linear and progressive course, such parallels between the development of notions and cognitive operations from elementary to higher levels of cognition can be traced between early and advanced stages of a given theory.

to be planned during his travels aboard the Beagle. As Anna Carolina Regner (2006) argued, during his juvenile years, Darwin aspired to make his contribution to science and gain recognition from his peers. Darwin was deeply intrigued by "the mystery of the mysteries," borrowing Charles Lyell's expression, as evident from his records on species change in the *Red Notebook* dating back to March 1837 (Sloan, 2019), or even earlier, during his journey. From this perspective, the first edition offers a purer expression of his ideas. It highlights how his quest for explanations for the origin of species was rooted in natural rather than supernatural causes. While Darwin introduced corrections and clarifications throughout the six editions,[2] the more significant alterations, involving changes in terminology and the addition of a chapter, intended to address contemporary critical responses.[3]

Historians and philosophers of science primarily hold a preference for the first edition. Jonathan Hodge suggested that this preference reveals their characteristic interest in investigating how was the knowledge of "Darwin's *Origin* in its own time and on its terms?"[4] On the other hand, scientists generally tend to read the book's last edition, driven by the question, "What parts of our knowledge today are owing to Darwin's Origin?" (Hodge, 2013, p. 2290). It can be assumed that sociologists of science also join scientists in reading the final edition, as their focus lies more on the impact of the contribution within the scientific community and society at large, as well as the driving forces behind its production and reactions of contemporaries and followers. It is important to acknowledge that this comparison may be somewhat artificial since both types of questions and their corresponding modes and sources of investigation can be of interest to any researcher (Martins, 1997, p. 1–3). However, this comparison serves as a heuristic tool to highlight the overall differences in interpretations by individuals from different backgrounds, providing valuable insights for interdisciplinary dialogue.

It is crucial to recognize that neither the act of "looking back" nor "looking forward" is inherently superior or inferior, and neither approach is right or wrong. As Michael Ruse noted, "reading a text is not a straightforward matter" (2013a). It is

[2]In the introduction of her online variorum edition, Barbara Bordalejo identified six types of changes: depersonalization, reinforcement, objectivization, clarification, updating, and semantic changes. This variorum identified every change made by Darwin between the six British editions from 1859 to 1872 and can be consulted on open platforms accessible at <http://darwin-online.org.uk/Variorum/Introduction.html>.

[3]In 1959, Morse Peckham published a canonical variorum of the six editions of Darwin's *Origin*. For in-depth thematic comparisons and commentary, works by Nelio Bizzo (1992, 2014, 2018, 2020) and Jim Endersby (2009) are recommended sources to explore.

[4]The first edition is finally available in reliable Portuguese translations coincidentally published in 2018. The volume edited by Edipro includes a careful review of the translation of biological terms by Nelio Bizzo, who also provides a preface and extensive notes that offer conceptual, epistemic, and historical contexts. On the other hand, the Ubu edition, while encountering some issues with biology terms, may be seen as closer to Darwin's style of writing. It includes additional texts, such as the publication with Wallace in 1858, the historical sketch, and the new seventh chapter, among others. These translations contribute to making the first edition more accessible to readers in the Brazilian context.

also evident that combining both perspectives can be a powerful tool in teaching and learning *science* and *about science*:

> Team teaching! You need to have both historians [and or philosophers] and scientists working together to present the ideas, not necessarily agreeing but playing off each other to bring out the relevant points and interpretations. (Ruse, 2013b, p. 2264)

The choice to focus on the first edition of Darwin's *Origin* does not diminish the value of the other five editions. Each edition carries the magical allure of being a primary source, offering distinct readings, sometimes disruptive but certainly potentially pluralistic. Exploring the uses and values for various readerships, including scientists, science students, biology teachers, and the general public is itself an important and worthwhile theme to deal with in contextual science education.

With that in mind, it's crucial to address the pressing question: What can we teach and learn from Darwin's *Origin*? The book holds immense educational value and can offer valuable insights and opportunities for learning, as highlithted in the next sections.

1.4 The Two Core Ideas of Evolution

What is truly worth knowing about the evolution of humans, nonhuman animals, plants, and all other living beings? Darwin's book relies heavily on two core ideas that form the foundation of his theory: common descent and natural selection (Radick, 2009). Is there anything else that is more important to consider for an informed understanding of biological evolution? When designing biology classes for specific audiences, it is essential for teachers to contemplate these types of questions. Instead of presenting an exhaustive list of facts, biology classes would hold greater significance if students were exposed to the fundamental underlying principles of the theories of biological diversity (Scheiner, 2010, p. 307). Common descent and natural selection provide a minimal explanatory framework and serve as a basis for future studies on evolution (Kampourakis & Zogza 2009, p. 1333).

Although teaching the concepts of common descent and natural selection may seem obvious or trivial, empirical research on the learning of evolution suggests otherwise.[5] Even among college students studying life sciences, providing a solid understanding of natural selection presents a challenge. It is crucial to address students' intuitive notions and preconceptions to promote conceptual change about natural selection and to improve explanatory coherence regarding evolution (Kampourakis & Zogza, 2009, p. 1314). Assessments of teaching learning sequences with episodes from the history of biology have shown promising results

[5] For some examples of the difficulties of a functional understanding of natural selection, see Anderson et al. (2002); Tidon and Lewontin (2004); Gregory (2008); Sinatra et al. (2008); Gregory (2009); Rosengren et al. (2012); Kampourakis and Nehm (2014); Cooper (2017); Bizzo (2019); Harms and Reiss (2019); and Oliveira and Cook (2019).

in enhancing the teaching of these two core ideas of evolution through an inquiry-based learning strategy and the use of diverse instructional materials, including primary sources.[6]

1.5 The Written Style and Argumentation

Darwin intended his book, *On the Origin of Species*, not only for his fellow natural philosophers but for a broader audience, with the aim of communicating the concept of evolution as a natural fact, along with its explanation through the theory of common descent and natural selection. In the final chapter, Darwin referred to the *Origin* as "one long argument," presenting his conception of answering the "big question" of species change. Unlike the earlier manuscript titled "Natural Selection," which he began writing in 1856 and included extensive tables of data and scholarly literature, these features and other elements were largely absent from the 1859 publication (Sloan, 2019).

The *Origin* does not adhere to the structure of a typical nineteenth-century scientific treatise, which would typically present a systematic sequence of principles, definitions, examples, and theory applications presented in a didactic manner. The novelty of the concept of evolution, which was embraced by only a limited number of naturalists during that time, necessitated an additional power of persuasion. Given the sensitivity of the theme and its stark contrast with the prevailing notion of special creation defended by natural theology, Darwin needed to employ convincing rhetoric. Furthermore, even for those who had entertained the idea of species transformation, Darwin aimed to demonstrate that the phenomenon had never been adequately explained.

The analysis of the argumentative strategy employed in Darwin's book presents significant potential for teaching evolution.[7] As an avid reader, Darwin was well-versed also in the emergent epistemology of science of the time. His engagement with the works of William Hershel and William Whewell led him to transcend the empiricist tradition of modern European science from the mid-sixteenth and early seventeenth centuries. In addition to being an astute observer and gifted

[6] At the Laboratory of the History of Biology and Teaching (LaHBE) at the Institute of Biosciences of the University of São Paulo, postgraduate research has been conducted on teaching evolution with an inquiry-based approach and using diverse strategies including primary sources, replication of historical experiments, narratives, and web games such as Silva (2013); Souza (2014, 2021); Jensen (2016); and Cortez (2018).

[7] In Brazil, unlike countries like the United States, the study of rhetorical language devices is primarily limited to Portuguese classes, typically at the secondary school level. This confinement creates the perception that these resources are only relevant within that specific context. However, if rhetorical devices were incorporated into the teaching and learning objectives across all disciplines, along with other metalinguistic aspects, students would be equipped to identify and analyze their usage in various texts, including scientific language. Such inclusion would be particularly valuable when engaging with science communication in both traditional and digital media.

experimentalist, Darwin's philosophical readings reveal him as an "indefatigable theorist" (Bradley, 2009). The structure of his argumentation becomes apparent by contrasting analyses such as those by Jonathan Hodge (1992, 2013) and Kenneth Waters (2009). As highlighted by Anna Carolina Regner (2006), Darwin was consistently aware of the inherent dependence of "explanation" on specific theoretical perspectives and assumptions throughout the book. Particularly, he relied on the comparison of different viewpoints as a fundamental strategy to construct and defend his own theory. He skillfully employed the strategy of approaching opposing perspectives, taking advantage of the clarity and broader scope of his vision to strengthen his argument.

1.6 How Science Works

The potential of the history of science to promote understanding of how science works dates back to Darwin's work (Jenkins, 1990). The idea took shape in pedagogical formulations in the mid-twentieth century, although accompanied by criticism and concerns about the associated difficulties (Pumfrey, 1991). The use of the history of science in science education gained momentum when it became explicitly linked to the epistemology of science, primarily based on the pedagogical construct Nature of Science (NOS). Especially from the 1990s, NOS research developed into a comprehensive "program of research," in Lakatos' sense. The confluence of principles from various domains such as constructivist pedagogy, contextualism in the history of science, and Science, Technology, and Society (STS) research contributed to the vitality of the program. Proposals for incorporating historical, philosophical, and sociological approaches into science curricula at all levels have multiplied in the literature.[8] NOS research has expanded into numerous branches and diverse approaches, becoming a recognized label in the typology of tools for contextual science education.

Initially, research on the nature of science gave more prominence to the intrinsic aspects of scientific inquiry processes, but it encompassed broader perspectives as well. In one of the widely recognized initiatives, Norman Lederman and collaborators proposed a set of seven metascientific elements, including sociocultural embeddedness (Fig. 1.1). Contrary to the argument put forth by some critics who claimed that the seven elements would be memorized rather than understood, it is worth remembering that Norman and Judith Lederman developed various

[8] Literature on NOS is extensive, and while acknowledging that the selection is subjective, here are some influential works: Jenkins (1990, 2013); Lederman (1992, 2019); Pérez et al. (2001); McComas (2002, 2020); Matthews (2012); Allchin (2013); Erduran and Dagher (2014); Lederman (2019). A special issue of the Brazilian journal *Filosofia e História da Biologia* (volume 8, number 3, 2013) focuses on episodes developed for high school education, available in Portuguese at <https://www.abfhib.org/revista/>. For Latin American works on the history, philosophy, and sociology of science in education, Prestes and Silva (2018) have compiled a collection.

Fig. 1.1 Posters with the seven aspects of the nature of science and eight aspects of knowledge of inquiry on the wall of Norman and Judith Lederman's teaching laboratory at the Illinois Institute of Technology, Chicago. Photo by the author, October 3, 2017

classroom inquiry-based activities to specifically address those NOS aspects (Flick & Lederman, 2006; Lederman et al., 2019.

The expansion of cultural studies of science has contributed to a more attentive treatment of the interaction between science and society. The focus of questions has shifted to issues such as why we should trust science and what aspects of science are worthy of trust. There is also an emphasis on avoiding scientificism and promoting a democratic balance between scientific authority and sociopolitical interests. Douglas Allchin's work exemplifies this shift, as he moved, in educational sets, from examining the *nature* of science to exploring the *reliability* of the "whole science" (Allchin, 2013). Inputs from philosophy, such as Wittgenstein's conception of "family resemblance," have also influenced the field. The model of the Family Resemblance Approach (FRA) to science education, developed by Gürol Irzik and Robert Nola and most prominently advocated by Sibel Erduran and Zoubeida Dagher (Erduran and Dagher, 2014), has gained significant traction in the science teaching community today.

It is worth noting that disputes within the field of science education reflect the clash between the "internalist" and "externalist" approaches that emerged in the history, philosophy, and sociology of science in the 1980s. Despite epistemic divergences, ongoing debates over prestige and dominance in the field, as well as lingering criticisms regarding unresolved challenges (Matthews, 2022, 2023), approaches to the nature of science (NOS) have marked a significant turning point in science education.[9] Regardless of the adopted NOS model, survey data consistently show that its integration into curricula came for good. And that it should be an explicit *content objective*. This necessitates intentional planning and assessment, with explicit and reflexive treatment in classrooms.[10]

NOS research focusing on evolution topics or Darwin's work is readily available.[11] Assessments conducted worldwide in teaching and learning scenarios related to evolution also provide support for the NOS approach. The results indicate that even when conceptual and epistemic understandings are achieved, worldview, religious, social, and cultural factors can persist as invisible barriers to acknowledging the existence of biological evolution (Deniz & Borgerding, 2018, p. 3).

1.7 Darwin's Epistemic Assumptions and Methods

When discussing the nature of science (NOS) within Darwin's book, it is important to address both the observational and conceptual dimensions of his scientific practices and the epistemic and ontological underlying them. Several topics of interest for understanding evolution emerge from this discussion, such as gradualism and naturalism:

> Darwin was clearly aware of the epistemic assumptions of his theory. It presupposes gradualism and naturalism as epistemological and ontological flags, as well as a view of evolution as a "natural" process of formation of new organic forms, which must be explained

[9] Comprehensive surveys that describe the historical developments and the variety of tendencies on the nature of science (NOS) include works such as Duschl (1994); McComas (2002, 2020); Matthews (2003); and Lederman and Lederman (2014).

[10] For the need of explicit and reflexive treatment of nature of science (NOS), see, for instance, Khishfe and Abd-El-Khalick (2002); Smith and Scharmann (2008); Hodson (2009); Rudge and Howe (2009); Duschl and Grandy (2013).

[11] In addition to previously mentioned works, other noteworthy contributions include Dennison (1993); Rudge et al. (2014); Cohen (2016); Lorsbach et al. (2019).

> by "natural" means, together with a non-essentialist view of species (he compares species to individuals). (Regner 2006, p. 59)

In his text, Darwin reveals a vision of nature as a system. As a natural philosopher, he observed the natural world as an integrated whole rather than isolated parts. One of his key methodological approaches, as noted by Anna Carolina Regner, is the interdisciplinary support that evidence from different areas provides to each other (Regner, 2006, p. 60).

Like Aristotle's studies on animals and plants, Darwin offers numerous examples of the reliability of knowledge he collected not only from scholars but also from professionals such as farmers and pigeon breeders. These individuals are quoted alongside botanists, geologists, anatomists, physicians, and others in *On the Origin of Species.*

Furthermore, until the end of the nineteenth century, there was no sharp distinction between science and philosophy. Darwin actively engaged in dialogues with philosophers treating them on equal grounds. He also acknowledged the influence of bishops, reverends, and fathers of the church, whose works he read or corresponded with regarding matters of the natural world. It is not merely a curiosity about Darwin's methods but rather an important consideration. Robert Richards highlights that even esteemed historians of science may misinterpret certain aspects of Darwin's theory due to biases shaped by present conceptions of science. He reminds us that during the mid-nineteenth century, the terms "philosophy" and "science" were used practically as synonyms. It was toward the end of the century that the term "science" gradually aligned more closely with its present understanding. In Richards's words:

> Toward the end of the century, the term 'science' would come ever closer to what we mean by the word, but through mid-century one would find 'philosophy' and 'science' used interchangeably – experimental philosophy or experimental science, moral philosophy or moral science. (Richards & Ruse, 2016, p. 114)

In addition to examining Darwin's methods, readers of the *Origin* encounter various procedural approaches employed in the elaboration and defense of his theory. These range from traditional methods such as experimentation to more innovative techniques like the use of metaphors and imagination in general (Regner, 2006, p. 60).

The notion of a comprehensive view of nature is not new. It can be traced back to Aristotle's studies on animals and the direct observation of nature, which was reintroduced into Western natural sciences initially in the thirteenth century and more significantly during the origins of modern science in the late sixteenth century. As modernity progressed, the sciences gradually became more segmented. This was seen in the division of natural history into biology and geology in the late eighteenth century, as well as in the institutionalization of specialized subfields within biology from Darwin's time until the present day. While there are clear social benefits to this deepening process in the sciences and the need to maintain it, the exclusive compartmentalization has also resulted in gaps and deficiencies, as evidenced by worldwide crises such as climate change. By exploring the history of biology and

Darwin's practice of science, we can steer away from succumbing to rigid methodological demarcations and encourage reflections on diverse ways of doing science today.

1.8 Structure of the Volume

This volume provides a chronological exploration of the development of evolutionary thought from the late eighteenth century until the present day. With a particular emphasis on Charles Darwin, the book is organized into four parts.

Part I. Transformation of Species, from the Beginning
In Part I of the book, the focus is on providing the conceptual and social context surrounding the development of evolutionary thought. Chap. 2 challenges the traditional view of biological evolution as solely a nineteenth-century achievement by examining the neglected inquiries into the history of evolutionary ideas. Pietro Corsi's research expands the scope of the investigation to include the contributions of eighteenth-century naturalists beyond the French and English languages, suggesting that the narrative of evolutionary theories can be reevaluated within broader historical contexts and more history of biology be done.

Chapter 3, authored by Gustavo Caponi, delves into the epistemic foundations and roots of the natural philosophy that ascended Darwin's learning, practice, and questioning. Specifically, Caponi focuses on two key concepts: the principle of the conditions of existence emphasized by Georges Cuvier and the unity of type highlighted by Étienne Geoffroy Saint-Hilaire. These concepts are seen as crucial elements that shaped Darwin's thinking and provided him with a framework for understanding the natural world. The author suggests that Darwin's work can be seen as a synthesis of these two opposing concepts and reconciles the unity of type with the principle of the conditions of existence. He proposes that nature acts as an austere demiurge, working with existing materials and modifying them to serve new functions rather than constantly creating entirely new structures. In this way, Darwin's perspective leans more toward Geoffroy's emphasis on unity and modification rather than Cuvier's emphasis on environmental adaptation.

Chapter 4 closes the section with a fresh look into Darwin's voyage on the Beagle and the sociocultural context of the time. Héctor Palma explores Darwin's anthropological interests, particularly his encounters with the Fuegians. The chapter deconstructs the anachronistic narrative that separates different fields of research during that period, bringing together human, nonhuman animals, and plants within the geographical and geological scenarios of Darwin's studies before he became an evolutionist. The sociopolitical climate of the time is also highlighted, especially in relation to the Beagle expedition and its impacts on Darwin's anthropological inquiries.

Part II. Constructing a Theory

Part II covers the period between Darwin's arrival from the Beagle voyage until the publication of *On the Origin of Species* in 1859.

In Chap. 5, Janet Browne examines Darwin's writing and record-keeping practices during his time on the Beagle voyage and the first years following his return to London. By examining the 1837–1838 *Notebooks on Transmutation* and the 1842 and 1844 *Essays*, Browne uncovers the connections between these materials and Darwin's thinking, as well as the scientific practices of nineteenth century. She argues that these practices contributed to the British colonial appropriation of information.

Chapter 6, authored by Kostas Kampourakis, delves into the alleged atheism of Darwin and challenges it based on evidence from Darwin's autobiography, intended for his family rather than publication, and personal correspondence. Darwin described himself as agnostic and someone who never explicitly rejected the existence of God. Kampourakis highlights the importance of this clarification as it reveals Darwin's conceptual shift from the notion of perfect adaptation of natural theology to the concept of relative adaptation of natural selection.

Chapter 7, written by João Cortese, expands on the discussion of the compatibility between Darwin's theory of natural selection and religious belief. Drawing upon Darwin's writings, Cortese argues that Darwin's statements can be seen as arguments regarding the reasonability of the existence of God, which can be related to theological arguments. By examining excerpts from Darwin's autobiography and letters, the chapter also sheds light on Darwin's responses to the *Vestiges of the Natural History of Creation* and his dialogue with the botanist and theologian Asa Gray.

In Chap. 8, Marsha Richmond explores an intriguing aspect of Darwin's scientific journey during his time on the Beagle expedition. She highlights Darwin's encounter with a peculiar barnacle species off the coast of Chile in 1835, which he named "Mr Arthrobalanus." Unlike the typical shelled species that attached themselves to rocks or ship hulls, this small creature lived "naked" in the crevices of seashells, piquing Darwin's curiosity. Around 10 years later, Darwin embarked on a comprehensive taxonomic project that lasted 6 years. He meticulously described and classified not only Mr. Arthrobalanus but also all known species of Cirripedes, both living and fossil specimens. This undertaking solidified Darwin's reputation as an eminent naturalist and revealed the connections between Darwin's meticulous taxonomic work and the foundational tenets of his groundbreaking theory of evolution as presented in *On the Origin of Species*.

In Chap. 9, written by Viviane Arruda do Carmo and Lilian Al-Chueyr Pereira Martins, the focus shifts to Alfred Russel Wallace and his ideas, examining their similarities and differences to Darwin's ideas expressed in the joint papers published in 1858. The analysis concludes that despite some differences in presentation and sequencing, the contributions of Wallace and Darwin are coherent and share crucial similarities. For instance, both Wallace and Darwin referred to the concept of the "struggle for existence" in nature, albeit Darwin did not use the same phrase.

Similarly, while Wallace did not use the term "natural selection," he alluded to a principle that carried the same connotation. Both Wallace and Darwin acknowledged that species initially exist as varieties, and they both recognized the principle of divergence. This analysis supports that both naturalists arrived at similar conclusions in 1858. The analysis underscores the parallel development of evolutionary thought by the two researchers and the convergence of their insights.

Chapter 10, written by Maria Elice Brzezinski Prestes, explores Darwin's inclusion of the "Historical Sketch" in the third edition of *On the Origin of Species*, addressing criticisms from his contemporaneous readers regarding the absence of historical context. The Sketch, however, poses challenges for today's readers due to its lengthy list of mostly unfamiliar authors. Additionally, the chronological presentation fails to effectively highlight the similarities and differences between these authors' ideas and Darwin's own. A research-based teaching exercise is introduced in Prestes's chapter to extract meaningful insights from Sketch. The exercise revealed Darwin's position that while some naturalists believed in species modification, none of them *simultaneously* defended the two fundamental principles of his theory: common descent and natural selection. The exercise also rectifies popular historical errors and re-evaluates the distinctions between Darwin's and Lamarck's theories. Viewing Sketch as a meta-scientific discourse encourages readers to explore the nature of science in Darwin's work.

Chapter 11, written by Bárbara Jiménez-Pazos, explores Darwin's linguistic strategies and provides a context for reading his main work. Considering the multiple aspects of the context, the author highlights the importance for motivated readers to engage with the original source, *On the Origin of Species*. However, many students, researchers, or interested readers often admit that the dense prose and complex explanations have discouraged them from continuing their reading. In this chapter, Jiménez-Pazos proposes a linguistic remedy by employing a semantic analysis of the most evocative and expressive parts of the work. That is to say, by carefully examining Darwin's sensitivity and expressiveness, particularly through his use of exclamatory passages that convey scientific-aesthetic emotions, such as admiration, passion, and respect for nature and its study. The author's aim is to awaken a similar fascination in the reader.

Part III. Spreading the New Theory to the World

Part III comprises an analysis of each chapter in the first edition of *On the Origin of Species*. In truth, the 12 chapters in this section correspond to the 14 chapters of the 1859 first edition of Darwin's book, as two original pairs of chapters are grouped together, each dealing with the same subject.

Every corresponding chapter assists the reader in identifying the content in Darwin's text by providing clues on *how* to approach them using tools from the history and philosophy of science. Depending on the specifics of each chapter in *Origin*, some contributors adopt a general approach, while others focus on specific topics. Certain chapters delve into aspects of the nature of science (NOS), whether from the observational, conceptual, or sociocultural dimension of nineteenth-century practices of science. Rather than presenting their own interpretations of Darwin's text, the contributors were encouraged to empower readers to discover their own ways of interpreting Darwin's words and navigating through the occasionally tumultuous waters of Darwin's extensive argument.

In Chap. 12, Gregory Radick gives an overall view of Darwin's Chapter I structuring it in two halves. The first examines different cases and the main causes of the variability of domesticated species. The second turns the subject from variation to selection. Additionally, Radick gives suggestions to help readers navigate the information-dense paragraphs of Darwin's text.

In Chap. 13, Roberto Rozenberg analyzes *Origin's* Chapter II, through the reading of an evolutionist, that is, in a more presentist perspective welcomed by science students with no background in the history of science; simultaneously, the different subjects are treated with caution to the pitfalls of anachronism.

In Chap. 14, Robert Richards preferred to address the richness of *Origin's* Chapter III by its two notable peculiarities, its puzzles and unexpected features.

In Chap. 15, Michael Ruse accounts for the overall structure of *Origin's* Chapter IV, facilitating the reader to identify not only the core ideas of Darwin's theory but also the secondary mechanism of sexual selection. Ruse calls attention too of the underlying economic notion of the division of labor and to the resulting model of evolution as a tree of nature.

Sander Gliboff highlights in Chap. 16 how much Darwin was attentive to philosophical and methodological issues of scientific practice as the title of *Origin's* Chapter V announces. Gliboff helps to identify and explore the multiple possible conceptions, patterns, and laws of variation raised by Darwin. The text has great potential ins science teaching, whether to deconstruct the supposed opposition between Lamarck and Darwin or to pose the challenges of current evolutionary development (evo-devo) and epigenetics.

Roberto Martins's Chap. 17 contributes with metascientific considerations that hold relevance to science education. In Chapter VI of *Origin*, Darwin addresses potential objections to his theory and offers responses to them. Martins emphasizes that these challenges emerged from his acquaintances' hesitance to accept Darwin's ideas and by published works presenting arguments against any natural explanation for species' origins, as well as Darwin's internal dialogue, initial doubts, and anticipation of criticism. Martins choose to outline the main difficulties and provide a more detailed examination of selected topics, analyzing Darwin's defense in those instances. Furthermore, he discusses some of the weak points in Darwin's line of reasoning from a diachronic perspective.

Chapter 18, authored by Nelio Bizzo and Lucas Marino Vivot, delves into *Origin's* Chapter VII on instinct. The authors explore how Darwin engages in a dialogue with the moral and philosophical beliefs prevalent among the Anglican elite of his time. He challenges the idea that nature reflects benevolence and compassion, highlighting morally repugnant behaviors observed in the natural world. Darwin questions whether instinct, as a form of natural behavior, is driven by an intelligent agent that contradicts benevolence. He argues that instead of continuous intervention, nature operates through fixed laws that lead to the advancement of organisms. This Darwinian interpretation suggests that repugnant behaviors in nature cannot be attributed to a cruel intelligent creator. These reflections anticipate the need for new theoretical frameworks in Anglican theology, which Darwin did not expect to be well-received. These insights have implications for the teaching and learning of biological evolution, particularly in reconciling moral, religious, and scientific perspectives.

P. Lorenzano wrote Chap. 19, related to *Origin's* Chapter VIII, where Darwin delves into hybridism. By examining the phenomenon, Darwin aims to explore the validity of the existence of a fundamental distinction between species and varieties, which forms an integral part of the broader “species problem.” Lorenzano's chapter deepens the historical context of this debate, tracing its origins from Linnaeus to Kölreuter and Gärtner, and analyzes the structure and arguments of Darwin's confronts this issue.

In Chap. 20, Charles H. Pence argues that Chapters IX and X of *Origin* signify a shift in the book's argumentative structure. While the earlier chapters address objections to natural selection, the tenth chapter explores the novel and unforeseen implications of evolutionary theory. Darwin examines domains such as extinction, biogeography, and embryology to indicate the potential of evolutionary theory. Notably, the fossil record serves as a pivotal point, offering both powerful objections and compelling explanations. In his chapter, Charles Pence analyzes Darwin's utilization of fossil evidence and investigates its unique role in different parts of his argument for evolution by natural selection, attributing its significance to the social and intellectual context of his time.

In Chap. 21, Tina Gianquitto addresses the two *Origin's* Chapters XI and XII on geographical distribution, showing how Darwin reinforces his evolutionary theory by highlighting the role of species distribution in supporting the idea of common ancestry. Gianquito also exposes other evolutionary concepts discussed there, including the influence of individuals on the future of a species, migration and dispersal processes, and the subsequent modification and proliferation of new forms. Seed dispersal emerges as a compelling explanation for the presence of related populations in geographically distinct areas. Furthermore, in the context of the book's publication in 1859, it serves as a rebuttal to the popular anti-evolutionary theory of separate centres of creation. This theory, linked to racist polygenist ideologies, justified slavery and perpetuated racial hierarchies. In contrast, Darwin's focus on seeds reveals a world interconnected by physiology, geography, and time.

In Chap. 22, Aldo Mellender de Araújo deals with Chapter XIII where Darwin emphasizes the importance of a natural system of classification based on both embryology and adult morphology. Darwin critiques previous methods of classification, such as affinity-based and geographic-based systems, as arbitrary and illogical. Instead, he proposes a genealogical approach, highlighting the affinities organisms share through inheritance from a common ancestor. By advocating for a classification system rooted in “descent with modification,” Darwin challenges traditional taxonomic hierarchies and calls for a natural system of classification based on genealogy.

In Chap. 23, Gerda Maisa Jensen, Bruno F. Lima, and Marcelo Monetti Pavani analyze the concluding Chapter XV of *On the Origin of Species*. Considering it an abstract within an abstract, the authors expose evidence of the changes Darwin made in the structure of the text based on correspondence with his acquaintances. Then they present and explain the main subjects discussed by Darwin. Additionally, they explore aspects of the nature of science (NOS) that make the chapter especially suitable for biology classes.

Part IV. Epilogue: What Came Next Was Extraordinary!
Part IV focuses on the task of presenting the comprehensive aspects of the exceptional research programs stemming from Darwin's theory. These programs play a crucial role in comprehending the pivotal transformations witnessed by the theory of evolution throughout the twentieth century and the initial two decades of the twenty-first century.

In Chap. 24, Susana Gisela Lamas delves into Modern Synthesis, a collaborative effort among researchers from various biological fields. The focus of this synthesis was to merge genetics data and theories with the concept of natural selection. Gisela Lamas examines the foundational principles of both Darwin's theory and the Modern Synthesis, shedding light on the consistencies and discontinuities that have influenced biological sciences for many decades.

Authored by Thierry Hoquet, Chap. 25 concludes this volume by providing an overview spanning from the Modern Synthesis era to the present day. Hoquet delves into the criticism from the fields that remained outside the scope of the 1930s and 1940s architectural unification, as well as the constraints that emerged in the last decades of the twentieth century. The future of Modern Synthesis as a body of knowledge hangs in the balance. Hoquet emphasizes that the ideal of unification has been a foundational aspiration from its inception, but whether this ideal will prevail or differing perspectives will shape the future is yet to be determined.

The epilogue holds significant importance in science education, not only for elucidating the distinct contributions of Charles Darwin to human knowledge but also for capturing the current dynamic state of the biological sciences. A comprehensive grasp of the historical trajectory of evolutionary thinking serves as a valuable foundation for shaping future developments, including meaningful engagement with the human sciences.

References

Allchin, D. (2013). *Teaching the nature of science: Perspectives & resources*. SHiPS.

Anderson, D. L., Fisher, K. M., & Norman, G. J. (2002). Development and evaluation of the conceptual inventory of natural selection. *Journal of Research in Science Teaching, 39*(10), 952–978.

Bizzo, N. M. V. (1992). Darwin on man in the "origin of species": Further factors considered. *Journal of the History of Biology, 25*(1), 137–147.

Bizzo, N. (2014). Prefácio. In C. Darwin (Ed.). *A Origem das Espécies* (Trad. of 1872 sixth edition by Carlos Duarte e Anna Duarte) (pp. 7–13). Martin Claret.

Bizzo, N. (2018). Prefácio, revisão técnica e notas. In C. Darwin (Ed.). *Sobre a Origem das Espécies por meio da seleção natural ou A preservação de raças favorecidas na luta pela vida* (Trad. of 1859 first edition by Daniel Moreira Miranda). Edipro.

Bizzo, N. (2019). What is the role of sound evidence in evolution education? A research program following students' narratives in Brazil and elsewhere. In A. Oliveira & K. Cook (Eds.). *Evolution education and the rise of the creationist movement in Brazil* (pp. 231–248). Lexington Books.

Bizzo, N. (2020). The history of science and science education: Tools for the practice and research in schools. In C. N. El-Hani, E. F. Mortimer, M. Pietrocola, & M. R. Otero (Eds.). *Science education research in Latin America* (pp. 426–456). Brill.

Bradley, B. S. (2009). Darwin's sublime: The contest between reason and imagination in *On the origin of species*. *Journal of the History of Biology, 44*, 205–232.

Browne, J. (2022). Reflections on Darwin historiography. *Journal of the History of Biology, 55*(3), 1–13.

Cohen, J. I. (2016). Exploring the nature of science through courage and purpose: A case study of Charles Darwin's way of knowing. *Springerplus, 5*, 1532.

Cooper, R. A. (2017). Natural selection as an emergent process: Instructional implications. *Journal of Biological Education, 51*(3), 247–260.

Cortez, E. P. M. (2018). *Descobrindo a seleção natural: uma proposta de ensino baseada na história da ciência*. Dissertação (Mestrado em Ensino de Ciências) – Universidade de São Paulo.

Deniz, H., & Borgerding, L. A. (Eds.). (2018). Evolutionary theory as a controversial topic in science curriculum around the globe. In *Evolution education around the globe* (pp. 3–11). Springer.

Dennison, R. (1993). Using Darwin's experimental work to teach the nature of science. *The American Biology Teacher, 55*(1), 5052.

Duschl, R. A. (1994). Research on the history and philosophy of science. In D. L. Gabel (Ed.). *Handbook of research on science teaching and learning: A project of the National Science Teachers Association* (pp. 445–465). MacMillan.

Duschl, R. A., & Grandy, R. (2013). Two views about explicitly teaching nature of science. *Science & Education, 22*, 2109–2139.

Endersby, J. (Ed.). (2009). Appendix 1, An evolving origin. In *Charles Darwin: On the origin of species*. Cambridge University Press.

Erduran, S., & Dagher, Z. R. (2014). *Reconceptualizing the nature of science for science education: Scientific knowledge, practices and other family categories*. Springer.

Flick, L. B., & Lederman, N. G. (Eds.). (2006). *Scientific inquiry and nature of science: Implications for teaching, learning, and teacher education*. Springer.

Gregory, T. R. (2008). Understanding evolutionary trees. *Evo Edu Outreach, 1*, 121–137.

Gregory, T. R. (2009). Understanding natural selection: Essential concepts and common misconceptions. *Evo Edu Outreach, 2*, 156–175.

Hall, T. S. (Ed.). (1970). *A source book in animal biology*. Cambridge University Press.

Harms, U., & Reiss, M. J. (2019). *Evolution education re-considered: Understanding what works*. Springer.

Hodge, J. (1992). Darwin's argument in the origin. *Philosophy of Science, 59*(3), 461–464.

Hodge, J. (2013). Darwin's book: On the origin of species. *Science & Education, 22*(9), 2267–2294.

Hodson, D. (2009). Making NOS teaching explicit and reflexive. In *Teaching and learning about science: Language, theories, methods, history, traditions and values* (pp. 77–112). Rotterdam.

Jenkins, E. W. (1990). The history of science in British schools: Retrospect and prospect. *International Journal of Science Education, 12*(3), 274–281.

Jenkins, E. W. (2013). The "nature of science" in the school curriculum: The great survivor. *Journal of Curriculum Studies, 45*(2), 132–151.

Jensen, G. M. (2016). *Charles Darwin e os peixes elétricos: História e natureza da ciência no ensino de ciências na Educação de Jovens e Adultos*. Tese (Doutorado em Ciências Biológicas, Biologia/Genética) – Universidade de São Paulo.

Kampourakis, K., & Zogza, V. (2009). Preliminary evolutionary explanations: A basic framework for conceptual change and explanatory coherence in evolution. *Science & Education, 18*, 1313–1340.
Kampourakis, K. (2010). Jim Endersby: Charles Darwin: On the origin of species. *Science & Education, 19*, 827–831.
Kampourakis, K., & Nehm, R. H. (2014). History and philosophy of science and the teaching of evolution: Students' conceptions and explanations. In M. Matthews (Ed.). *International handbook of research in history, philosophy and science teaching* (Vol. 1, pp. 377–399). Springer.
Khishfe, R., & Abd-El-Khalick, F. (2002). Influence of explicit and reflective versus implicit inquiry-oriented instruction on sixth graders' views of nature of science. *Journal of Research in Science Teaching, 39*(7), 551–578.
Lederman, N. G. (1992). Students' and teachers' conceptions of the nature of science: A review of the research. *Journal of Research in Science Teaching, 29*(4), 331–359.
Lederman, N. G., & Lederman, J. S. (2014). Research on teaching and learning of nature of science. In N. G. Lederman & S. K. Abell (Eds.). *Handbook of research on science education* (Vol. II, pp. 600–620). Routledge.
Lederman, N. G. (2019). Contextualizing the relationship between nature of scientific knowledge and scientific inquiry: Implications for curriculum and classroom practice. *Science & Education, 28*, 249–267.
Lederman, J., Lederman, N., Bartles, S., & Jimenez, J. (2019). An international collaborative investigation of beginning seventh grade students' understanding of scientific inquiry: Establishing a baseline. *Journal of Research in Science Teaching, 56*, 486–515.
Lorsbach, A. W., Meyer, A. A., & Arias, A. M. (2019). The correspondence of Charles Darwin as a tool for reflecting on nature of science. *Science & Education, 28*, 1085–1103.
Martins, L. A-C. P. (1997). *A teoria cromossômica da herança: Proposta, fundamentação, crítica e aceitação*. Tese (Doutor em Ciências, área de concentração de Genética e Evolução) – Universidade de Campinas.
McComas, W. F. (Ed.). (2002). *Nature of science in science instruction: Rationales and strategies*. Kluwer.
McComas, W. F. (Ed.). (2020). *Nature of science in science instruction: Rationales and strategies*. Springer.
Matthews, M. (2003). The nature of science and science teaching. In B. J. In Fraser & K. G. Tobin (Eds.). *International handbook of science education. Part two* (pp. 981–1000). Kluwer.
Matthews, M. (2012). Changing the focus: From Nature of Science (NOS) to Features of Science (FOS). In M. S. Khine (Ed.). *Advances in nature of science research* (pp. 3–26). Springer.
Matthews, M. (2022). Thomas Kuhn and science education: Learning from the past and the importance of history and philosophy of science. *Science & Education*. https://doi.org/10.1007/s11191-022-00408-1
Matthews, M. R. (2023). Cultural studies in science education: A philosophical appraisal. *Cultures of Science*. https://doi.org/10.1177/20966083231173721
Moulton, F. R., & Schifferes, J. J. (1957). *Autbiografia da Ciência* (Trad. português de Mário Henrique Leiria). Sociedade de Intercâmbio Cultural Luso-Brasileiro. 2 vols.
Oliveira, A., & Cook, K. (2019). *Evolution education and the rise of the creationist movement in Brazil*. Lexington Books.
Pérez, D. G., Montoro, I. F., Alís, J. C., Cachapuz, A., & Praia, J. (2001). Para uma Imagem não deformada do trabalho científico. *Ciência & Educação, 7*(2), 125–153.
Piaget, J., & García, R. (1983/2008). *Psicogénesis e Historia de la ciencia*. Siglo XXI.
Prestes, M. E. B., & Silva, C. C. (Eds.). (2018). *Teaching science with context: Historical, philosophical, and sociological approaches*. Springer.
Pumfrey, S. (1991). History of science in the national science curriculum: A critical review of resources and their aims. *British Journal for the History of Science, 24*, 61–78.
Radick, G. (2009). Is the theory of natural selection independent of it's history? In J. Hodge & G. Radick (Eds.), *The Cambridge companion to Darwin*. Cambridge University Press.
Regner, A. C. K. P. (2006). A Polêmica Darwin versus Mivart: Uma lição em refutar objeções. *Filosofia e História da Biologia, 1*, 55–89.

Reznick, D. N. (2009). *The origin then and now: An interpretive guide to the origin of species*. Princeton University Press.

Richards, R., & Ruse, M. (Eds.). (2016). *Debating Darwin*. Chicago University Press.

Rosengren, K. S., Brem, S. K., Evans, E. M., & Sinatra, G. M. (Eds.). (2012). *Evolution challenges: Integrating research and practice in teaching and learning about evolution*. Oxford University Press.

Rudge, D. W., Cassidy, D. P., Fulford, J. M., & Howe, E. M. (2014). Changes observed in views of nature of science during a historically based unit. *Science & Education, 23*, 1879–1909.

Rudge, D. W., & Howe, E. M. (2009). An explicit and reflective approach to the use of history to promote understanding of the nature of science. *Science & Education, 18*, 561–580.

Ruse, M. (Ed.). (2009). *Philosophy after Darwin: Classic and contemporary readings*. Princeton University Press.

Ruse, M. (2013a). Teaching the classics: The origin of species as a case study. *Science & Education, 22*, 2255–2265.

Ruse, M. (Ed.). (2013b). *The Cambridge encyclopedia of Darwin and evolutionary thought*. Cambridge University Press.

Scheiner, S. (2010). Toward a conceptual framework for biology. *The Quarterly Review of Biology, 85*(3), 293–318.

Silva, T. T. (2013). *Darwin na sala de aula: Replicação de experimentos históricos para auxiliar a compreensão da teoria evolutiva*. Dissertação (Mestrado em Ensino de Ciências) – Universidade de São Paulo.

Sinatra, G. M., Brem, S. K., & Evans, E. M. (2008). Changing minds? Implications of conceptual change for teaching and learning about biological evolution. *Evo Edu Outreach, 1*, 189–195.

Sloan, P. (2019). Darwin: From *origin of species* to *descent of man, The Stanford encyclopedia of philosophy* (Summer 2019 Edition) (E. N. Zalta, Ed.). https://plato.stanford.edu/archives/sum2019/entries/origin-descent/

Smith, M. U., & Scharmann, L. (2008). A multi-year program developing an explicit reflective pedagogy for teaching pre-service teachers the nature of science by ostention. *Science & Education, 17*, 219–248.

Souza, R. A. L. d. (2014). *A viagem de Wallace ao Brasil: uma aplicação da história da ciência no ensino de biologia*. Dissertação (Mestrado em Ensino de Ciências) – Universidade de São Paulo.

Souza, R. A. L. d. (2021). *A viagem de Wallace ao Brasil: uma aplicação da história da ciência no ensino de biologia*. Dialética.

Tidon, R., & Lewontin, R. C. (2004). Teaching evolutionary biology. *Genetics and Molecular Biology, 27*(1), 124–131.

Waters, C. K. (2009). The arguments in the origin of species. In R. Richards & M. Ruse (Eds.). *The Cambridge companion to the "origin of species"* (pp. 120–143). Cambridge University Press.

Maria Elice Brzezinski Prestes researches the history and philosophy of biology, including its approach to Science Education. She focused on the generation of living beings, especially on the modes of observation and experiments in the seventeenth and eighteenth centuries. Currently, she is developing studies on Darwin's and Aristotle's works. She is a Sênior Professor at the Department of Genetics and Evolutionary Biology of the Institute of Biosciences of the University of São Paulo, where she coordinates the Laboratory of History of Biology and Teaching (LaHBE). She is a former president of the Brazilian Association of Philosophy and History of Biology (ABFHiB) and the current editor of the Brazilian journal *Filosofia e História da Biologia*.

Part I
Transformation of Species, from the Beginning

Chapter 2
Debates About Life's Origin and Adaptive Powers in the Early Nineteenth Century

Pietro Corsi

Abstract In the "Historical Sketch" he published in the third edition of *On the Origin of Species* (1861), Darwin famously argued that "the great majority of naturalists believe that species are immutable productions and have been separately created." Commentators have implicitly endorsed this self-serving view since then. In fact, since the last decades of the eighteenth century, writers about nature and practitioners of the "system of nature" genre had engaged in discussions on the origin of life and the limits and extent of its capability to adapt to changing environments. By the early decades of the nineteenth century, several highly popular French encyclopedias and dictionaries published entries devoted to discussing spontaneous generations or the thesis that varieties were incipient species. A handful of such articles endorsing transformism caught Darwin's attention, though their authors were ignored or snubbed in the "Historical Sketch."

2.1 Introduction

When in the spring of 1800, Lamarck announced to the pupils attending his class at the Muséum National d'Histoire Naturelle that he now endorsed species mutability and spontaneous generation, he would have been deeply disconcerted to learn that he had earned a place among the precursors of Darwin – as historians should be. Though the hunt for precursors has been epistemologically and historically debunked, considerable energy has been spent in evaluating whether someone, somewhere, had "really" been a precursor of Darwin. Authors as diverse as the Latin poet Titus Lucretius Carus (ca. 94–54 BC), the diplomat and writer Benoît de Maillet (1656–1738), or the French naturalist Georges-Louis Leclerc, count Buffon (1707–1788), or indeed Darwin's own grandfather, Erasmus (1731–1802) have been enshrined in the hall of fame of evolution.

P. Corsi (✉)
The University of Oxford and EHESS, Paris, France
e-mail: pietro.corsi@history.ox.ac.uk

M. Elice Brzezinski Prestes (ed.), *Understanding Evolution in Darwin's "Origin"*, History, Philosophy and Theory of the Life Sciences 34,
https://doi.org/10.1007/978-3-031-40165-7_2

True, all the authors mentioned above played an important role in the debates on life and its history that interested so many readers throughout the eighteenth and nineteenth centuries. They raised or suggested questions leading authors throughout Europe to explore the possibility that organisms kept adapting to changing environments and that a common origin for all life forms could be envisaged. Some argued that species could indeed change in response to changed environmental conditions, but the relevant theoretical context and implications had little to do with Darwin. Around 1800, a rich population of readers shared a common fascination for all-encompassing views of nature, often called systems of nature, which enthralled philosophers and naturalists, theologians and politicians, civil servants and compilers of dictionaries and encyclopedias since the early eighteenth century. It is within this broader context that debates on life at the time have to be placed (Corsi, 2018) – a context that was far from dominated by crude fixist and creationist prejudices.

When Lamarck gave his introductory speech in 1800, he was commenting on, and to some extent adapting to, wide-ranging contemporary debates on systems of nature. At the same time, he fought hard to distance himself from the population of writers about nature, filling scores of publications with their personal views of the history of the Universe, the Earth and Life. It could even be argued that the relatively few "professional," full-time, naturalists active in 1750–1850 experienced great difficulty convincing the cultivated public that they were the exclusive depositary of expertise and authority. The so-called amateurs continued to carry the day.[1] Yet, it should be stressed that the line of demarcation between "professional" and amateur or between specialized publications and works directed to the lay public was not as sharply drawn as historians have assumed.

In this paper, I would like to briefly call attention to a number of French authors and publications catering for readers seemingly insatiably thirsty for new systems of nature and life. My contribution aims at complementing important work done, for instance, on German debates on nature during the decades spanning the eighteenth and the nineteenth centuries and their impact on the development of Darwin's vision of nature. I will call attention to authors, readers, and editorial ventures that rarely play a role in our reconstruction of contemporary debates about organic change. Within this rich literature, I will focus on conversations concerning the origin of life, of species, and the power and limits of life's capacity to adapt to changing geoclimatic circumstances.[2] In a final section, I will briefly consider if and how Darwin reacted to some of the authors and themes discussed in this paper.

[1]For a masterly discussion of broadly "evolutionist" views of living nature debated in French literary circles, see Wanlin (2018).

[2]Robert Richards has strongly argued for the influence of German *Naturphilosophie* on Darwin's work: Richards (1992, 2002). Another source is Gliboff (2008).

2.2 Systems of Nature and Theories of Life

Among the neglected features of European debates about nature, a place of prize is occupied by the reception and legacy of Buffon's works. Buffon was not the first to write about the history of the Earth and its inhabitants (Rossi, 1984; Rudwick, 2005; Poole, 2010). He was certainly the most successful and attracted a wide and varied readership. Dailies, periodicals, and encyclopedias relentlessly summed up the contents of Buffon's works. Needless to say, encyclopedias were as expensive (albeit sold in instalments) as his books, but they were at least a good investment. They informed and commented on the literary productions everyone had to have an opinion on. Encyclopedias and journals allowed the social and cultivated élites to take part in salon conversations, from nation's capitals down to provincial towns. This composite population of readers were the public publishers relied on to embark upon costly and often dangerously ambitious editorial plans: readers who longed to be informed but also amused and thrilled.

In view of the widespread interest in the mineralogy of the last decades of the eighteenth century – after all, Buffon himself, as always careful to ride the wave of consent, published his own *Histoire naturelle des Minéraux* (1783–1788) – most comments on the last work Buffon published, the *Époques de la Nature*, dealt with the chronology of the formation of the various mineral components of the planet. Yet, from within the discussions focusing on mineralogy, there emerged substantial comments on the existence and properties of forms of living matter Buffon had called *molécules organiques* and, more generally, on the history of life on Earth. Buffon insisted that, for instance, the scores of mammals he had described could be reduced to a smaller number of prototypes that gave rise to local adaptations – jackals, dogs, and wolfs belonged to the same family, the descendant of primitive *souches* (strains). Buffon was no evolutionist. He insisted that only animals capable of generating viable offspring belonged to the same species; the fact that dogs and wolfs could also reproduce indicated that they belonged to the same *souche*. He believed that omnipresent organic molecules were spread around the surface of the globe and in the depth of the seas; each species was made possible by the existence of a distinctive *moule intérieure*, an internal mold capable of appropriating the organic molecules and of arranging them according to the architecture of the mold. Combinations of organic molecules were also responsible for the spontaneous generation of parasites or infusoria (Farber, 1972; Sloan, 1979; Hoquet, 2005).

It is important to emphasize that the fashion for systems of nature was not limited to the Buffonian tradition. The temptation to engage in drafting omni-comprehensive systems of nature allured even naturalists who did not like Buffon and, in later years, spent considerable energy fighting the fashion for systems. The list comprised even young Georges Cuvier, later in life, the archenemy of speculations in natural history. As he wrote to his friend Christoph Heinrich Pfaff (1773–1852) in 1788:

> I think that one should carefully investigate the relations of all existing beings with the rest of nature and show especially their part in the economy of this great Whole. In doing this work, I would like to start from the simplest things, for example, water and air, and after having

> spoken of their influence on the whole, to pass little by little to the compound minerals; from these to the plants, and so on, and that at each step one would seek exactly the degree of composition, or, which amounts to the same thing, the number of properties that this degree presents more than the preceding one, the necessary effects of these properties, and their utility in creation. (Marchand 1858, pp. 70–71)

With due distinctions, even the famous programmatic lecture Carl Friedrich Kielmeyer (1765–1844) delivered on 11 February 1792, sketching a research program for the life sciences that marked German and European biology for decades to come, could be read within debates on competing systems of nature.[3]

2.3 The Abbot and the Jesuit

Of the many and varied comments, Buffon's *Époques* elicited, I will briefly touch upon the polemical exchange between Jean-Louis Giraud Soulavie (1752–1813), an abbot from the Vivarais region (today's Ardèche), and his nemesis, the former Jesuit and catholic polemist Augustin Barruel (1741–1820). Soulavie combined the critique of the mineralogical tenets of Buffon with his own interpretation of the successive appearance of life forms. He devoted a long section of the first volume of his *Histoire naturelle de la France Méridionale* (1780–1784) to a critique of Buffon's doctrines concerning the successive formation of mineral substances and the strata. Soulavie also commented on the succession of species, putting forward "a chronological history of fossil and living animals, established upon incontrovertible facts." He contrasted Buffon's *Époques* with a succession of organic "ages," starting with the time when all organisms were totally different from the ones living today, up to man, the most recent appearance on Earth (Soulavie, 1780–1784, v. 1, p. 346).

Soulavie was convinced that the Earth was several million years old. It is not equally clear whether he thought the successive animal populations he discussed had been transformed into different species due to environmental change or whether, in Buffonian terms, the populations that appeared in successive ages were the result of the "degeneration" of original *souches* (strains). The latter appears to be Soulavie's conviction:

> The multiplication of families is, properly speaking, nothing but a multiplication of external forms. (...) With men and seashells, as with all the families of the organised world, the soil, food, the temperature of the climate, etc., slowly altered the primitive form established by the Author of their existence. (Soulavie, 1780–1784, v. 1, p. 356)

---

[3] K.F. Kielmeyer, "Über die Verhältnisse der organischen Kräfte unter einander in der Reihe der verschiedenen Organisationen, die Gesetze und Folgen dieser Verhältnisse"; the manuscript lecture circulated widely throughout Europe. Analyses of the work of Kielmeyer can be read in Coleman (1973) and Bersier (2005).

Though Soulavie claimed that he had followed the history of Creation narrated in the book of Genesis, he was eventually asked by the Académie des Sciences to suppress the incriminated section from the first volume, which was duly reprinted.

Soulavie's nemesis was Augustin Barruel, a former Jesuit destined to European fame as the fiercest denouncer of the crimes of the French Revolution. The *Mémoires pour servir à l'Histoire du Jacobinisme* Barruel started publishing in 1796 quickly became a European bestseller. Back in 1780, he had been allowed by Soulavie to see a manuscript draft of his work containing estimates of the age of the Earth, calculated in hundreds of million years. Barruel was in the process of publishing the first volume of his own work, *Les Hélviennes*, a sarcastic tirade against the *Philosophes* and their anti-Christian follies. He hastily added a small opening pamphlet of 60 pages (pp. xiii–lxxiii) when the volume was already printed, entirely devoted to the demolition of Soulavie.

One recurrent theme in Barruel's apologetic diatribes was the contention that philosophical materialists were the new Lucretius, predicating the eternity and self-sufficiency of matter in order to dispense with the Creator. Soulavie, a personal acquaintance (they were officiating in the same diocese), had really angered Barruel. A fellow priest doubting the Mosaic narrative of creation, hinting that the 6 days had, in fact, covered a period of millions of years, was providing weapons to the infidel camp.

It is revealing of the climate of the 1780s that Barruel failed to convince his superiors to join in the crusade against Soulavie. Indeed, the latter sued Barruel in a civil court, arguing that his critic had abused his trust by selectively publishing draft paragraphs which contained his speculations and thought experiments, not his firmly held convictions (Terrinière, 1785). His Bishop did not share Barruel's conviction that the *Histoire naturelle de la France Méridionale* represented a threat to revealed religion and advised caution (Mazon, 1893, v. 1, pp. 31–50; Aufrère, 1952).

Barruel's *Les Hélviennes* have very rarely been commented upon and never in relation to debates on the systems of nature put forward by eminent representatives of French Enlightenment thought. The former Jesuit was no bigot, nor was he against scientific undertakings. He was a good mathematician and kept abreast of physics and astronomy.

The mockery of Soulavie was followed by equally flippant surveys of theories of nature and life published during the eighteenth century: Barruel summed up a century of debates for readers of the 1780s. What deserves particular mention is the care with which he had read the works of his enemies: the caricature of their position Barruel offered was fairly detailed. He also listed the different theories put forward to account for the origin of life and the succession of organisms. Barruel discussed at length Buffon's doctrine of organic molecules and his *Époques*; De Maillet's *Telliamed*, and the latter's view that all organisms living on land or in the air were originally born out of specific seeds resting in the waters of the oceans; and Jean-Baptiste-René Robinet's (1735–1820) *De la Nature* (1761), including the latter's view that the Earth, the moon, planets, and stars were living organisms, and that all matter was in fact alive.

Only Diderot was credited with the idea that all animals descended from a single prototype through processes of adaptive modifications. This view, first put forward in *Pensées sur l'interprétation de la Nature* (1754, pp. 33–36), had been inspired by passages in Buffon's *Histoire Naturelle, générale et particulière*. In the article "L'âne" (IV, 1753, pp. 381–383), the naturalist discussed the structural resemblances between allied species, suggesting further exploration of analogies between higher categories of classification. He reached and immediately rejected the hypothetical conclusion that all animals probably descended from a single original organism. Diderot was quite taken by the idea. He borrowed the term "prototype" Buffon had used to indicate the type of each species and applied it to the hypothetical ancestor of all living organisms.

Barruel rightly insisted that the majority of the authors he pilloried had not considered as viable the idea that species could change into different ones – change only affected external features, not the "essence" of the species. As he put it in his abrasive polemical prose:

> With Buffon, new species could well form when the old ones stopped eating organic molecules; but a monkey was never anything else than a monkey, and man is not afraid of becoming a mouse. With Telliamed, we were, it is true, pikes, salmons; but we do not fear to be them again. With Robinet, I don't even know what I was; but the moon only begets a moon, and each thing remains in its kind. With Diderot, I say, what could he not have been? What should he not become again? Greetings to the prototype animal. I am his most humble servant and yours: but don't talk to me about it anymore! (Barruel, 1784–1788, v. 1, pp. 255–256)

It is noteworthy that to Barruel Robinet – an author almost invariably relegated to the periphery of eighteenth-century studies – was an opponent enjoying the same status as Diderot, de Maillet, or Buffon (Murphy, 1976; Bourdin, 1992). Equally revealing is the fact that to a cultivated catholic apologist like Barruel, active in the 1780s, the (to his eyes) dangerous speculations on the history of nature were making inroads into provincial social and intellectual circles, reaching representatives of the lower clergy such as Soulavie. As Barruel aptly put it, "Tout le monde veut lire des systèmes." As we are going to see, within a decade, systems of nature insisting on the autonomous powers of natural processes to bring about variously adapted organisms all along the history of the Earth became popular and widely debated throughout contemporary French and European societies. They thrived despite rebuke from leading representatives of institutional scientific bodies and accusations of leading to atheism and political radicalism.

2.4 Organic Molecules, Matter, and the Origin of Species

Not everyone among contemporaries and successors of Buffon was prepared to accept organic molecules. In the *Entretien entre d'Alembert et Diderot* (only published in 1830), the latter famously argued that minerals, too, could be absorbed by living organisms and contribute to their nourishment and growth. A later die-hard materialist such as Jean-Claude De la Métherie (1743–1817, Delamétherie since 1793), editor of the widely read *Journal de Physique*, also maintained that material elements and material laws were responsible for all forms of organization: from crystals to embryos. Indeed, a universal law of crystallization had given rise to mineral formations in the primaeval ocean, spontaneous marine generations, and a limited number of ancestral animal and vegetable prototypes. These differentiated with time and adapted to the directional change of the Earth's emerging surface. Even the fecundated eggs developed into embryos and new individuals, thanks to a fitting implementation of the universal law of crystallization. Delamétherie (wrongly) equated his law of crystallization with the concept of the *nisus formativus* Johann Friedrich Blumenbach (1752–1840) had put forward to explain the goal-directed growth of the embryo (Delamétherie, 1799, p. 8).

As far as he was concerned, after 1800, Lamarck vigorously argued that organic molecules or living matter were a logical and physical impossibility. Life emerged from particular assemblies of very soft material molecules and fluids, establishing *orgasme vital*, a tension among molecules rendering them capable of diastole-systole-like contractions and reacting to external stimuli. Such contractions created the condition for an elementary fluid dynamic within the barely organized molecular combinations. Biology was, to him, a branch of terrestrial physics, dealing with fluid dynamics acting within molecular arrangements defined by membrane boundaries. This did not make simple living organisms independent from the laws regulating material phenomena: simple organisms did, in fact, exist only because environmental fluids (heat, light, water, air, etc.) were penetrating and traversing them, thus originating and maintaining vital motions. Only with increased levels of organization did living beings generate and maintain their own internal, still wholly material fluid dynamics.

As we have already stressed, a plurality of authors elaborated a plurality of doctrines. If the position defended by Delamétherie was on the whole close to the teaching of Buffon, with the only exception of rejecting organic molecules, others abandoned the search for prototypes of existing organisms and put organic molecules, and living matter in general, to good use to develop alternative explanations. Some argued that living beings were materially produced in the location they were found, thanks to combinations of living molecules favored by the physicochemical nature of the soil.

Toward the end of the eighteenth century, Jean André Deluc (1727–1817), a collaborator to Delamétherie's *Journal de Physique*, firmly opposed theories of the spontaneous generation then current and Delamétherie's doctrine of the production of prototypes for several Phila through crystallization. Deluc's *Remarques sur l'origine des Êtres organisés*, dated December 1795, also conveyed his awareness that some naturalists and members of the public believed organisms had originated through spontaneous generations in the locations where they were found to thrive. As Barruel before him, Deluc adopted the literary format of correspondence, in his case with a young gentleman who avidly consumed systems of nature and took most of his information from the *Journal de Physique*: the conversation, in other words, was aimed at winning over the cultivated reader, not fellow naturalists. When discussing the flora and fauna of continents, Deluc's correspondent addressed the problem head-on:

> Why then the species that are common to the two hemispheres were not spontaneously born following the same Law of Nature, while conditions were favourable to its action? Why then, generally speaking, plants, animals and men – in other words, all the species of organised beings – were not born in the very location they are indigenous of, and all descended from the first individual of each species? (Deluc, 1798, pp. 392–393)

Deluc denied such a possibility and reiterated his belief in a unique act of creation. However, he admitted that geoclimatic change could alter a species' external appearance. A fully blown theory of the local production of living organisms was put forward in the very early 1800s (among others) by Jean-Baptiste Fray-Fournier (1764–1835), a firm believer in spontaneous generations daily produced by combinations of active molecules present in infusions. Fray-Fournier was no atheist nor materialist and often referred to nature as God's creation. One of his chosen examples was the flora of the high Pyrénées: many of the region's trees could not be found anywhere else on Earth. Hence, they were clearly the result of the development of seeds spontaneously generated locally. No need for constantly adapting prototypes when the Creator had providentially endowed nature and matter with the power to fill every corner of the Earth with suitable forms of life.

In 1805, Fray performed his experiments of spontaneous generations at Arcueil in front of the great chemist Claude Louis Berthollet (1743–1817), who was not convinced – probably, Fray surmised, because the famous chemist's eyes were not sufficiently trained to see the extremely small, newly generated combinations of living molecules (Fray, 1817, p. 9). He had more success with Pierre Jean Georges Cabanis (1757–1808). The famous doctor became a supporter and endorsed Fray in a lengthy footnote printed in the second edition of his extremely successful *Rapports du Physique et du Moral de l'homme* (1805, v. 2, pp. 301–302). Echoing a tradition going back to Lucretius, Cabanis speculated that if today we only witness the spontaneous generation of very simple organisms, it was still possible that in the past, under more favorable circumstances, more complex organisms owed their origin to the appropriate combination of living molecules.

Fray, a supplier to the army, followed French troops throughout Europe and was stationed for several years in the German states under Napoleonic hegemony. He forged close links with several German medical and scientific colleagues, to the point that the book summing up his views, the *Essai sur l'origine des corps organisés et inorganisés*, was firstly printed in Berlin and Leipzig in 1807 and was often referred to by leading representatives of the German biomedical community (Fray, 1807, with excerpts appearing in Fray (1810); Gruithuisen (1809); Treviranus (1811), pp. 75–76). His work was deemed sufficiently important and dangerous to deserve a highly critical chapter made up of lengthy quotations in French in John Barclay's *An Inquiry Into the Opinions, Ancient and Modern, Concerning Life and Organization* (1822, pp. 126–141).

In a seminal article published in 2005, Nicolaas Rupke has called attention upon a number of German-speaking authors who, starting in the 1810s, had proposed an alternative between strict creationism and transformism. They argued, albeit with different overtones and argumentative styles, that organisms could be directly produced by physicochemical agency in the locations where they thrive. Not just barely structured spontaneous generations but indeed highly organized animals, men included. It would be interesting to explore the relationship (if any) between the authors studied by Rupke and the earlier French debates we have summarily sketched above. Indeed, Fray could represent a key instance of direct communication between the French post-Buffonian debates on living matter and organic molecules and analogous conversations being carried on in several German scientific capitals (Rupke, 2005).

A final instance, among several, of speculations on "continuous creations," occurring locally and emanating from molecular combinations, is represented by the many articles and book sections Jean-Baptiste Bory de Saint Vincent (1778–1846) devoted to promoting his views on the matter and on *matière agissante* in particular. In a chapter of his 1804 *Voyage dans les quatre principales îles des mers d'Afrique* Darwin discussed with Joseph Dalton Hooker in 1844–1845, Bory sketched a history of the surface of the Earth emerging from a primitive ocean, in which the first forms of life had originated through spontaneous generation (1804, v. 3, pp. 123–171). New lands kept emerging even in the contemporary world due to the action of volcanoes. Pin-head islands lost in the seas hosted a great variety of cryptogams and very simple animals, almost identical everywhere:

> We must surely admit the possibility of modern creations and even of future ones, taking place on some future points on the surface of the Earth, on the occasion of the gathering of sufficient circumstances. (Bory, 1804, v. 3, p. 161)

Bory thought he had found incontrovertible proof of the relatively recent production even of highly organized animals specific to isolated locations. Mauritius had hosted the now extinct Dodo, a defenseless member of the Columbidae family – the bird had literally been eaten up by mariners. The Dodo was found only on Mauritius and nowhere else: irrefutably, the Dodo was a local, recent production, as were the 15 races of humans he described in later works.

Bory spent years experimenting with the microscope and claimed to have repeatedly witnessed the combination of molecules of active matter and green matter (produced at the surface of ponds), giving birth to organisms intermediate between plants and animals. He never expanded upon the probable mechanism leading to the development of higher organisms. He only stated, echoing debates summed up above:

> Life and vegetation have always started in the same way; beings have been formed, according to the elements offered by each locality; temperature and other causes have modified a small number of primitive species; species that are born again and again, and move on to different states, the more they depart from the original form of the type; their first

degradations will take on well-defined forms, and they will reproduce as constant species. (Bory, 1804, v. 3, pp. 168–169)[4]

Bory's doctrine of continuous creations well illustrates the way in which, in the early decades of the nineteenth century, a debate started with Buffon and Diderot took on new dimensions and adapted to the curiosities and interests of a much-expanded readership of natural history works. Narratives of voyages and the global dimension of natural history observations or the issue of polygenism increasingly attracted the attention of cultivated readers.

2.5 Theories of the Earth and the Limits of Species Change

Throughout the second half of the eighteenth century and well into the nineteenth, the joint heritage of Lucretius and *Telliamed* was perceptible in authors reflecting on the history of the Earth and of life. The Lucretian theme of a primordial overproduction of organized beings through spontaneous generation, destined for the greatest part to be destroyed because of structural imperfection or weakness, kept emerging in writings by Julien Jean Offray de La Mettrie (1709–1751) or Diderot, down to geologists such as Philippe Bertrand (1739–1811) or Ami Boué (1794–1881). Often, the narrative of the way in which primordial organisms adapted to changing conditions echoed the description *Telliamed* tendered of how a fish stranded by powerful waves or winds on a very humid marsh, unable to regain the sea, slowly turned its fins into feet and became a terrestrial animal. Adapting the Lucretian theme of massive extinctions to his own purposes, De Maillet commented that most marine organisms stranded on land were destined to die. However, "If one hundred thousand have perished in contracting the Habitude, yet if two have acquired it, they are sufficient to give Birth to the Species" (De Maillet, 1748, pp. 223–224). In the first edition of his *Principes de la Philosophie naturelle* (1778, pp. 113–114), and again in the second one (1787, pp. 360–362), Delamétherie described in analogous terms how the monkey-like early humans left the trees and adjusted to live on the ground. They adopted a permanent vertical posture, which in turn entailed a cascade of related anatomical and behavioral change (Stoczkowski, 1995). Needless to say, Delamétherie did not share many of De Maillet's assumptions: his monkey-man had not emerged from the seas in cold northern regions, as *Telliamed* had maintained, but lived on land, in a tropic-like climate. Still, adaptationist narratives inspired by *Telliamed* continued to find an echo in the writings of several practitioners of the theory of the Earth genre well into the nineteenth century.

During the last decades of the eighteenth century, only a limited number of writers fully endorsed the view that life unfolded through repeated processes of

[4]Bory was also the editor of the *Dictionnaire Classique d'histoire naturelle*, 1822–1831. Analyses of his work can be found in Ferrière (2006, 2009).

adaptation (most famously Diderot). Yet, at the dawn of the nineteenth century, in several European countries, naturalists and practitioners of the systems of nature genre multiplied imaginative descriptions of adaptive change to explain species formation. Erasmus Darwin recruited elephants and seagulls to drive the point home and listed sexual characters or defense traits to insist on the capacity of organisms to modify their external form. Lamarck relied on birds' legs, shortening or elongating according to need, and only on three occasions in several thousand pages, and in passing, he mentioned the neck of the giraffe, for which he earned centuries-long mockery. Needless to say, Darwin's idea that a bear swimming for miles with his mouth open could become a whale-like animal has been discreetly ignored (Darwin Correspondence Project, 1985–present). A militant anti-materialist naturalist as Julien-Joseph Virey looked for instances of adaptive change (to him, providentially arranged by the Creator) even in the beak of ducks, where appropriately prolonged nervous terminations endowed the animal with the capacity to search for food in muddy waters.

I have discussed elsewhere the examples of adaptation listed in 1801 by Johann Christian Rödig (1772–1863), an author Cuvier despised and historians have simply written off. It is appropriate to insist here on the interest of Rödig, a citizen of Saxony, as a witness to the spread of the fashion for systems of nature in provincial settings throughout Europe. A reader of Lucretius and Telliamed, but also of Lavoisier and Laplace, Rödig developed his system of nature along lines reminiscent of Cuvier's ambitious work plan, from chemical elements to life. He saw a progressive development of organisms from simple spontaneous generations through increasingly complex structures, a process in which habits and needs played a key role. Rödig described how a group of hippos lost at sea could turn into whale-like animals (they could revert to their original form if stranded on land) and how an American squirrel (*Glaucomys volans*) jumping from tree to tree sustained by a membrane joining the front and rear legs was clearly on its way to become a bat-like organism. Finally, the Italian Giuseppe Gautieri (1769–1833), who in the very early 1800s enjoyed the conversation of Goethe and Schelling in Jena, invested his energies in producing his own system of nature and completed his narrative of living organisms by expanding upon – among others – the process that gave the giraffe its long neck (Gautieri, 1805).

Virey elaborated his sophisticated adaptive explanations well before Lamarck embarked upon rather crude descriptions of how birds living in marshes could elongate their legs not to get wet; Erasmus Darwin explained how the trump of the elephant got elongated well before Lamarck evoked the stretching neck of the giraffe; Rödig composed his work (1800–1801) when Lamarck's introductory lecture had not yet been printed; and Gautieri did not mention Lamarck's early transformist works (though he was keen to parade his wide-ranging reading list) and, anyway, his description of the cascade of anatomical consequences stemming from the stretching of the neck of the giraffe was to a certain extent more sophisticated than the cursory mention proffered by his French colleague (Corsi, 2005). Darwin hinted that Erasmus had preceded Lamarck, indeed, as many others Erasmus had never heard of.

To ask whether the choice of authors I have listed were truly evolutionists or proto-evolutionists (and some indeed were, in France as well as elsewhere in Europe), let alone precursors of Darwin, misses the point completely. They all exemplify the wider dimensions of debates on living organisms and their differential power of adaptation at the surface of the Earth and throughout its long history. The conversation involved a varied population of writers and readers who bought their works: full-time "professional" naturalists were not the loudest nor the most listened to voices.

It also needs to be stressed that during the early decades of the new century, some naturalists considered that varieties could become true species without experiencing any anxiety concerning the dogma of the creation of specific forms. Thus, for instance, the ultra-conservative Leopold von Buch (1774–1853) took for granted that plants blown by winds on the other side of a towering mountain like the Pic of Tenerife in the Canary Islands could generate local varieties. In due time, the varieties could become true species, that is, could not produce new individuals with the original species. When Darwin dotted that Galapagos species descended from an immigrant ancestor undermined the belief in the stability of species, he was clearly talking for himself. To von Buch, and many contemporary writers of natural history, the phenomenon simply indicated the adaptative flexibility species could display in particular circumstances.

2.6 Encyclopedias and Dictionaries

Over the years, I have repeatedly called attention on Virey, as representative of a type of authors supplementing their income and building a reputation out of prolific authorial commitments. In particular, I have insisted *ad nauseam* that the success of the *Nouveau dictionnaire d'histoire naturelle* he contributed to (first edition 1803–1804, second edition 1816–1819) should alert us to the active role the reading public played in the diffusion of broad issues in natural history, the question of species among others. Following the success of the *Nouveau dictionnaire*, the editors of competing ventures tried to offer their own choice of leading articles covering issues debated in salons or among the cultivated public. Successive French encyclopedias offered updates on the literature on species, in spite of snubs and open rebuke from a handful of authoritative representatives of institutional science. Readers' continuing interest in systems of nature that included accounts of the origins and transformations of life kept subscriptions up: the demand very much conditioned the offer.

It is also important to stress that authors writing for editorial enterprises directed to the general public were not amateurs, though some indeed were. The divide authoritative, professional science vs amateur improvisations does not reflect the real state of affairs as the nineteenth century progressed. The cohort of scientific personnel contributing to encyclopedias included professional paleontologists (e.g., Pierre Théodore Virlet d'Aoust, 1800–1894), medical men (e.g., Achille Requin,

1803–1854), naturalists employed in various public and private natural history collections, and members of the Académie des sciences. It should be recollected that even Étienne Geoffroy Saint-Hilaire exploited his contacts within publishing houses to obtain favorable mention of his own theories and contributed himself highly polemical articles to collective works. His son Isidore Geoffroy Saint-Hilaire (1805–1861) started his prestigious institutional career by contributing to various encyclopedic ventures (Corsi, 2011).

Since I have discussed elsewhere a few such editorial ventures and their authors, I will limit my analysis to insist again on selected features of Virey's contributions to the *Nouveau dictionnaire*. A prolific and fast writer, Virey was the ideal author several encyclopedias hired for their leading articles. An omnivorous reader, he spent his life in libraries and at his desk, compiling extracts from a great variety of sources (Corsi, 1986). It is unfortunate that borrowings were rarely acknowledged. Yet, Virey almost invariably added his own original reflections on the sources at hand. The way in which he summed up the discussions on the prototype Buffon had unwittingly started and Diderot had amplified is a case in point. Though a rich collection of Diderot's writing was finally available thanks to the 1798 edition by Jacques-André Naigeon (1735–1810), or that works by La Mettrie and even *Telliamed* had been reprinted in the late 1790s, by 1803–1804 these were names editors of the *Nouveau dictionnaire* felt it was prudent to leave out of the printed page. Thus, no names were dropped when Virey rewrote in terms of current classification categories Buffon's (rejected) hypothesis that all living beings descended from a unique primordial form. In the entry "Corps organisés" (1803), Virey summed up the question in his own terms:

> Is it not a great presumption that these families are only nuances emanating long ago from the same source which we call class today? For the families of living beings are to the class what the present species are to the family. Consequently, these so-called primitive species will only be varieties originating from the class. But as the same reason which subsisted for the families, still subsists for each class; that is, as the classes are linked together by common bonds of analogy, we are led to think that in fact nature has created in each kingdom of living beings, only one original form which was the primitive and common trunk from which the various branches of the present species did spring. (Virey, 1803, v. 6, p. 268)

In words reminding analogous hypotheses formulated by Erasmus Darwin, Virey suggested that nature had created a single primordial "germ" for all animals and all plants, though it was possible that the two germs were, in fact, varieties of a single primitive living form. The conclusions Virey drew from his discussion of the filiation of beings was destined to be taken up by several successive encyclopedic publishing ventures:

> Thus, a single germ, by developing successively, creating a great number of similar individuals, will have seen them modify themselves little by little in the long space of centuries, and by the influence of climates, temperatures, &c., turn into species more or less similar to each other. These will have been further modified by the succession of ages as they have experienced the long and profound influences of all that surrounds them, and as they have mixed with each other. These mixtures, these variations, these species, will go on becoming subdivided; for one day, let there be no doubt, what we regard as varieties, will become a species which will in turn have its varieties. (Virey, 1803, v. 6, p. 268)

2.7 To Darwin's Doorsteps

The view that varieties were incipient species, and that, given enough time, organisms and their descendants would keep splitting up, found expression in several French encyclopedias up to the 1840s. The article "L'espèce" (1844) Fréderic Gérard published in the *Dictionnaire universel d'histoire naturelle* (1841–1849) edited by Charles Dessalines d' Orbigny (1806–1876) contained a long discussion of the difficulty naturalists experienced to distinguish between species and varieties, guided by the assumption:

> I am thus convinced, with Lamarck, Poiret [Jean-Louis, 1758–1834] and Geoffroy, that varieties become Species, and that it is in this way that new Species are formed which throws so much hesitation and uncertainty in the science. (Gérard, 1841–1849, p. 447b)

The article caught Darwin's eye.

Four decades had gone by since Virey had put forward a similar claim in less than a page of the *Nouveau dictionnaire*. Reflecting increasing specialization even in encyclopedias directed to the general public, in 1844 Gérard substantiated his thesis through pages and pages of concrete examples illustrating the difficulties naturalists experienced in deciding which organism counted as a variety or a species. It was the listing of examples that caught Darwin's eye. Gérard was not a professional naturalist but gained his living as a translator at the Ministry of War; he complemented his income by contributing to a variety of editorial enterprises. A handful of Gérard's articles for the *Dictionnaire Universel* provided a spirited literature review of the debate on species. Contrary to Virey, Gérard was a committed and coherent transformist, eclectically marshalling doctrines put forward by Lamarck and Geoffroy Saint-Hilaire, but also by Robinet, Bory de Saint-Vincent, and Poiret. Gerard paid great attention to German authors such as Karl Friedrich Burdach (1776–1847), who were available in French language editions. In spite of the scorn some academic Parisian naturalists and Hooker poured on Gérard (a radical socialist, by the way), the Belgian conservative politician and geologist D'Omalius d'Halloy hosted an article by Gérard in the bulletin of the Belgian Academy of Sciences. The negative opinions of a few naturalists active in France or in the British Isles should not be taken by historians as the final verdict: after all, D'Orbigny's *Dictionnaire Universel* was on the whole a competent and appreciated collection.

Before recalling Darwin's encounter with Gérard, it is useful to mention that on 25 December 1844 Darwin had written to Joseph Dalton Hooker (1817–1911) a typically ambiguous letter, trying to understand what his friend thought of Bory's contention that isolated volcanic islands presented a preponderant number of polymorphous organisms while at the same time making clear that he did not trust an author he had nevertheless referred to quite often during the Beagle voyage: "I grieve there is no better authority for Bourbon, than that stupid Bory: I presume his remark that plants, on isolated Volcanic isl^{ds}. are polymorphous (i.e., I suppose, variable?) is quite gratuitous." Indeed, Hooker answered, contrary to Bory's statement, the flora of Saint Helen, for instance, contained very few polymorphous plants. Yet, Hooker was less severe than Darwin and acknowledged that Bory had acquired some merit in natural history.

With Gérard, Darwin's ambiguity was even more marked. As it was common practice, dictionaries were printed in instalments, often sold as independent pamphlets. The French algologist Camille Montagne (1784–1866) had sent a copy of Gérard's article "L'espèce" to William Jackson Hooker, Joseph Dalton's father, and the latter had passed it on to Darwin, who received it in early June 1845: "I am particularly glad to see & have sent to endeavour to get one." Darwin was clearly taken by Gérard, marked passages of interest and sent the pamphlet back to Hooker in August 1845, asking him not to erase the pencil marks he had used to highlight passages. Early in September, Darwin received from his friend a letter well known to Darwin scholars. In it, the demolition of Gérard turned into a denunciation of people who talked about species without having seen and studied thousands and thousands of specimens. The best botanists who had done so, Hooker explained, believed in the reality of species: "Quoting instances by tens or hundreds of variations in individual species is nothing new, few have an idea of the labor required to establish or destroy a species of a mundane genus" (Hooker to Darwin, 4–9 September 1845). Darwin took the criticism to heart, feeling it was directed at him as well.

The interesting feature of this late 1845 exchange is that by late October to early November, Darwin received his own copy of the pamphlet, once again thanks to Hooker's good offices. In spite of Hooker's view that heaping instances of variations served no purpose, this was precisely what Darwin was interested in. Darwin asked his friend to send back the copy he had marked: "Could you lend me sometime, your former copy that I may transpose my marks (or rather exchange copies) as I do not want the trouble of looking it over again. I shall be glad to see the other pamphlets; though I do not expect much, if they are by Gérard" (Darwin to Hooker, 5 or 12 November 1845). To make sure his friend got the message, an equally contorted request was sent a few days later: "I must sometime beg your copy of l'Espece to copy my marks, as I by no means want to wade through so poor a performance again" (Darwin to Hooker, 6 November 1845).

Like Bory, Gerard was severely judged by Darwin. Still, he felt the two authors had raised questions he had to investigate. So, Bory was stupid and Gerard poor. Yet, Darwin had taken good notice of what they had written and sounded out the true expert, Hooker, on how far they could be trusted. In spite of the severe judgment passed by his friend, Darwin insisted to get back his annotated copy of Gérard while appearing to despise the author as worthless. He also bought the other "pamphlets" the French author had contributed to the *Dictionnaire Universel*, for instance, on "Génération spontanée" and "Géographie zoologique."[5] The materialist and transformist views Gérard enthusiastically promoted cost him his collaboration to the *Dictionnaire*, as he bitterly complained. In the "Historical sketch" Darwin prefaced to the third edition of *On the Origin*, there is no mention of Gérard, in

[5] For several references to Gérard in Darwin's notebooks on works to read or read, and abstracts of his readings, see J. van Whye, ed., (http://darwin-online.org.uk)

spite of the fact that the French author had been the occasion for an exchange of views with Hooker he could not possibly have forgotten. Darwin also kept rather vague on some of the French authors he had read, to mark his distance from a literature his learned friends found wholly unacceptable. In the *Historical Sketch*, he told readers that he had learned through reading Alexandre Godron (1807–1880) *De l'Espèce* (1859) that Bory probably disbelieved the fixity of species, as if he had never heard his name. How Darwin read Lamarck and Bory, Gérard and Poiret still awaits a comprehensive and informed critical assessment.[6]

2.8 Conclusion

As I often alluded to, French dictionaries, encyclopedias, and periodicals of the decades 1800–1840 have been rarely studied (to employ a euphemism), and the question of their diffusion in Europe and elsewhere never hinted at: in spite of the fact that French was the language of scientific, literary, and philosophical exchanges before starting to partially lose ground to the German and English languages during the 1840s. Sample findings support the view that the scientific literature published in French we alluded to circulated widely. To quote some instances, when a so-called precursor of Darwin, the picturesque and picaresque Constantin Rafinesque (1783–1840), maintained that varieties were incipient species, he was quoting from Virey's *Nouveau Dictionnaire* – he owned a copy of the dictionary and kept traveling with it in his trunk. In a successful textbook, *Prodromo della storia naturale generale e comparata d'Italia*, the Italian Francesco Costantino Marmocchi (1805–1858) printed without acknowledgment the rather sophisticated evolutionary entry "Animal" Achille Requin (1803–1854) had contributed to the *Encyclopédie nouvelle* edited by Pierre Leroux and Jean Raynaud. Requin, eventually a Professor of Medicine and the teacher of Jean-Martin Charcot, contributed one of the most interesting surveys of the question of species printed in early nineteenth century France and Europe, originally inspired by Lamarck and Geoffroy Saint-Hilaire. I have recently shown how a transformist article inspired by Geoffroy Saint-Hilaire and printed in *Le Globe*, a French newspaper, on 1 April, 1829, was translated and printed in the *New Edinburgh Philosophical Journal* in less than 2 weeks (Corsi, 2021). Even German authors such as Friedrich Tiedemann, Burdach, Gottfried Reinhold Treviranus, and Johannes Peter Muller circulated in the British Isles and elsewhere in the Western World in French translation.

The tradition of debates on living organisms and their capacity to adapt to changing circumstances I have tried briefly to sketch clearly indicates the need to complement the customary concentration on Darwin or Lamarck with systematically taking into account populations of authors, editorial ventures, and ideas vigorously

[6]Johnson (2019, pp. 324 and 329) wrongly claims that Darwin had never read a line of Bory and that Poiret did not believe in species change.

circulating throughout the European cultural space. It is equally clear that it is unprofitable – to say the least – to continue to engage in highly selective, anachronistic historical reconstructions centered on a handful of issues and an even smaller number of authors. Systematic perusal of periodicals, encyclopedias, and dictionaries is dramatically enlarging the list of issues the cultivated reading public was interested in, highlighting the way in which readers' demand (or curiosities, if you prefer) very much conditioned the offer. The success of many such editorial ventures, but also the success of the *Vestiges of the Natural History of Creation*, the "Victorian Sensation" Jim Secord has brilliantly reconstructed (Secord, 2000), and indeed the success of Darwin, Huxley, Haeckel, and Spencer were not due to élite culture opening the eyes of the masses but to members of the intellectual élites finally talking a language the reading public were prepared to listen to – at times, for reasons like Empire, misogynism, social subordination, and racialism we tend to eliminate from our historical reconstructions.

References

Aufrère, L. (1952). *Soulavie et son secret: Un conflit entre l'actualisme et le créationnisme. Le temps géomorphologique*. Hermann.

Barruel, A. (1784–1788). *Les Helviennes ou Lettres Provinciales Philosophiques*, 2 vols.

Bersier, G. (2005). Visualizing Karl Friedrich Kielmeyer's organic forces: Goethe's morphology. *Monatschefte, 1*, 18–32.

Bory, J.-B. de. (1804). *Voyage dans les quatre principales Îles des mers d'Afrique*, 3 vols. Buisson.

Bourdin, J.-C. (1992). Le "profonde serieux" de Robinet selon Hegel. *Studies on Voltaire and the Eighteenth Century, 302*, 409–437.

Coleman, W. (1973). Limits of the recapitulation theory: Carl Friedrich Kielmeyer's critique of the presumed parallelism of earth history, ontogeny, and the present order of organisms. *Isis, 64*, 341–350.

Corsi, P. (1986). Julien–Joseph Virey, le Premier Critique de Lamarck. In S. Atran et al. (Eds), *Histoire du concept d'espèce dans les sciences de la Vie*, (pp. 181–192). Fondation Singer–Polignac.

Corsi, P. (2005). Before Darwin: Transformist concepts in European natural history. *Journal of the History of Biology, 38*, 67–83.

Corsi, P. (2011). The revolutions of evolution: Geoffroy and Lamarck, 1825–1840. *Bulletin du Musée d'Anthropologie préhistorique de Monaco, 51*, 97–122.

Corsi, P. (2018). "Systèmes de la nature" and theories of life: Bridging the eighteenth and the nineteenth centuries. In J. Riskin (Ed.), *Early adventures in evolution* (Republics of Letters 6, issue 1). https://arcade.stanford.edu/rofl_issue/volume-6-issue-1

Corsi, P. (2021). Edinburgh Lamarckians? The Authorship of Three Anonymous Papers (1826–1829). *Journal of the History of Biology, 54*, 345–374. https://doi.org/10.1007/s10739-021-09646-5

Darwin Correspondence Project. (1985–present). *The Whale Bear*. https://www.darwinproject.ac.uk/letters/darwins-works-letters/rewriting-origin-later-editions/whale-bear

Delamétherie, J.-C. (1799). Discours introductif. *Journal de Physique, 48*, 3–99.

Deluc, J. A. (1798). *Lettres sur l'histoire physique de la terre*. Nyon.

De Maillet, B. (1748/1750). *Telliamed: Or, discourses between an Indian philosopher, and a French missionary, on the diminution of the sea, the formation of the earth, the origin of men and animals*. Osborne.

Farber, P. (1972). Buffon and the concept of species. *Journal of the History of Biology, 5*, 259–284.
Ferrière, H. (2006). *Bory de Saint-Vincent, militaire naturaliste entre Révolution et Restauration*. Université Paris 1 Panthéon-Sorbonne.
Ferrière, H. (2009). *Bory de Saint-Vincent: l'évolution d'un voyageur naturaliste*. Syllepse.
Fray, J.-B. (1807). *Nouvelles expériences extraites d'un manuscrit qui a pour titre: Essai sur l'origine des Corps Organisé et Inorganisés*. Louis Quien/Grieshammer/Nicole.
Fray, J.-B. (1810). Nouvelles expériences extraites d'un manuscrit qui a pour titre: Essai sur l'origine des Corps Organisé et Inorganisés. *Journal de Physique, 70*, 427–429.
Fray, J.-B. (1817). *Essai sur l'origine des Corps Organisé et Inorganisés, et sur quelque phénomènes de Physiologie Animale et Végétale*. Mme Ve Courcier.
Gautieri, G. (1805). *Slancio sulla genealogia della Terra e sull'origine dei Vermi abitanti le interiora degli Animali*.
Gérard, F. (1844). L'espèce. In C. D. d'Orbigny (Ed.), *Dictionnaire Universel d'histoire Naturelle*. d'Orbigny.
Gliboff, S. (2008). *H. G. Bronn, Ernst Haeckel, and the origins of German Darwinism: A study in translation and transformation*. MIT Press.
Gruithuisen, F. V. (1809). Ueber die chemischen und dynamischen Momente bey der Bildung der Infusorien mit einer Kritik der Versuche des Herrn Fray. *Journal für die Chemie, Physik und Mineralogie, 8*, 511–547.
Hoquet, T. (2005). *Buffon, histoire naturelle et philosophie*. Champion.
Johnson, C. N. (2019). *Darwin's historical sketch: An examination of the "preface" to the origin of species*. Oxford University Press.
Marchand, L. (Ed.). (1858). *Lettres de Georges Cuvier à C. M. Pfaff*. V. Masson.
Mazon, A. (1893). *Histoire de Soulavie (naturaliste, diplomate, historien)*, 2 vols. Fischbacher.
Murphy, T. (1976). Jean Baptiste René Robinet: The career of a man of letters. *Studies on Voltaire and the Eighteenth Century, 150*, 183–250.
Poole, W. (2010). *The world makers: Scientists of the restoration and the search for the origins of the Earth*. Peter Lang.
Richards, R. (1992). *The Meaning of Evolution. The morphological construction and ideological reconstruction of Darwin's theory*. Chicago University Press.
Richards, R. (2002). *The Romantic Conception of Life: Science and Philosophy in the Age of Goethe*. Chicago University Press.
Sloan, P. (1979). Buffon, German biology, and the historical interpretation of biological species. *The British Journal for the History of Science, 12*, 109–153.
Soulavie, J.-L. G. (1780–1784). *Histoire naturelle de la France méridionale*. 7 vols. J.-F. Quillau.
Rossi, P. (1984). *The Dark Abyss of Time: The History of the Earth & the History of Nations from Hooke to Vico*. Chicago University Press.
Rudwick, M. J. S. (2005). *Bursting the Limits of Time. The Reconstruction of Geohistory in the Age of Revolution*. University of Chicago Press.
Rupke, N. (2005). Neither creation nor evolution: The third way in mid-nineteenth century thinking about the origin of species. *Annals of the History and Philosophy of Biology, 10*, 143–172.
Secord, J. A. (2000). *Victorian sensation: The extraordinary publication, reception, and secret authorship of* Vestiges of the Natural History of Creation. University of Chicago Press.
Stoczkowski, W. (1995). Le Portrait de l'ancêtre en singe: L'Hominisation sans évolutionnisme dans la pensée naturaliste du XVIIe Sicèle. In R. Corbey & B. Theunissen (Eds.), *Man, Ape, Apeman: History of an idea, 1600–1900* (pp. 141–155). Pitecanthropus Centennial Congress.
Terrinière, L. d. l. (1785). *Pièces Relatives à l'Histoire naturelle de la France Méridionale*. Grangé.
Treviranus, L. C. (1811). *Beyträge zur Pflanzenphysiologie*. Dieterich.
Virey, J-J. (1803). Corps organisés. In: *Nouveau dictionnaire d'histoire naturelle*, 36 vols. Abel Laone.
Wanlin, N. (Ed.). (2018). *Evolutions. Littérature et sciences au XIXe siècle: Une Anthologie*. Classiques Garnier.

Pietro Corsi is Professor Emeritus of the History of Science, the University of Oxford. He has been conducting research work and has taught at Cambridge, Harvard, and Université Paris 1. He is also Director of Research Emeritus at the EHESS, Paris. He has published several studies on European evolution theories before Darwin, on science and religion in Victorian England, and on the relations between science and the state in nineteenth-century Italy.

Chapter 3
The Darwinian Not Too Strictly Balanced Arrangement Between Cuvier and Geoffroy Saint-Hilaire

Gustavo Caponi

Abstract In 1859, Darwin proposed a solution for the tension between unity of type and conditions of existence that had shaped the growth of pre-evolutionary morphology. The theory developed in *On the Origin of Species* contemplated and integrated both ideas: the unity of type explained by common descent and the arrangement to the conditions of existence explained by natural selection. That integration, however, was not a presupposition of the argument developed there but rather its most important achievement. The compatibility between the requirements of the principle of the conditions of existence stressed by Georges Cuvier and the Unity of Type highlighted by Étienne Geoffroy Saint-Hilaire was, in fact, the main tour de force of Darwin's work and maybe its main key of reading. However, despite this, it can also be said that the synthesis that Darwin proposes for the antithesis between unity of type and conditions of existence ends up tipping the balance a little in Geoffroy's favor. He led us to consider nature as an austere demiurge that tended to work with the same materials, modifying them indefinitely in virtue of the new functions that came to them in virtue of different circumstances and minimizing the production of new structures.

3.1 Introduction

In 1837, in the third volume of his *History of the Inductive Sciences*, William Whewell (1837, pp. 456–457) mentioned the antagonism between two "schools of physiologists." One was those who, denying the doctrine of final causes, grounded their analyses of animal morphology on the theory of the unity of composition proposed by Étienne Geoffroy Saint-Hilaire. The other was those naturalists who accepted the doctrine of final causes as condensed in "the principle of the *conditions of existence*" proposed by Georges Cuvier (Whewell, 1837, p. 472). And many years

G. Caponi (✉)
Philosophy Department, Federal University of Santa Catarina, Florianópolis, SC, Brazil

M. Elice Brzezinski Prestes (ed.), *Understanding Evolution in Darwin's "Origin"*, History, Philosophy and Theory of the Life Sciences 34,
https://doi.org/10.1007/978-3-031-40165-7_3

later, Edward Stuart Russell insisted on this opposition when, in *Form and Function*, sustained:

> [. . .] the contrast between the teleological attitude, with its insistence upon the priority of function to structure, and the morphological attitude, with its conviction of the priority of structure to function, is one of the most fundamental in Biology [. . .]. (Russell, 1916, p. 78)

He also said that Geoffroy and Cuvier could be the paradigmatic representatives of the two possible solutions to the dilemma. But, by continuing to speak of the dilemma, preserving the opposition raised by Whewell, Russell was undermining the solution Darwin had proposed for this tension between the unity of type and the conditions of existence.

As it can be read in *On the Origin of Species*, those ideas were contemplated and integrated into the theory presented there, the unity of type explained by common descent, and the arrangement of the conditions of existence explained by natural selection:

> It is generally acknowledged that all organic beings have been formed on two great laws – Unity of Type and the Conditions of Existence. By unity of type is meant that fundamental agreement in structure, which is quite independent of their habits of life. In my theory, unity of type is explained by unity of descent. The expression of conditions of existence, so often insisted on by the illustrious Cuvier, is fully embraced by the principle of natural selection. (Darwin, 1859, p. 206)

That integration, however, had not been a presupposition of the argument developed there but rather its most important achievement. Compatibility between the requirements of the principle of the conditions of existence and the unity of type, explained by common descent and natural selection, was, in fact, the leading *tour de force* of *On the Origin of Species*. And stated in the first pages of the work:

> In considering the Origin of Species, it is quite conceivable that a naturalist, reflecting on the mutual affinities of organics beings, on their embryological relations, their geographical distribution, geological succession, and other such facts, might come to the conclusion that each species had not been independently created, but had descended, like varieties, from other species. Nevertheless, such a conclusion, even if well founded, would be unsatisfactory, until it could be shown how the innumerable species inhabiting this world have been modified, so as to acquire that perfection of structure and coadaptation which most justly excites our admiration. (Darwin, 1859, p. 3)

The unity of type, the general similarity of structure that exists between the different groups of living beings, shown by comparative anatomy and comparative embryology, together with the evidence of paleontology and biogeography, suggested the possible common descent of living beings. It was in the common descent that the unity of type and the evidence of paleontology and biogeography found their explanation. However, therein lies the problem, the postulation of any mechanism capable of deriving different forms of living things from a single primitive form should also explain how that process could occur in accordance with the multiple conditions that the different kinds of living things should meet for making their existence possible. And the mechanism that was able to fulfill that theoretical requirement was natural selection. It produces a diversification of forms while generating beings whose forms, to a certain extent, seem to be the result of a negotiation between Cuvier and Geoffroy. This certainly gave Darwin's thesis a lot of force because, at the time of the publication of *On the Origin of Species*, these two French naturalists were still the main references in studying the natural history of organized beings. Undoubtedly, establishing a synthesis between the views of both naturalists was quite an achievement for Darwin.

3.2 Cuvier and the Principle of the Conditions of Existence

The canonical formulation of the principle of the conditions of existence is the one Cuvier gives in *Le Règne Animal* of 1817:

> Since nothing can exist unless it meets the conditions that make its existence possible, the different parts of each being must be coordinated in such a way as to render possible the total being, not only in itself but also in relation to those beings that surround it. (Cuvier, 1817, p. 6)

But, to correctly understand how Cuvier understood the functional requirements for the organic forms derived from this principle, it is necessary to consider that, by "conditions of existence," Cuvier understood something that is not the same as what we, following Darwin, understand today (*cf.* Russell, 1916, p. 34; Caponi, 2008, p. 41). In our way of thinking, the conditions of existence have to do, above all, with the demands of the struggle for existence: they are conditions of life that are related to the ecology of each living being (Limoges, 1970, p. 40). In turn, Cuvier had a more physiological than ecological perspective (*cf.* Reiss, 2009, p. 19). According to him, and as Russell (1916, p. 34) explained it, "the very condition of existence of a living being, and part of the essential definition of it, is that its parts work together for the good of the whole"; and, therefore, Cuvier's inquiries were centered, almost exclusively, "on the adaptations of function and organ within the living creature" (Russell, 1916, p. 34). For that, it is even possible to say that Cuvier's perspective, decidedly organicist, is best enunciated in that corollary of the principle of the conditions of existence which is the principle of the correlation of organs (*cf.* Coleman, 1964, p. 67):

> Every organized being forms a whole, a unique and closed system, in which all the parts correspond to each other, and converge to the same definitive action by a reciprocal reaction. (Cuvier, 1992 [1812], p. 97)

It is understandable, nonetheless, that Darwin preferred to invoke the principle of the conditions of existence. The reference to the environment that Cuvier makes therein seems to be aligned with the image of living beings that emerges from *On the Origin of Species*. Moreover, the Darwinian perspective suggests that biological structures are insidiously adapted to the pressing demands of the struggle for existence (*cf.* Caponi, 2020). Every detail of structure can be, then, either the response to a threat from the environment or the resource to take advantage of an opportunity offered by that environment, and such an idea seems to be better considered by the principle of the conditions of existence than by the principle of the correlation of organs. However, in the analyses of organic forms effectively developed by Cuvier, and especially in his paleontological reconstructions, references to the environment of living beings have always been very generic and even marginal.

Cuvier's studies, it is true, include some general references to the habitat of organisms. But, as Marjorie Grene pointed out, they did not go much beyond generalities such as "birds in the air, fish in the sea" (Grene, 2001, p. 188). That is to say, "fishes live in water – that's why they have gills, not lungs; birds live in the air – that's why they have light, air-filled bones" (Grene & Depew, 2004, p. 135). But, as Marjorie Grene (2001, p. 188) stresses, what mattered to Cuvier "first and foremost was the integrated, harmonious coordination of all the parts, each functioning to produce a functioning whole." Cuvier thought that for an organism to be able to respond to the demands of its environment, it must first be an organizationally possible structure. And that is why, from such a perspective, the anatomical structures that define how the animal links to its environment and its food sources were considered as the mere corollary of its internal organization. As far as this is concerned, and maybe contrary to Stephen Jay Gould's interpretation (*cf.* Gould, 2002, p. 294), in Cuvier, there is nothing similar to Darwinian "adaptationism." Cuvier was not an "externalist." For him, in understanding organic forms, the internal organizational coherence of the living being was more relevant than the contingencies of the environment.

A carnivore's stomach, said Cuvier (1805, p. 55) in his *Leçons d'Anatomie Comparée*, requires teeth and claws suitable to such a diet; and fish physiology, of course, is only viable in an aquatic environment. Nevertheless, as far as Cuvier's perspective is concerned, it is not necessary to go much further in these correlations between organization and environment (*cf.* Limoges, 1970, p. 40). In general, as far as a physiologist is concerned, to understand the functioning of the digestive and respiratory apparatus of any organism, it is sufficient to correlate these organ systems with "the nature of the molecules" that he must assimilate "either by respiration or by feeding" (Foucault, 1970, p. 68). But, although today these generalities may seem too obvious to have any significant theoretical impact, the actual results of Cuvier's investigations showed that if these principles of analysis were taken into account and applied with rigor and detail in morphological studies, the understanding of organic forms could progress as never before.

Considering identifiable but previously ignored organizational requirements, the comparative study of functional correlations between organs, and organ systems, of different kinds of animals would allow us to understand the variances of structure that those shown by those kinds of organic beings. This certainly contributed to the development of physiology. But, besides that, the postulation of a functional necessity that excluded any contingent nexus between the organizationally relevant parts of living beings was crucial for the development of Paleontology. Cuvier (1810, p. 330) argued that, given the organic correlations established by comparative anatomy, it would be possible not only to "recognize an animal by a single bone, by a single part of bone" but even to reconstruct such an animal, with geometrical certainty, departing just from that single fragment. If each part of an organism "stands in a necessary relation to all the others," then, and to a certain extent, "the whole can be inferred from any one of those parts and vice versa" (Cuvier, 1801, p. 68). Thus, under the perspective of this organicist functionalism, the reconstruction of a complete fossil from a fragment could be seen as the incontrovertible result of a physiological calculation of characters (Cuvier, [1812] 1992, p. 100).

In actual research, things were less impressive than reconstructing a complete animal from a single isolated tooth. However, Cuvier could justify this by appealing to the still imprecise and fragmentary character of our knowledge of the laws that ruled organic correlations. Moreover, despite these limitations, we already knew enough about these correlations to advance gradually in our paleontological investigations. For instance, the fragments of an herbivore's dentition could lead us to infer a particular conformation of the animal's digestive system, and the presence of the remains of horns and hooves could confirm that inference. In addition, this same dentition could also indicate that the animal under examination must have been a placental mammal. Another form of dentition could have indicated that the animal to be reconstructed was a carnivorous marsupial, which would also allow us to know something about the shape of its pelvis. Thus, the study of fossils was definitively removed from the fabulous domain, where any bizarre creature could have a place, and it was unquestionably placed in the field of positive science.

3.3 Geoffroy and the Unity of Type

The thesis that articulated all of Étienne Geoffroy Saint-Hilaire's morphological studies was that from mollusks to man, the configuration of all animals, despite the notorious variety of their forms, basically obeys a single plan of composition (Russell, 1916, p. 53; Schmitt, 2006, p. 214). The expression he used was "unity of organic composition" (*unité de composition organique*) (Geoffroy Saint-Hilaire, 1822, p. xxxiv). In his relatively early *Memoire sur les rapports naturelles des Makis-Lemur*, that idea, which in English came to be designated by the expressions "unity of plan" (Whewell, 1837, p. 456) and "unity of type" (Darwin, 1859, p. 206), was already clearly enunciated (Le Guyader, 1998, p. 35):

> It would seem that nature has confined herself within certain limits, and has formed all living beings on a single plan, essentially the same in its principle, but which she has modified in a thousand ways in all its accessory parts. If we consider particularly one class of animals, it is especially there that her plan will appear evident to us: we shall find that the various forms under which she brought each species into existence, all derive from one another; being sufficient for her to change certain proportions of the organs to adapt them to new functions, or to extend or restrict their uses. [...] The pouch of the marsupial females, a skin fold of great depth; the trunk of the elephant, an excessive prolongation of its nostrils; the horn of the rhinoceros, a considerable pile of hairs adhering to each other, and so on. Thus, the forms, in each class of animals, however, varied they may be, are all, finally, organs common to all: nature refuses to employ new organs. Thus, all the differences, even the most essential, which affect each family belonging to the same class, derive only from another arrangement, from a complication, from a modification, in short, of these same organs. (Geoffroy Saint-Hilaire, 1796, pp. 1–2)

This thesis, it is true, had not lacked respectable precedents (*cf.* Perrier, 1884, pp. 95–96). Isidore Geoffroy Saint-Hilaire (1847, pp. 142–145), son and foremost biographer of Étienne, counted eight authors who, in some way, had already enunciated it, glimpsed it, or presupposed it before his father proposed it in 1796. They were Aristotle, Pierre Belon, Isaac Newton, Buffon, Félix Vicq-D'Azir, Johann Gottfried Herder, Johann Wolfgang Goethe, and Philippe Pinel. But we must be cautious. We must also not get carried away by the "myth of the precursor." In a sense, it can be said that all these authors had glimpsed the idea of the unity of plan. But not all of them did so with clarity, and none succeeded in giving it a methodologically operational form, as Étienne Geoffroy Saint-Hilaire was able to do. And to achieve this, Geoffroy distinguished and formulated the two fundamental elements that came together in the idea of the unity of composition: the theory of analogs and the principle of connections.

In the theory of analogs is the assumption that the materials of which animals are composed are always the same (Geoffroy Saint-Hilaire, 1822, p. xxxi). However, for the correct understanding of such an idea, it is necessary to consider that with the word "materials," Geoffroy – as Goethe realized ([1832] 1837, p. 177) – alluded neither to the organs that are integrated into an animal nor to the tissues of which those organs are made. For Geoffroy, materials were the parts of which an organ or structure is composed. Thus, the materials of the skull are just the bones from which the skull is formed; the same can be said about the sternum or any other complex bony structures. Therefore, the complete anatomical description of any vertebrate's sternum or skull had to identify all those parts, at least at some stage in developing that structure. "The prediction to which this truth leads us," said Étienne Geoffroy Saint-Hilaire (1818, p. xxxii), is that "we shall always find, in each family, all the organic materials which we had perceived in another." It can be said, in this sense, that the theory of analogs functioned as a principle of parsimony that compelled us to assume that nature always works with a limited repertoire of materials. And our anatomical descriptions had to admit that "Geoffroy's razor," without denying that these materials could present different configurations and perform other functions.

Naturalists, said Étienne Geoffroy Saint Hilaire (1818, p. xxii) in the preliminary discourse to the first volume of the *Philosophie Anatomique*, must accept that "an organ, varying in its conformation, often passes from one function to another." They, he then went on to say, can verify this by following "the forefoot both in its various uses and in its numerous metamorphoses": by seeing it "successively applied to flying, swimming, jumping, running, etc.; being here a useful tool for searching, there a hook for climbing, elsewhere defensive or offensive weapons; or even becoming, as in our species, the principal organ of touch, and, consequently, one of the most effective means of our intellectual faculties" (Geoffroy Saint Hilaire, 1818, pp. xxii–xxiii). Geoffroy, as explained by Russell (1916, p. 305), "held that nature formed nothing new, but adapted existing materials of organization to meet new needs." And, with the principle of connections, to this parsimony of materials, Geoffroy would add a constancy in the way in which these materials could be disposed of (Fischer, 1993, p. 58): "an organ is altered, atrophied, annihilated, rather than transposed" (Geoffroy Saint Hilaire, 1818, p. xxxx).

However, to properly appreciate the impact of Geoffroy's anatomical philosophy, it is necessary to perceive that, to a certain extent, his theoretical and methodological principles were only turning explicit and giving due importance to presuppositions that, in a way, were already implicit in morphological studies carried out since the end of the eighteenth century. Assumptions that, at times, it is true were not rigorously respected and that from Geoffroy onward could be followed with greater rigor and greater coherence. Thus, a possible conflict of epistemological hierarchy between them and other methodological principles also followed by naturalists became evident. Such was the case of the functional correlations that Cuvier advocated as being more important than the morphological invariants emphasized by Geoffroy. His anatomical philosophy, we can also say, was not the product of the speculation of a visionary; instead, it was rooted in the actual practice of comparative anatomy. And that is why its impact on the development of such studies was unavoidable. Geoffroy brought to light something that the naturalists already glimpsed, thought vaguely and confusedly, in their morphological investigations (*cf.* Schmitt, 2006, p. 248). But he also showed, by the infallible rhetoric of concrete results, that the consistent application of these assumptions could produce important discoveries.

3.4 Conditions of Existence *vs.* Unity of Type

To understand the conflict that could arise between Cuvier's and Geoffroy's perspectives, it may be more helpful to think about paleontology than comparative anatomy itself. If the postulates of Geoffroy's anatomical philosophy are applied with minimum rigor, it is necessary to accept that the reconstruction of a fossil has to start before the functional correlations between the parts may be considered from the morphological restrictions imposed by the theory of analogs and by the principle of connections. A starting point implies admitting that the functional requirements on which the Cuvierian perspective would lead us to focus could only be satisfied within the framework of morphological constraints alluded to by these principles on which Geoffroy based his analyses. Cuvier ([1812] 1992, p. 97), on the other hand, argued as if paleontological reconstructions could be based on purely functional correlations (Caponi, 2008, p. 59). Some of these were very obvious, such as the correlation that a large and heavy humerus must correlate with the size and robustness of the acromion, clavicle, scapula, and other shoulder parts. And others could be more complex, as in the case of this one proposed by Cuvier himself:

> If the intestines of an animal are organized to eat exclusively fresh meat, its jaws must be constructed to devour the prey, its claws to grasp and tear it; its teeth to tear and divide the meat; the whole system of its organs of movement to pursue and reach it; its organs of sense to see it from afar, and it is even necessary that nature has placed in its brain the necessary instinct to know how to hide and to set traps for its victims. (Cuvier, [1812] 1992, pp. 97–98)

Thus, if we find fragments of a large carnivore jaw exhibiting saber-toothed tiger-like fangs, that will allow us to know a lot about the other organ systems of that animal: we will infer that it had claws and intestines consistent with a carnivorous diet, and we will also infer that that animal had a skull, and a neck, large and strong enough to support that jaw and those canines, making them work. We shall also infer something of the size of its humerus from the remainder of the scapula we find along with the jaw (*cf.* Cuvier, [1812] 1992, p. 100). But, more than how reasonable all this may seem, it is also necessary to admit that there is implicit a presupposition akin to Geoffroy's perspective. A presupposition without which all this reasoning makes no sense. Because it is evident that where there is a humerus, there is an acromion, a clavicle, and a scapula, it is also obvious that these elements keep certain relative positions among themselves similar to those they possess in other known species.

Not so many years after the death of Cuvier and Geoffroy, Thomas Huxley ([1856] 1898, pp. 433–435) would say that, for reconstructing a fossil, it was mandatory to consider both functional and morphological correlations – but for Geoffroy, the morphological correlations were the most important, the first to be considered. The reason was that such structural correlations determined the set and the relative position of the elements, pieces, or materials whose functional correlation had to be established at a second moment. If one assumes that an already unearthed humerus must be functionally correlated with a scapula to be found, one already presupposes that if there is a humerus, there is a scapula. Not to mention that one must also assume that if there is a left humerus and scapula, there must be a right humerus and scapula. From the perspective of anatomical philosophy, the theory of analogs and the principle of connections defined the conditions which were prevalent over the requirements foreseen in the principle of correlation of organs proposed by Cuvier.

It is true that Cuvier could think that these morphological constants, such as "where there is humerus, there is the scapula," always had a functional explanation to be found (*cf.* Guillo, 2003, p. 158; Amundson, 2005, p. 56). One of the conclusions that close the first volume of the *Histoire Naturelle des Poissons* stated this idea very well: "if there is a similarity between the organs of fishes and those of other classes, this can only be so insofar as there is a similarity of function" (Cuvier & Valenciennes, 1828, p. 406). But then, again, Geoffroy Saint-Hilaire (1829, p. 24) – who liked to quote this statement as a clear indication that Cuvier did not understand the real point of the question and that he was still imprisoned in a naive finalist perspective (Piveteau, 1961, p. 491) – might reply that what was at stake was not the simple global similarity of forms, but rather the identity and relative position of the materials, or parts, of which the anatomical structures were composed.

In this regard, Geoffroy could allude to the hand of the bats. This structure was composed of the same materials that made up the hand of a monkey (Geoffroy Saint-Hilaire, 1829, pp. 14–15), even if its function was more similar to that performed by the entire forelimb of some birds (Geoffroy Saint-Hilaire, 1829, p. 12). However, and this was not unimportant either, how, in bats and birds, this function was performed also seemed to be constrained by the composition of each structure (Geoffroy Saint-Hilaire, 1829, p. 13). Nor was it irrelevant that in many birds, the wing did not serve for flight: although its composition, and not its form, was the same as that of flying birds. Moreover, if we compare the hand of the monkey with the hand of the bat, or the hand of the pterodactyl with that of the lizard (*cf.* Cuvier, 1809, p. 434), we shall note analogies of structure, "homologies" it will be said later, without analogies of function. Let's compare the bat's wing with the hand of the pterodactyl and the forelimb of the lark as a whole. We shall see greater analogies of function between them than between the hand of the monkey and the hand of the bat, between the hand of the pterodactyl and that of the lizard, and between the arm of the lark and that of the penguin. Nevertheless, between the hands of the monkey and the bat, or between the hands of the pterodactyl and the lizard, and between the arms of the lark and the penguin, there are some quite remarkable structural resemblances.

Therefore, a direct and necessary correlation between the analogy of form and the analogy of function is impossible. On the contrary, function always comes after structure because of the restrictions that the theory of analogs and the principle of connections lead to foresee for all animals in general, or at least for each family or genus of animals. This, which even Cuvier himself had to accept, even if only tacitly, when testing the determination of a particular fossil, was one of the teachings of the anatomical philosophy that had the most significant impact, both in the development of comparative anatomy that preceded the emergence of evolutionism and in the evolutionist perspective itself. However, given that the functional requirements emphasized by Cuvier could not be ignored either, a pragmatic and eclectic attitude prevailed in natural history immediately before Darwinism, in which a morphological analysis would be considered successful to the extent that it showed that the forms under study conformed both to the requirements of the unity of type and to those of the conditions of existence (*cf.* Guillo, 2003, p. 160).

3.5 An Evolutionist Arrangement Between Cuvier and Geoffroy

Almost repeating something already said by Buffon (1753) in "The donkey," in his *Critique of Judgment*, Kant also dared to suggest that unity of type insinuated common descent:

> So many genera of animals share a certain common schema on which not only their bone structure but also the arrangement of their other parts seems to be based; the basic outline is admirably simple but yet was able to produce this great diversity of species, by shortening some parts and lengthening others, by the involution of some and the evolution of others. Despite all the variety among these forms, they seem to have been produced according to a common archetype, and this analogy among them reinforces our suspicion that they are actually akin, produced by a common original mother. (Kant, [1790] 1987, §80, p. 418)

However, even recognizing that these facts offered to "the mind a ray of hope, however, fain, that in their case at least we may be able to accomplish something with the principle of natural mechanism," Kant considered that, ultimately, this was impossible: "the technic that nature displays in organized beings" was something beyond the reach of explanations based on "forces governed by mechanical laws" as those involved in "crystal formations" (Kant, [1790] 1987, §80, pp. 418–419).

However, although strictly speaking, the theory that Darwin was to propose did not offer mechanical laws, it achieved this explanatory goal by postulating a purely natural process (natural selection) which, in addition to being governed by purely natural factors, could also account for the "technique of nature" and the fact that living beings are something close to what Kant called the "organized products of nature" (Kant, [1790] 1987, §66, p. 376).

In a conceptual flip-flop almost as unexpected and surprising as the maneuver of a conjuror, but not doubtful or illegitimate, the difficulty that the functional adequacy of organic structures posed to the thesis of common descent was thus erected as the key to explaining the diversification of living forms from a common ancestor: natural selection produces, preserves, and reformulates this adequacy, generating, almost simultaneously, this diversification. But it is important not to overlook that Darwin's proposed mechanism of transformation cannot be limited to operating on the internal correlation of organic parts and on those generic and obvious correlations between organs and the environment on which Cuvier's analyses were centered. Inevitably, natural selection has to go beyond, producing an adjustment between living beings and the demands of the environment much more thorough than that foreseen by pre-Darwinian natural history. Urged and pressured by a struggle for life, whose intensity, and even whose existence, had not been recognized by a natural history based on the idea of a natural economy in equilibrium, natural selection cannot limit itself to operating on the internal functional organization of living beings and their most obvious correlations with the physical environment (*cf.* Limoges, 1970). In addition, it will end up rewarding any modification of structure or behavior that, regardless of its organizational importance (Guillo, 2007, p. 80), confers on its bearers some advantage, however minimal it may be, in the struggle for existence.

The theory of natural selection makes it possible to overcome the difficulty that the adjustment of living beings to their conditions of existence poses to the evolutionist thesis. But this is achieved by assuming that those obvious correlations between form, function, and environment, which previous natural history already recognized, are the result of a process that, besides preserving and improving such correlations, also generates conformations and instincts whose necessity could hardly be derived from anything similar to the mere laws of correlation of parts postulated by Cuvier. Conformations and habits, or instincts, whose raison d'être must be sought in the demands and opportunities that the environment poses to living beings. It is precisely there that lies the difference between the terms in which Darwin initially posed his problem and the terms in which he finally solved it. The problem statement is, so to speak, Cuvierian, and it points to what any naturalist of the beginning of the nineteenth century could understand by "conditions of existence" (*cf.* Whewell, 1837, p. 472). But its resolution leads to understanding these conditions in a significantly different way: more as the adequacy of organic profiles to environmental requirements than as the mutual functional correlation of the parts. Thus, although Darwin continues to use the expression conditions of existence, also invoking Cuvier, the fact is that he ends up giving the term a meaning that is no longer Cuvierian. Darwin, as Russell (1916, p. 239) emphasized in *Form and*

Function, uses the expression "conditions of existence" as equivalent to conditions of life derived from the contingencies of the struggle for existence (*cf.* Darwin, 1859, p. 127).

In addition, besides answering the question of the conditions of existence, giving feasibility to the thesis of common filiation, the postulation of natural selection as a diversifying mechanism of living forms also allows accounting for a finding that was central to Geoffroy's anatomical philosophy. The adaptation of organic structures to the different conditions of existence of living beings was reduced to changes of forms occurring in a repertoire of structures whose elements were constant, even if their forms could vary significantly, and whose relative positions remained constant. This is so because natural selection, as follows from what Darwin (1859, p. 111) called the "principle of divergence," was a mechanism that, by modifying pre-existing structures in virtue of the demands of the struggle for existence, also diversifies them (Kohn, 2009). Natural selection does not produce changes in only one direction. Otherwise, phyletic diversification would not have occurred as it did. But this diversification of forms also has as its starting point a pre-existing form. It does not result from the advent of new structures: it arises from the modification and, eventually, the iteration of previous structures – pre-existing structures that change in form and function.

All this ratified the postulates of Geoffroy's anatomical philosophy: the unity of type and constancy of morphological elements modified under the diversity of conditions of existence. The forelimb of vertebrates, consistently with its same parts changed in the most diverse ways under the different functions that this limb could perform, was, of course, the most precise illustration of these principles of the anatomical philosophy. But, having explained this by appealing to that exigent and merciless process which was natural selection, Darwin had shown that this subjection of living forms to the morphological constraints evidenced in the unity of type was not in absolute conflict with the exigencies of the conditions of existence. On the contrary, to explain that all living forms were modifications of a single fundamental type, it was necessary to accept that this resulted from a process of form-derivation governed by factors that, in a certain way, resulted from these conditions of existence and obliged to satisfy them in a much more complex way than Cuvier had imagined. However, despite this, it can also be said that the synthesis Darwin proposes for the antithesis between the unity of type and conditions of existence ends up tipping the balance a little in Geoffroy's favor. And this is so not only because it is based on an evolutionary point of view that Geoffroy Saint-Hilaire (1833), unlike Cuvier, had already dared to sustain (*cf.* Caponi, 2011a).

Once, returning to the antagonism between Cuvier and Geoffroy that was pointed out by Whewell, Thomas Huxley ([1864] 1893a, p. 86) said that "the apparently divergent teachings of teleologists and morphologists were reconciled by the Darwinian hypothesis" (see also: Huxley, [1878] 1893b, p. 223). However, I dare say, in that reconciliation, the teleological perspective advocated by Cuvier had to yield on a crucial point: the analogy of form did not have to be explained by the analogy of function. On the contrary, it was better to explain it by common descent (Darwin, 1859, p. 206); and although this link between the unity of type and common descent was not a presupposition of Geoffroy's anatomical philosophy, that was not an evolutionary theory – the association between both notions put the unity of type in a place of preeminence. And that was so because the association between the unity of type and common descent came to be a critical key for phylogenetic reconstructions (Sober, 2008, p. 265; Caponi, 2011b, p. 48), and, consequently, the unit of type became, as Geoffroy wished, the obligatory initial reference for any study of form.

3.6 Darwin's Tactful Materialism

Apparently, Darwin would contradict what I am saying; for, after having referred to the unity of type and the conditions of existence as being the two great principles that govern the configuration of living beings, he says that according to his point of view: "the law of conditions of existence is the higher law; as it includes, thought the inheritance of former adaptations, that of unity of type" (Darwin, 1859, p. 206). Furthermore, in some literature not so recent but still highly regarded, it has been insisted on corroborating this, showing Darwin closer to Cuvier than Geoffroy. That was made by Stephen Jay Gould (2002, pp. 253–254) in *The Structure of Evolutionary Theory* and later endorsed by Ron Amundson (2005, pp. 102–103) in *The Changing Role of the Embryo in Evolutionary Thought*. And what Darwin said in this respect seems to confirm that this was the idea he had of his position. However, if we follow the line of analysis proposed by Edward Stuart Russell (1916, p. 305) in *Form and Function*, we may be led to think that, beyond Darwin's personal opinion, the approach to morphology that objectively derives from *On the Origin of Species* tends more toward Geoffroy's side than toward Cuvier's side.

The great research program outlined in the pages of Darwin's book is the complete reconstruction of the "tree of life" (*cf.* Darwin, 1859, pp. 484–485): that genealogical classification of all terrestrial life forms, extant and extinct, which, attending to morphological, biogeographical, and paleontological evidence, would also give us a phylogenetic representation of the evolutionary history of all those forms (Bowler, 1996, p. 7; Waters, 2003, p. 127; Caponi, 2011b, p. 3). And if functional requirements, already transformed by Darwinism into contingencies of the struggle for existence (Russell, 1916, p. 239), were to be cited in these phylogenetic studies, it would only be to explain morphological particularities understood as states derived from the same previous form (Caponi, 2011b, p. 81). In that framework, morphological similarities that only respond to functional similarities, such as those between the wing of the bat and the wing of the pterodactyl, could not be ignored, of course. Still, for evolutionists, the first and most important thing to do in those cases was to identify the structures from which each wing could be considered a derivation or modification (Caponi, 2011b, p. 70). The morphological convergence produced by the similarity of function had to be regarded as secondary to a divergence whose starting point was a previously defined repertoire of parts.

From this point of view, the evolutionary history of organic structures was the history of the changes of form of these structures resulting from the changes of functions they underwent in virtue of the different conditions of existence they had to attend to. Felix Anton Dörhn (1975, p. 60) explained this idea by appealing to the principle of change of function. But, in *On the Origin of Species*, Darwin (1859, p. 454) had already alluded to this change of functions, to which he resorted in a very clear way in *The Various Contrivances by which Orchids are Fertilized by Insects* (Darwin 1877, pp. 283–284). And it can be said that Dörhn's principle is nothing but the evolutionist formulation of one of the most immediate corollaries of the theory of analogs: functions can multiply, but not the elements that perform them. In a sense, by applying this principle enunciated by Döhrn, the evolutionists only continued the path initiated by Geoffroy. As Russell pointed out:

> The evolutionists followed Geoffroy rather than Cuvier. They laid great store by homological resemblances, and dismissed analogies of structure as of little interest. They were singularly unwilling to admit the existence of convergence, or parallel evolution, and they held very firmly the distinctively Geoffroyan view that nature is so limited by the unity of composition that she can and does form no new organs. (Russell, 1916, p. 305)

That is to say, Darwinian evolutionists tended to respect Geoffroy's razor: they considered that nature was an austere demiurge that tended to work with the same materials, modifying them indefinitely in virtue of the new functions that came to them in virtue of different circumstances and minimizing the production of new structures. But, in doing so, Darwin's followers did no more than heed what followed from the theses outlined in *On the Origin of Species*. By adopting that parsimony, they remain loyal to Darwin. However, and similar to what had already happened with Owen, in a natural history not entirely determined to break with natural theology, as was the case of the English natural history where Darwinism was called to irrupt, the approach with Cuvier that Darwin essayed was more comfortable than the association with the Robespierrean materialism of that impenitent follower of Napoleon Bonaparte who had been Étienne Geoffroy-Saint Hilaire.

References

Amundson, R. (2005). *The changing role of the embryo in evolutionary thought*. Cambridge University Press.

Bowler, P. (1996). *Life's splendid drama*. Chicago University Press.

Buffon, G. (1753). L'Ane. In G. Buffon (Ed.), *Histoire Naturelle Générale et Particulière* (Tome IV, pp. 377–405). L'Imprimerie Royale.

Caponi, G. (2008). *Georges Cuvier: Un Fisiólogo de Museo*. UNAM-LIMUSA.

Caponi, G. (2011a). Compensation and correlation of the organs in Étienne Geoffroy-Saint Hilaire's transmutationism. In J. Martínez-Contreraz & A. P. de Leon (Eds.), *Darwin's evolving legacy* (pp. 100–110). Siglo XXI.

Caponi, G. (2011b). *La Segunda Agenda Darwiniana*. Centro Lombardo Toledano.

Caponi, G. (2020). The Darwinian turn in the understanding of biological environment. *Biological Theory, 17*, 20–35.

Coleman, W. (1964). *Georges Cuvier: Zoologist*. Harvard University Press.

Cuvier, G. (1801). Extrait d'un Ouvrage sur les Espèces de Quadrupèdes dont on a Trouvé les Ossements dans l'Intérieur de la Terre Adresse aux Savants et aux Amateurs des Sciences. *Magasin Encyclopédique*, 7me année (tome premier), 60–82.
Cuvier, G. (1805). *Leçons d'Anatomie Comparée* (Tome I). Baudouin.
Cuvier, G. (1809). Sur Quelques Quadrupèdes Ovipares Fossiles Conservés dans les Schistes Calcaires. *Annales du Muséum d'Histoire Naturelle, 13*, 401–437.
Cuvier, G. (1810). *Rapport Historique sur les Progrès des Sciences Naturelles depuis 1789 at sur leur État Actuel*. L'Imprimerie Impériale.
Cuvier, G. (1817). *Le Règne Animal Distribué d'après son Organisation* (Tome I). Deterville.
Cuvier, G. (1992). *Recherches sur les Ossements Fossiles de Quadrupèdes* [Discours préliminaire, 1812]. Flammarion.
Cuvier, G., & Valenciennes, A. (1828). *Histoire Naturelle des Poissons* (Tome I). Levrault.
Darwin, C. (1859). *On the origin of species*. Murray.
Darwin, C. (1877). *The various contrivances by which orchids are fertilized by insects* (2nd ed.). Murray.
Döhrn, F. (1975). *Der Ursprung der Wirbelthiere und das Princip des Functionswechsles*. Engelman.
Fischer, J.-L. (1993). L'Anatomie Transcendante et le Concept de Récapitulation chez Étienne Geoffroy Saint-Hilaire. In P. Mengal (Ed.), *Histoire du Concept de Récapitulation* (pp. 55–68). Masson.
Foucault, M. (1970). La situation de Cuvier dans l'Histoire de la Biologie. *Revue d'Histoire des Sciences et leurs applications, 23*(1), 49–60.
Geoffroy Saint-Hilaire, É. (1796). *Memoire sur les Rapports Naturelles des Makis-Lemur*. L'Imprimerie du Magasin Encyclopédique.
Geoffroy Saint-Hilaire, É. (1818). *Philosophie Anatomique des organes respiratoires*. Baillière.
Geoffroy Saint-Hilaire, É. (1822). *Philosophie Anatomique des Monstruosités Humaines*. Chez l´ Auteur.
Geoffroy Saint-Hilaire, É. (1829). *Cours de l'Histoire Naturelle des Mammifères*. Pichon & Didier.
Geoffroy Saint-Hilaire, É. (1833). *Le Degré de Influence du Monde Ambiant pour Modifier les Formes Animales* (pp. 63–92). *Mémoires de l'Académie Royale de Sciences de Institut de France*.
Geoffroy Saint-Hilaire, I. (1847). *Vie, Travaux et Doctrine scientifique d'Étienne Geoffroy Saint-Hilaire*. Bertrand.
Goethe, J. (1837). Principes de Philosophie Zoologique par M. Geoffroy-St-Hilaire (suite et fin). In J. Goethe (Ed.), *Oeuvres d'Histoire Naturelle* (pp. 160–182). Cherbuliez. (Original work published 1832).
Gould, S. (2002). *The structure of evolutionary theory*. Harvard University Press.
Grene, M. (2001). Darwin, Cuvier and Geoffroy: Comments and questions. *History and Philosophy of the Life Sciences, 23*(2), 187–211.
Grene, M., & Depew, D. (2004). *The philosophy of biology: An episodic history*. Cambridge University Press.
Guillo, D. (2003). *Les Figures de l'Organisation*. PUF.
Guillo, D. (2007). *Qu'est-ce que l'Évolution?* Ellipses.
Huxley, T. (1893a). Criticism on the origin of species. In I. T. Huxley (Ed.), *Darwiniana* (pp. 80–106). Appleton. (Original work published 1864).
Huxley, T. (1893b). Evolution in biology. In I. T. Huxley (Ed.), *Darwiniana* (pp. 187–226). Appleton. (Original work published 1878).
Huxley, T. (1898). On the method of paleontology. In M. Foster & R. Lankester (Eds.), *The scientific memoirs of Thomas Henry Huxley* (Vol. I, pp. 432–444). Macmillan. (Original work published 1856).
Kant, I. (1987). *Critique of judgment*. Hackett. (Original work published 1790).
Kohn, D. (2009). Darwin's keystone: The principle of divergence. In M. Ruse & R. Richards (Eds.), *The Cambridge companion to the origin of species* (pp. 87–108). Cambridge University Press.

Le Guyader, H. (1998). *Geoffroy Saint-Hilaire: Un Naturaliste Visionnaire*. Belin.
Limoges, C. (1970). L'Économie Naturelle et le Principe de Corrélation chez Cuvier et Darwin. *Revue d'Histoire des Sciences et leurs applications, 23*(1), 35–45.
Perrier, E. (1884). *La Philosophie Zoologique avant Darwin*. Félix Alcan.
Piveteau, J. (1961). Anatomie Compare des Vertébrés. In R. Taton (Ed.), *La Science contemporaine I: le XIXe siècle* (pp. 485–501). PUF.
Reiss, J. (2009). *Not by design*. California University Press.
Russell, E. (1916). *Form and function*. Murray.
Schmitt, S. (2006). *Aux Origines de la Biologie Moderne*. Belin.
Sober, E. (2008). *Evidence and evolution*. Cambridge University Press.
Waters, K. (2003). The arguments in the origin of species. In J. Hodge & G. Radick (Eds.), *The Cambridge companion to Darwin* (pp. 116–137). Cambridge University Press.
Whewell, W. (1837). *History of the inductive sciences* (Vol. III). Parker.

Gustavo Caponi Born in Rosario, Argentina, in 1961, Gustavo Caponi attained his Philosophy degree at the National University of Rosario at the beginning of 1984. In 1992, he obtained his Ph. D. in Logic & Philosophy of Science at the State University of Campinas (UNICAMP), Brazil. Since 1993, he has taught at the Federal University of Santa Catarina (UFSC), Brazil. In February 2015, he became a Full Professor (Professor Titular) at the Philosophy Department. He is currently a researcher at the Brazilian National Council for Scientific and Technological Development (CNPq) and in the Philosophy Graduate Program of UFSC. His areas of work are Philosophy and the History of Biology. In addition, he was visiting scholar and visiting professor in various institutions, such as the REHSEIS team at Paris VII (September 1999 and August 2000), the EHESS of Paris (2004), the National University of Colombia (2004), the Université de Bourgogne (2007), and the IHPST of Paris I-Sorbonne (2011). He also taught at the Universidad Autónoma Metropolitana de México (UAM) (2014). Furthermore, he was visiting scholar at the University of Buenos Aires (March to August 2017) and at the IHPST of Paris I (September 2017 to February 2018). He published up to two hundred papers in collective works and academic journals and authored several books: *Georges Cuvier: un fisiólogo de Museo* (Universidad Nacional Autónoma de México: Mexico, 2008); *Buffon* (Universidad Autónoma Metropolitana: Mexico, 2010); *La segunda agenda darwiniana: contribución a una historia del programa adaptacionista* (Centro Lombardo Toledano: Mexico, 2011); *Função e adaptação na Biologia contemporânea* (Editora 34: São Paulo, 2012); *Réquiem por el centauro: aproximación epistemológica a la la Biología Evolucionaría del Desarrollo* (Centro Lombardo Toledano: Mexico, 2012); *Leyes sin causa y causas sin ley en la explicación biológica* (Universidad Nacional de Colombia: Bogotá, 2014); *El darwinismo de Ameghino* (Universidade Federal de Santa Catarina: Florianópolis, 2017); *Determinismo y organización: fundamentos y límites del programa de Claude Bernard* (Universidad Nacional de Colombia /Universidad El Bosque: Bogotá, 2018); and *Evolucionismo* (Editora Fibra, São Paulo 2021).

Chapter 4
An Amazing Journey: Darwin and the Fuegians

Héctor A. Palma

Abstract Between 1826 and 1836, His Britannic Majesty sent two maritime expeditions to South America with different objectives. In the first one (which reached Tierra del Fuego), the young Captain Robert FitzRoy, in command of the *Beagle*, embarked on four Fuegians. The first three Fuegians were taken aboard as hostages to force the other Fuegians to return a whaling boat they had stolen; the fourth was taken aboard under different circumstances. Finally, FitzRoy changed his mind and decided to take them to England to teach them English, religion, and some trades and return them to their homeland. Unfortunately, one of them died as soon as they arrived. The other three were repatriated on the second Southern Hemisphere voyage around the world between December 1831 and October 1836, with Charles Darwin as the naturalist on board. Darwin deals extensively and in detail in his voyage diary, describing the Fuegians with whom he established a certain relationship and the dramatic and complex process of repatriation. This lesser-known part of Darwin's voyage has generated multiple opinions and controversies for almost two centuries. This article analyzes the facts from the available direct sources and interprets them differently from the repeated in the historiographical tradition.

4.1 Introduction

On a sunny January morning in 1833, a British vessel sails alongside a smaller boat through one of the Southern Channels of the Tierra del Fuego. Through screams and smoke, the natives of the area quickly communicate with each other about the novelty, and dozens of canoes, with hundreds of them, emerge from observing the peculiar event. Curious and friendly, somewhat aggressive, others watch the smallest boat approaching the shore with three Fuegians (two men and one woman) returning to their homeland after almost 3 years of absence. To the surprise of their

H. A. Palma (✉)
National University of San Martín (UNSAM), Buenos Aires, Argentina
e-mail: hpalma@unsam.edu.ar

M. Elice Brzezinski Prestes (ed.), *Understanding Evolution in Darwin's "Origin"*, History, Philosophy and Theory of the Life Sciences 34,
https://doi.org/10.1007/978-3-031-40165-7_4

compatriots, who receive them almost naked, these three Fuegians dressed in European clothes with short hair, speak in English, and bring with them porcelain tea sets, bed linens, hats, and dresses. This unique scene, pretty well known, is only a tiny part of a larger story headed to oblivion at that hostile southern tip of South America.

It was common for European vessels to take onboard natives from visited regions for different purposes (Taladoire, 2017), which were, in most cases, neither noble nor innocent. Nevertheless, we still remember this case, especially because the naturalist on board would later become one of the most influential scientists of the modern world: Charles Darwin.

4.2 The *Adventure* and *Beagle* Expeditions

Between 1826 and 1836, His Britannic Majesty sent two sea expeditions to the Southern Hemisphere. During the first one (between May 1826 and October 1830), under Captain Philip P. King's (1791–1856) command, four Fuegian natives were brought aboard to England. Unfortunately, one died upon arrival, and the other three were repatriated during the second voyage (between December 1831 and October 1836) under Robert FitzRoy's (1805–1865) command. For the first voyage, two vessels were used, the *Adventure* and the *Beagle*, and for the second voyage, only the latter was used, which had been previously improved.[1]

The European power expeditions to several lands of the world, with military, commercial, and also scientific objectives, as well as their explorers' accounts, dating back to the sixteenth century or even earlier. But it was from the eighteenth century that the journey diary started to include a systematic report on fauna, flora, geology, and human groups' observation, as was the case with Darwin.

The experiences of the two extensive voyages were published in London in 1839 in three volumes under the title *Narrative of the Surveying Voyages of His Majesty's Ships Adventure and Beagle (1826–1836).* Volume I (in this paper identified as "a"), on the first expedition, was authored by Captain King, although much of it was written by FitzRoy. As in such edition, FitzRoy's accounts and thoughts are clearly distinguishable by being enclosed in quotation marks; the present chapter refers to them as authored by him – notwithstanding the obligatory citation as King's publication. Volume II (in this paper identified as "b"), on the second expedition, written by FitzRoy, also includes an extensive Appendix. Volume III (in this paper identified as "c"), written by Darwin, was later republished separately and under different titles in 1839, 1845, and 1860.

The double expedition is part of an extensive series of British journeys and reflects the expansion strategy designed and developed throughout the nineteenth

[1] In both cases minor vessels were temporarily included for specific tasks.

century (the "Imperial Century") with the already known outcome (see Powell, 1993). The objectives of the Beagle mission were:

> The object of the expedition was to complete the survey of Patagonia and Tierra del Fuego, commenced under Captain King in 1826 to 1830 – to survey the shores of Chile, Peru, and of some islands in the Pacific – and to carry a chain of chronometrical measurements round the World. (King et al., 1939c, p. 1; see also King et al., 1939b, chapter II, for full instructions)

At the beginning of the twentieth century, the British Empire dominated about 25% of the world's population and 20% of the Earth's land area, apart from other forms of diplomatic and commercial domination.

4.3 Four Fuegians on Their Way to London

On January 29, 1830, while the Beagle was conducting a coastal survey in the Tierra del Fuego, accompanied by some men, the master was sent to Cape Desolation in a whaleboat to look for a harbor and make some measurements. After no news from them for several days, only three men returned "in a clumsy canoe, made like a large basket, of wicker-work covered with pieces of canvas, and lined with clay, very leaky, and difficult to paddle" (King et al., 1939a, p. 392). They warned that the whaleboat had been robbed by the natives and that the master and the rest of the crew remained in that place.

Then, the captain sailed a boat with men, supplies, and equipment to the incident scene and began an obsessive boat search that lasted for many days. The sailors could only find boat pieces and other items from it in the hands of different groups of natives. As the days went by, FitzRoy wrote:

> I became convinced that so long as we were ignorant of the Fuegian language, and the natives were equally ignorant of ours, we should never know much about them, or the interior of their country; nor would there be the slightest chance of their being raised one step above the low place which they then held in our estimation. (King et al., 1939a, p. 405)

To force them to give back the boat, they captured a group of Fuegians and held them hostage, but almost all escaped. Only three children remained on board. Two of them were sent back to shore, and only one girl, about 8 years old, remained:

> [. . .] she seemed to be so happy and healthy, that I determined to detain her as a hostage for the stolen boat, and try to teach her English. (King et al., 1939a, p. 409)

That girl, named *Yokcushlu*, was of the *Alacaluf* group ("alikhoolip," in FitzRoy's writing). They named her Fuegia Basket to remember the basket-like canoe they use and "by which we received intelligence of the loss of our boat" (King et al., 1939a, p. 411). She was the first native aboard.

On March 3, while some crewmen were busy building a new boat, some Fuegians came near. The captain strived to scare them away by ordering a gun to be fired over their heads. But he thought that if he could keep:

> [. . .] one of these natives on board, there would be a chance of his learning enough English to be an interpreter, and that by his means we might recover our lost boat, I resolved to take the youngest man on board [. . .]. (King et al., 1939a, p. 410)

He was named *El'leparu*, also an *Alacaluf*, about 26 years old, and was renamed York Minster (a nearby island name). He was the second native taken aboard.

Some days later, they saw two canoes they were chasing, as they were sure those who had stolen the boat were in them. However, they finally captured only one native. FitzRoy does not mention his original name, but he calls him Boat Memory, an *Alacaluf* of about 20 years old.

The *Beagle* continued with the intended tasks, and a few months after the incident with the stolen boat, the fourth native was brought to the ship, but under different circumstances. Finally, on May 11, three canoes carrying natives eager to exchange items intercepted the *Beagle*:

> We gave them a few beads and buttons, for some fish; and, without any previous intention, I told one of the boys in a canoe to come into our boat, and gave the man[2] who was with him a large shining mother-of-pearl button. The boy got into my boat directly, and sat down. Seeing him and his friends seem quite contented, I pulled onwards, and, a light breeze springing up, made sail. Thinking that this accidental occurrence might prove useful to the natives, as well as to ourselves, I determined to take advantage of it. [. . .] 'Jemmy Button,' as the boat's crew called him, on account of his price, seemed to be pleased at his change, and fancied he was going to kill guanaco, or wănăkāye, as he called them – as they were to be found near that place. (King et al., 1939a, p. 444)

He was named *Orundellico*, from the Yamana or Yagan group, about 14 years old, the fourth Fuegian aboard. Now the previously stolen boat obsession, FitzRoy's initial plan of holding the natives' hostage, gave place to a most ambitious project: to take them to England, to teach them English, religion, and some trades. As a nineteenth-century man, FitzRoy did not hesitate to get approval for such a project, tearing the Fuegians from their culture and beloved ones, without knowing where they were going, for what, and for how long. He did not even express any qualms about the little girl, who would be far from her family at such a young age.

4.4 Observations About the Fuegians

Direct testimonies of Darwin and FitzRoy about the Fuegians should be treated with historical and contextual caution.

There is no doubt that they were formed out of imperialistic and racist European prejudices, which were also sustained by the science of the time.[3]

[2]Later, Jemmy explained the man was one of his uncles.

[3]The science of anthropology with its characteristic and unmistakable racist character, was emerging by them with the creation of the first expert societies and journals, such as *La Société ethnologique de Paris* (1839), followed by *The American Ethnological Society* in New York (1842) and *The Ethnological Society of London* (1843). In 1859, by Paul Broca's initiative, appeared the *Société d'Anthropologie de Paris*, as a spin-off of the previous *Société de biologie* (Fletcher, 1882, p. 141).

In addition, although there are many derogatory and negative comments about the Fuegians, others are more benevolent and even laudatory, especially regarding the four people involved in this story. Thus, a strategic and intentional selection of quotations could help prove opposing theses easily. A balance between naive historicism, where everything is fair in the name of context and historical peculiarity, on the one hand, and an out-of-time interpretation with current values and categories, on the other, is required – as presented and defended in this chapter.

The testimonies above deal with some common topics about the supposed inferior and savage condition of the natives: the nudity/precarious clothing, the painting on their bodies and faces, their animal-like appearance, their tendency to steal, the controversial question of cannibalism, and the supposedly inferior and primitive language. Let us have a look at passages reflecting some of these considerations.

The voyage journals have many mentions of natives' savagery. FitzRoy states, for instance: "They are creatures barely superior to the brute creation" (King et al., 1939b, p. 6).

Darwin expressed in a similar way when he observed a group of Fuegians in Good Success Bay for the first time:

> [. . .] I could not have believed how wide was the difference between savage and civilized man [. . .]. One of our arms being bared, they expressed the liveliest surprise and admiration at its whiteness, just in the same way in which I have seen the ourang-outang do at the Zoological Gardens. (King et al., 1939c, p. 205)

Yet, in another one, he wrote:

> Viewing such men, one can hardly make oneself believe that they are fellow-creatures, and inhabitants of the same world. (King et al., 1939c, p. 208)

And a few months later:

> I believe, in this extreme part of South America, man exists in a lower state of improvement than in any other part of the world. (King et al., 1939c, p. 230)

But there are also different opinions:

> I do not think that our Fuegians were much more superstitious than some of the sailors. (King et al., 1939c, p. 215)

In his other fundamental book, the *Descent of Man*, Darwin also expressed in an ambivalent way:

> He who has seen a savage in his native land will not feel much shame, if forced to acknowledge that the blood of some more humble creature flows in his veins. (Darwin, 1871, vol. II, p. 404)

And also:

> [. . .] I was incessantly struck, whilst living with the Fuegians on board the "Beagle," with the many little traits of character, shewing how similar their minds were to ours [. . .]. (Darwin, 1871, vol. I, p. 232)

There is a noticeable difference between the general observations about the Fuegians and the more individual and personalized descriptions of the natives who were part of the journey.

Comments on Fuegia and York are scarce. And as regards York, they are almost always negative (although Darwin was a little more benevolent than FitzRoy) not only on his characteristics but also on his relationship with the Englishmen and the other Fuegians. On the other hand, for Boat Memory, FitzRoy only has words of praise. Finally, Jemmy seems to have been the favorite. There is evidence that he had established a caring relationship with his teachers in England, with Darwin, with some members of the *Beagle* crew, and above all with FitzRoy.

When FitzRoy refers to his treatment of the natives, he shows himself as cautious and careful. Several times he points out he had tried to avoid situations that seemed to lead to moments of conflict with the natives (e.g., King et al., 1939b, Chapter I). He also mentions that he had found himself pushed to show one's strength, especially with those who did not yet know the power of firearms, to dissuade them from undertaking any attack with tragic consequences. However, he narrates at least one extremely violent event while searching for the stolen boat, when a native was shot dead.

The idea that natives were thieves was part of the usual considerations. Although FitzRoy attributes thefts to the Fuegians' fascination with certain items, it was something quite common among the natives themselves. The most potent or most numerous groups used to steal whatever they could get from the weakest. Jemmy's people were constantly afraid that the *Oens* (natives also named Selk'nam, guanaco hunters but not navigators) would come down to rob them, which was quite common.

Cannibalism or anthropophagy was a war ritual practice widely known by the Europeans and attributed to conquered peoples throughout the world, sometimes with credible documentation and references, sometimes with little evidence (Harris, 1991). European culture considered that practice a typical trait of primitive peoples, together with human sacrifice. Dozens of travelers' stories helped to install this belief. The too-much painted and naked Fuegians in a hostile land, thousands of miles from the "center" of the world, with very basic technology, with incomprehensible – to the European – customs and languages, were the suitable candidates to be blamed for cannibalism. Human zoos had spread out in Europe by the second half of the nineteenth century, attracting attention and provoking fascination in public that those exposed human beings were cannibals (see Blancel et al., 2002). About the subject, FitzRoy said:

> The acts of cannibalism occasionally committed by their countrymen, were explained to me in such terms, and with such signs, that I could not possibly misunderstand them [. . .]. (King et al., 1939b, p. 2)

Darwin also gives a similar version, and both narrate that in ties of food shortage, they used to eat the eldest women of their tribes:

> The different tribes when at war are cannibals. From the concurrent, but quite independent evidence of the boy taken by Mr. Low, and of Jemmy Button, it is certainly true, that when pressed in winter by hunger, they kill and devour their old women before they kill their dogs: the boy, being asked by Mr. Low why they did this, answered, "Doggies catch otters, old women no." This boy described the manner in which they are killed by being held over smoke and thus choked; he imitated their screams as a joke, and described the parts of their bodies which are considered best to eat. Horrid as such a death by the hands of their friends and relatives must be, the fears of the old women, when hunger begins to press, are more painful to think of; we were told that they then often run away into the mountains, but that they are pursued by the men and brought back to the slaughter-house at their own fire-sides! (King et al., 1939c, p. 214)

However, there are very different considerations on this issue. Fitz Roy says that Jemmy

> [...] in telling this horrible story as a great secret, seemed to be much ashamed of his countrymen, and said, he never would do so – he would rather eat his own hands. (King et al., 1939b, p. 183)

FitzRoy also comments that Jemmy never wanted to eat land birds since, as he said, "they eat dead men". He did not even dare mention his own dead friends.

Likewise, Lucas Bridges, an expert in Fuegians' culture and idiosyncrasy, vehemently denies that they had that custom and assures us they rejected even the idea of eating animals that might have previously eaten human flesh, such as vultures. He also speculates that the Europeans would have shaped that theory because of their own prejudices and a mutual misunderstanding of language (Bridges, 1952, p. 25).

Another persistent belief supported the poverty and gutturality of the Fuegians' language. Darwin said, "The language of these people, according to our notions, scarcely deserves to be called articulate" (King et al., 1939c, p. 205). This opinion has to be contextualized for a period when the absence of linguistic tools, prejudices of the time, and the fact of being faced with an absolute strange language. Today, we acknowledge that the language was far from being a poor or simple one, as Bridges indicated more than 70 years ago:

> Yagan or Yamana – English dictionary, written by my father comprises no less than thirty-two thousand words and inflections, which could have been considerably increased keeping the right language. (Bridges, 1952, p. 27)

4.5 Scientific Description of the Fuegians

In addition to numerous descriptions of the Fuegians' traits, characteristics, and customs from various sources, physiognomic and phrenological studies are undertaken on Jemmy, Fuegia, and York in the *Narrative* Appendix. Physiognomy was the name of a discipline of science used to detect people's character and skills in facial traits. Phrenology was a method to identify, through a craniological map, people's mental, intellectual, and moral capacities, which were thought to be located in different areas of the brain.[4]

The three Fuegians' results indicate the following.

Yokcushlu, a female, 10 years of age:

- Strong in attachment.
- If offended, her passion's strong.
- A little disposed to cunning, but not duplicity.
- She will manifest some ingenuity.
- She is not at all disposed to be covetous.
- Self-will, at times very active.
- Fond of notice and approbation.
- She will show a benevolent feeling when able to do so.
- Strong feelings for a Supreme Being.
- Disposed to be honest.
- Rather inclined to mimicry and imitation.
- Her memory good of visible objects and localities, with a strong attachment to places in which she has lived.
- It would not be difficult to make her a useful member of society in a short time, as she would readily receive instruction.

Orundellico, a Fuegian, aged 15:

- He will have to struggle against anger, self-will, animal inclinations, and a disposition to combat and destroy.
- Rather inclined to cunning.
- Not covetous; not very ingenious.
- Fond of directing and leading.
- Very cautious in his actions: but fond of distinction and approbation.
- He will manifest strong feelings for a Supreme Being.
- Strongly inclined to benevolence.
- May be safely intrusted with the care of property.
- Memory, in general, good; particularly for persons, objects of sense, and localities.
- To accustomed places he would have a strong attachment.
- Like the female, receiving instruction readily, he might be made a useful member of society; but it would require great care, as self-will would interfere much.

El'leparu, about 28:

- Passion's very strong, particularly those of an animal nature; self-willed, positive, and determined.
- He will have strong attachment to children, persons, and places.
- Disposed to cunning and caution.
- He will show ready comprehension of things and some ingenuity.
- Self will not be overlooked, and he will be attentive to the value of property.
- Very fond of praise and approbation and of notice being taken of his conduct.
- Kind to those who render him a service.
- He will be reserved and suspicious.
- He will not have such strong feelings for the Deity as his two companions.
- He will be grateful for kindness, but reserved in showing it.

---

[4]The analysis of observable and measurable traits of human characteristics used to set hierarchies (racial or other) between superior and inferior individuals was noticeable throughout the nineteenth and early twentieth centuries. In addition to those mentioned: the various forms of craniometry, or, in the last third of the century, criminal anthropology and biotypology (see Gould, 1993).

- His memory, in general, good: he would not find natural history, or other branches of science, difficult, if they can be imparted to him; but, from possessing strong self-will, he will be difficult to instruct and will require a great deal of humoring and indulgence to lead him to do what is required.

(King et al., 1939b, Appendix, p. 147)

4.6 In England

Immediately after they arrived in England, FitzRoy arranged a second inoculation; the first had been in Montevideo to protect the Fuegians from the contagion that had proved fatal to other natives from distant lands when they were brought to Europe. Two days later, they were taken to a:

> [. . .] quiet farmhouse, where he hoped they would enjoy more freedom and fresh air and at the same time run less risk of infection than in a populous port city, that would stimulate their curiosity. (King et al., 1939b, p. 33).

Soon after, in November 1830, Boat Memory contracted smallpox and died shortly after that. FitzRoy took advantage of Reverend W. Wilson's offer (of the Church Missionary Society) to receive the Fuegians into his parish, in Walthamstow, near central London. Rev. Wilson promised, "[...] that he would talk to the master of the Infant School about taking them into his house, as boarders and pupils" (King et al., 1939b, p. 10). FitzRoy would pay for their room and board, the schoolmaster's work, and unforeseen expenses.

During the year they spent in London, the Fuegians learned the basics of Christianity, English, and some skills and trades. In addition, they received valuable items and tools as gifts to take back home to improve their lives in such hostile regions. However, they also received rather frivolous gifts that showed the ignorance of British people not only about idiosyncratic and cultural differences but also about those natives' country of origin. They even had a friendly encounter with King William IV and his wife Adelaide, who gave Fuegia a hat and money, according to FitzRoy (King et al., 1939b, p. 12).

The truth is that when they were ready to embark, extra boats were required to carry all those gifts to the *Beagle*:

> [. . .] clothes, tools, crockery-ware, books, and various things which the families at Walthamstow and other kind-hearted persons had given. In the small hold of the Beagle, it was not easy to find places for the stowage of so many extra stores [. . .]. (King et al., 1939b, p. 16).

The sailors themselves were aware of this absurd situation and made fun of it:

> [. . .] at the expense of those who had ordered complete sets of crockery-ware, without desiring that any selection of articles should be made. (King et al., 1939b, p. 16)

4.7 Coming Home

When they arrived in Tierra del Fuego, the captain wanted to disembark the natives with their respective groups, that is, Jemmy in one region, Fuegia, and York in the other, but York told FitzRoy that he would rather live with Jemmy along with his now fiancée, Fuegia. They reached Woollya (a deep bay to the west of what is now Navarino Island), chose a free space to set up a camp, posted guards, and made all the necessary arrangements to receive the visit of hundreds of natives.

At first, only a few arrived, but soon, hundreds of natives reached the site. FitzRoy wrote that they camped some distance away for his relief, so he did not have to make any further effort to prevent thefts. As a result, the group of about 30 men could devote themselves to cutting wood from a relatively distant forest, preparing the land for a vegetable garden, and building the three Fuegian tents (called "wigwam"), one for Matthews, one for Jemmy, and the other for York and Fuegia. It could take an hour to set one up, and they were only used for a couple of days. However, these three tents were stronger and sturdier, so it took several days to complete the task.

While the activities to complete the settlement were in progress, the area filled with canoes and Fuegians, and Jemmy began to get annoyed with his compatriots' queries:

> [. . .] a deep voice was heard shouting from a canoe more than a mile distant [. . .] upon a repetition of the shout, exclaimed "My brother!" He then told me that it was his eldest brother's voice, and perched himself on a large stone to watch the canoe, which approached slowly, being small and loaded with several people. [. . .] When it arrived, instead of an eager meeting, there was a cautious circumspection which astonished us. [. . .] Animals when they meet show far more animation and anxiety than was displayed at this meeting. [. . .] Jemmy was evidently much mortified, and to add to his confusion and disappointment, as well as my own, he was unable to talk to his brothers, except by broken sentences, in which English predominated. After a few minutes had elapsed, his elder brother began to talk to him; but although Jemmy understood what was said, he could not reply. York and Fuegia were able to understand some words, but could not or did not choose to speak. (King et al., 1939b, p. 209)

Darwin also repeatedly records the dramatic situation of forgetting their native language. There is a great deal of written material regarding asymmetric relationships across cultures. Anthropology has widely dealt with it, especially since the late nineteenth century. In those days, there was little left to know about the geography of the Earth or exotic native groups, and the European colonial expansion had reached its height. There was no uniform pattern of these always uneven processes: plain and straightforward extermination of cultures in some cases, different degrees of survival in others, domination, and resistance, along with alteration or adaptation of native cultures after their contact with exotic cultures. The issue of Jemmy is previous to the dramatic and fast process where many peoples from Southern Patagonia have simply vanished. But the fact that they were coming back to their homeland dressed in strange clothes and that they had forgotten part of their mother tongue and knew very little of the foreign one, is an unquestionable symbol of acculturation (Berry, 2006). Thus, no culture is imposing itself on another; there is just a loss.

As the settlement organization went smoothly, FitzRoy decided to complete some pending west surveys, leaving the three Fuegians and Reverend Matthews – who felt confident of his task – in their new home. When he returned a few days later, everything seemed fine, but Matthews negatively evaluated his experience with the Fuegians. He recounted that he had been constantly mistreated and harassed and feared for his life. Therefore, FitzRoy decided to embark Matthews on the *Beagle*, hoping for another opportunity in the future.

4.8 Last Encounter with Jemmy

In March the following year, FitzRoy returns to Woollya on the *Beagle*, and there he has the last and touching encounter with Jemmy:

> He was naked, like his companions, except a bit of skin about his loins; his hair was long and matted, just like theirs; he was wretchedly thin, and his eyes were affected by smoke. We hurried him below, clothed him immediately, and in half an hour he was sitting with me at dinner in my cabin, using his knife and fork properly, and in every way behaving as correctly as if he had never left us. He spoke as much English as ever, and, to our astonishment, his companions, his wife, his brothers and their wives, mixed broken English words in their talking with him. [...] Jemmy gave a fine otter skin to me, which he had dressed and kept purposely; another he gave to Bennett. Next morning Jemmy shared my breakfast, and then we had a long conversation by ourselves; the result of which was, that I felt quite decided not to make a second attempt to place Matthews among the natives of Tierra del Fuego. [...] I cannot help still hoping that some benefit, however slight, may result from the intercourse of these people, Jemmy, York, and Fuegia, with other natives of Tierra del Fuego. Perhaps a ship-wrecked seaman may hereafter receive help and kind treatment from Jemmy Button's children; prompted, as they can hardly fail to be, by the traditions they will have heard of men of other lands; and by an idea, however faint, of their duty to God as well as their neighbour. That Jemmy felt sincere gratitude is, I think, proved by his having so carefully preserved two fine otter skins, as I mentioned; by his asking me to carry a bow and quiver full of arrows to the schoolmaster of Walthamstow, with whom he had lived; by his having made two spear-heads expressly for Mr. Darwin; and by the pleasure he showed at seeing us all again.
>
> As nothing more could be done, we took leave of our young friend and his family, every one of whom was loaded with presents, and sailed away from Woollȳa. (King et al., 1939b, p. 323)

Darwin describes the reunion and farewell with Jemmy in similar terms and concludes:

> I do not now doubt that he will be as happy as, perhaps happier than, if he had never left his own country. Everyone must sincerely hope that Captain Fitz Roy's noble hope may be fulfilled, of being rewarded for the many generous sacrifices which he made for these Fuegians, by some shipwrecked sailor being protected by the descendants of Jemmy Button and his tribe! (King et al., 1939c, p. 229)

4.9 After the Beagle

The lives of the protagonists of our story took different directions. Darwin and FitzRoy increased their profound differences over time, especially after the former published his theory of evolution. None of them ever saw Jemmy again. Nevertheless, over the years, Darwin became one of the most prominent scientists of modern times. FitzRoy remained in history for his voyages, his meteorology contributions, and his performance as governor of New Zealand.

Almost nothing is known about York's life, but that other natives murdered him. However, as regards Fuegia, there is more information.

Darwin collected an account from Captain Sulivan when he was devoted to the exploration and study of the Islas Malvinas. He says he heard a seal hunter telling that, probably in 1842, while he was on the western side of the Strait of Magalhães, he was stunned when he saw a savage woman who spoke in English. Undoubtedly, it was Fuegia. She was also seen in Ushuaia in 1873 and by Captain Giacomo Bove while commanding the "Argentine Southern Scientific Expedition" in 1882. Finally, Lucas Bridges had the opportunity to see her on her deathbed in 1883, surrounded by the love of her daughter and her second husband.

However, the story of Jemmy continued to be tragically linked to the English people. Between 1848 and 1851, a retired British officer called Allen Gardiner, committed to religious preaching in Patagonia and who had founded the *Patagonian Mission* (from now on, "the Mission") in 1844, tried to find Jemmy. Still, he and all the expeditioners died in the *Beagle* area (Bridges, 1952; Hazlewood, 2000).

In October 1854, the Mission sent a large vessel named Allen Gardiner to accomplish the failed intention of its founder. Unfortunately, due to a storm, the ship had to veer to Keppel Island (one of the Islas Malvinas), where the missionaries found an excellent refuge and decided to settle there. A year later, in November 1855, the Gardiner arrived in Woollya to fulfill the second part of Gardiner's plan. There they found Jemmy and, as it is told, everyone was astonished when he stood in front of the ship's commander – William Parker Snow – and spoke a few words in English. Snow offered him to move with his family to Keppel, but Jemmy, who by this time had two wives and several children, declined the invitation, and the vessel returned to Keppel (Bridges, 1952).

Soon after, the Mission, under Reverend George Despard's spiritual guidance, got several Yagans to travel to Keppel. Jemmy finally agreed to travel with his family in 1857. However, he quickly felt annoyed by the missionaries' instructions and expressed his desire to return to his land. A few days later, he was taken back to Woollya. His relationship with these Englishmen was not as he remembered it from his youth. According to the Mission reports, the Englishmen began to understand the Yahgan language, and the natives learned English. Encouraged by those achievements, the Mission decided in October 1859 to take the final step: to settle a mission in Woollya. However, nothing happened as expected (Bridges, 1952; Hazlewood, 2000).

4.10 Woollya Massacre

When the natives, returning from Keppel Island, were preparing their bundles to disembark, one of the crewmembers reported that some possessions had been stolen, and the stolen items were found in the natives' bundles. The situation became tense with those natives; one of them attacked Captain Fell. Later, it also affected those who were waiting in the surrounding area, including Jemmy Button, who was "the most insolent with his endless and insatiable demands and his bad temper when he was not pleased; there is no doubt that he had been extremely spoiled during the previous visits of the Mission," (Bridges, 1952, p. 36). Despite these drawbacks, they succeeded in building a shed within a week and held their first church service in the Tierra del Fuego on November 6, 1859. However, as soon as the service began, the Englishmen were attacked and murdered by many natives. Cole, as mentioned above, terrified on the boat, could only watch the shocking scene, hide, and flee. Strangely, they spared his life, and he lived among the natives for about 3 months until he was rescued.

The Woollya massacre had caused a strong repercussion in London. Although most people attributed those events to the savagery of the Fuegians, others, such as the former captain of the Gardiner, William Snow, were very critical of the Mission. They took the Fuegians to Keppel Island, and with the excuse of instructing them in more civilized practices, they took them for farm work and other tasks; they were ripped from their own lands to enslave them. Nevertheless, Rev. Snow expressed the need to respect the natives' will even if one was convinced that such an action did not represent their own good.

A few years later, in 1884, Jemmy died from one of those epidemics that began to annihilate the Fuegians.

4.11 History Review

Much has been written about the Fuegians' case: that they were taken hostage; that it was a cultural experiment developed by FitzRoy – some people have described it as a "philanthropic capture" (Penhos, 2018); that they were subjected to an acculturation process without the slightest respect for their idiosyncrasy; that it was part of the imperial plan to promote communication with the natives to get the control and dominance of different regions of the planet; that it was in the end, a sample of the empire arrogance that overlooked the rights of those people they simply considered inferior; and that it was part of a giant sequel of kidnappings perpetrated by European seamen. Thus, all those assertions may be factual in some way, but, at the same time, if taken in isolation, they are just partial, plain, and linear. A full understanding of those facts requires a comprehensive evaluation, with no politically correct or out-of-time passions and with no naive readings, at the same time (Palma, 2021). A plausible reinterpretation faces two fundamental limitations.

First, the available direct sources, the words of FitzRoy and Darwin, are tinged with all the beliefs, prejudices, and scientific knowledge of the time. However, they can be considered honest and transparent. In reading them, we do not perceive a trace of cynicism, double, subliminal, or insidious messages. On the other hand, the absence of direct testimonies from the natives means that the experiences, feelings, reactions, and deep motivations of the natives can only be captured by us from a certain human empathy (precarious and insecure in terms of our cultural distance).

But there is another limitation: out of available direct sources, it is possible to defend different interpretations of past events by applying a more rigorously methodological selection of quotations with their opinions.

Darwin and FitzRoy's texts show at different times some ambiguity or change of mind about the temper of the natives. Still, they also alternate between optimism and pessimism about the possibility of "civilizing" the natives. In continuity with other researches that have been conducted, this one found that such tensions and ambiguities result from these men who (with their significant and irreconcilable differences, which increased over time) are living the profound European tensions of the nineteenth century: sincere and well-meaning but ethnocentric; convinced of humanity union but unable to understand the reason why there is such a perceived enormous difference with the Fuegians; persuaded by the benefit they believe they are offering to the natives, but without cultural or scientific tools to understand the reason why their enterprise is doomed to failure; and aware of their own cultural superiority (and responsibility) compared to the rest of the peoples, but unable to understand the respect for other cultures and idiosyncrasies, and a lot less to critically consider the imperial plan they are taking part in, with the necessary objectivity or awareness (see Palma, 2021).

Perhaps the tension between the racist Darwin (as could not be otherwise for a nineteenth-century European citizen) and the always profound and combative antislavery and abolitionist Darwin is like the tip of an iceberg that hides the rest of the tensions, ambiguities, and doubts. According to Gould (1993), he was not a supporter of irremediable inequality like biological determinists. On the contrary, he was a meliorist, meaning that he was convinced that the world tends to improve and that human activity and effort can contribute. At the same time, he had a paternalistic attitude toward the so-called inferior peoples. This blend of meliorism and paternalism can lead to very scornful opinions of these people and the confidence that they will improve. Indeed, the meliorist may wish to eliminate cultural traditions and be cruel and inflexible in his lack of esteem for differences. Still, he sees the savages both as primitive due to their social circumstances and as beings capable of improving. As a matter of fact, of Westernizing or, more properly, Europeanizing, FitzRoy, who strictly rejected slavery, took part in this paternalistic attitude when he took the Fuegians on board in an attempt to "Westernize" them.

Now let's review some topics that, due to over-repetition, have been installed in the on-the-subject narrative.

Several elements show that this was an important episode and not just another case in an extensive series of kidnappings perpetrated by European expeditions worldwide. First, the extensive passages, entire chapters at times, that both FitzRoy and Darwin dedicated to the Fuegians in the *Narrative*. Second, FitzRoy was willing to pay not only for their education but also for the expensive repatriation voyage. The third is the enormous expenditure of men and time to build and organize the mission in Woollya. Fourth, the treatment they received was also peculiar. Although they were subjected to a painful and unfair acculturation process, the British Royal couple received the Fuegians in an atmosphere of certain respect (at least according to the available accounts). Finally, although they became a constant peculiarity for Londoners who knew of their arrival, they were not exhibited as attractions or exotic phenomena at fairs or in human zoos, which unfortunately began to spread gradually in Europe but were discreetly placed in boarding schools.

It has been repeated time after time that the Fuegians have merely been held as hostages. It is true that FitzRoy first expressed his intention to take some natives hostage, but at least four questions can be made out of this interpretation of the facts. First, after that initial intention, FitzRoy changed his mind until he finally decided to take them to England for his civilizing project. Second, it would make no sense to take hostages to England. Third, only the first three were embarked according to the initial intention of forcing the boat's return, but Jemmy was boarded more than 3 months later and in a different context. Fourth, according to FitzRoy's testimony, the Fuegians seemed pleased on board. Although FitzRoy's version might be doubtful, the truth is that the four Fuegians did not escape. And it is also true that other natives who were forcibly taken on board at the initial moment of the boat search, though under guard, quickly escaped.

Another element that has been installed due to too much repetition is that FitzRoy "bought" a native for a mother-of-pearl button.[5] FitzRoy's account of this episode is very brief and somewhat ambiguous (still, his words can be reinterpreted given the communication difficulties between British people and the Fuegians and the cultural and idiosyncratic differences). What Bridges asserts: "[...] no Indian would have sold his son not even for the Beagle itself with everything it contained on board" (Bridges, 1952, p. 22). They were exchanging different objects for fresh fish with the Fuegians, and in that context, the mother-of-pearl button was part of the items given, but it never constituted the payment for the boy who got aboard the boat. They were simply successive events in time but did not take part in a causal chain. These quite familiar exchange situations, where fish were thrown from the boats toward the vessel and pieces of cloth, nails, and some beads from the vessel to the canoes, were quite chaotic and spontaneous to confirm that there was an agreement by which a boy was worth a button. The crew's over-repetition of this forced, scornful, and equivocal interpretation of the situation, plus FitzRoy's account in the same sense, helped install it in history.

[5]This version comes from FitzRoy's own account and is later repeated by Darwin (King et al., 1939b, c). It was later taken up by many authors such as Huxley and Kettlewel (1965) and Stone (1981).

4.12 Jemmy, a Man from Nowhere

It is certain that out of the three Fuegians that returned to the Tierra del Fuego, the outstanding case that undoubtedly motivates us to keep mentioning them is Jemmy's, even after almost 200 years. He returns again and again for nearly 30 years to occupy the core of a tragedy despite, or maybe just for that reason, his place in a cultural limbo from where he never managed to get out.

Hundreds of anonymous Fuegians or natives from other regions were habitually embarked by Europeans. But Jemmy Button has a first and last name from the very beginning. Although fictional and scornful, it is true. But Jemmy Button (who is no longer even Orundellico) is a figure built and rebuilt by third parties with no self-intervention, without having a say. However, there is a large combination of factors that interact in his figure: FitzRoy's missionary-religious-evangelizing plan in tension with his racial and ethnocentric prejudices; his frustrated desires and voluntarism; the emotional impact of meeting the savages of the Tierra del Fuego on the young Darwin; interests and political strategies of the British Empire and the *South American Mission Society* in South America; and the re-readings of social scientists who, with a strategic selection of sites from the original sources, show, now with first and last name, the systematic oppression and annihilation of the original peoples. But Jemmy is also someone who has lost much of his mother tongue and could not acquire good English. Everyone says something about Jemmy, or they make him say things, but we are not sure of what he has said. Who is Jemmy, apart from the stories we have reviewed? Definitely, he has not been the mythological and romantic "noble savage" some people believed to see, overwhelmed by his naivety by the voracious empire. Nor has he probably been the savage with no limits or scruples that others perceived. But who is Jemmy Button?

Jemmy was deeply astonished by what he had seen and learned during his trip to London. For him, that was a strange and stunning world where he was attended and, in a certain sense, cared for and appreciated. However, an impossible-to-overcome tear took hold of young Jemmy as he came to know that world to which he did not belong and would never belong but which attracted him. A tear deepens as the leap into another culture occurs in his adolescence in search of his identity and place in the world. Jemmy returned as a young man from nowhere who had forgotten much of his native language but taught his people a few English words (King et al., 1939a, b, c).

He tried to convince FitzRoy and the Englishmen on several occasions that his people were wonderful and full of virtues. We can easily infer from his words; however, reality often annoyed him and fustigated his own as a way of showing an apology, like when someone needs to apologize to a stranger for what a little boy could have done. The last time he met Fitz Roy on the *Beagle*, his brother Tommy, fearing that Jemmy would not leave the ship, shouted his name in English. Would he always make his people call him "Jemmy Button"? What effect would have had his English on his people and the fact that – more than 20 years later – those strangers would have come asking for him by his (false) name? It is quite probable that Jemmy has built his identity trapped in that shocking youth experience that definitely took him away from his former life but did not give him another. Might he have told his compatriots stories (real or fabled) and wonders of his remarkable journey and those people he had met, those new and far away friends? Might he have told them that he used to treat them like old comrades? Might he have told them that he talked as equals with the highest chief of the foreigners?

More than 20 years after this youthful experience, Jemmy felt powerful when these strangers came back looking for him and called him by his English name. However, Jemmy cooperated little and reluctantly with those he did not know from his past and who had neither the intentions nor the human qualities FitzRoy and his other former friends used to have. He and the other Fuegians were now expected to work as servants and stay in the same place for a long time. He expected special, warm, and exclusive treatment from these Englishmen, as he had received more than 25 years ago, but this was not the case. He felt disappointed and humiliated and could no longer tell his people fantastic stories about his power over the Englishmen. The conditions for tragedy were in motion until it finally occurred. Perhaps an accumulated resentment, a sense that he had not been recognized as he deserved, and his failure to stop his people might have led him to support the course of such brutal events, if only by omission.

His belief that the Englishman respected and recognized him was perhaps the only way to survive the violent acculturation process he had been subjected to. But that fantasy had blown into a thousand pieces in Jemmy's mind.

As in ancient Greek tragedies, everything was triggered as was expected. The personages inevitably found themselves on opposite sides. There were no gods involved here – as in the classic format– but in any case, the condition for the final tragedy had begun to unfold almost 30 years before, silently and inevitably in that young boy who, trapped in his cultural limbo, had ceased to be what he used to be forever but never came to be what he imagined he could.

4.13 The Journey in History

After leaving the Tierra del Fuego, they sailed north across the Pacific Ocean and returned to England, crossing the Indian Ocean and passing through Southern Africa. Over time, Darwin became one of the most influential scientists in modern history. The theory of evolution not only permanently changed biology but also triggered the most important cultural, philosophical, and anthropological revolution that can be attributed to a scientific theory.

The voyage on the *Beagle* was the best opportunity for Darwin to train his mind scientifically. It flooded him with perturbing questions about the natural world and provided him with an enormous amount of biological, geological, and paleontological information, as well as the fossils and specimens he collected in different regions. The interaction with different types of human societies, which sowed in a nineteenth-century Englishman so many doubts about the origin and evolution of humanity, enriched and strengthened his vision and interests. The finds in Punta Alta (Argentina), Patagonia, and the Galápagos Islands would later become essential pieces of the evolutionary puzzle. In his own words:

> The voyage of the Beagle has been by far the most important event in my life, and has determined my whole career [. . .]. I have always felt that I owe to the voyage the first real training or education of my mind; I was led to attend closely to several branches of natural history, and thus my powers of observation were improved, though they were always fairly developed. [. . .] The glories of the vegetation of the Tropics rise before my mind at the present time more vividly than anything else; though the sense of sublimity, which the great deserts of Patagonia and the forest-clad mountains of Tierra del Fuego excited in me, has left an indelible impression on my mind. The sight of a naked savage in his native land is an event which can never be forgotten. (Darwin, 1892, p. 61)

References

Berry, J. W. (2006). Acculturation: A conceptual overview. In M. H. Bornstein & L. R. Cote (Eds.), *Acculturation and parent-child relationships: Measurement and development* (pp. 13–30). Lawrence Erlbaum.

Blancel, N., et al. (2002). *Zoos humains*. La Découverte.

Bridges, L. (1952). *El último confín de la tierra*. Emecé.

Darwin, C. R. (1859). *On the origin of species by means of natural selection or the preservation of the favored races in the struggle for life*. J. Murray.

Darwin, C. R. (1871). *Descent of man, and selection in relation to sex*. J. Murray.

Darwin, F. (Ed.). (1892). *Charles Darwin: His life told in an autobiographical chapter, and in a selected series of his published letters*. J. Murray.

Fletcher, R. (1882). Paul Broca and the French School of Anthropology. In *The saturday lectures, delivered in the lecture-room of the U. S. National Museum under the Auspices of the Anthropological and Biological Societies of Washington in March and April 1882* (pp. 113–142). D. Lothrop & Co./Judd & Detweiler.

Gould, S. J. (1993). *The mismeasure of man*. W.W. Norton Company.

Harris, M. (1991). *Cannibals and kings: Origins of cultures*. Vintage Books.

Hazlewood, N. (2000). *Savage: Survival, revenge and the theory of evolution*. Sceptre.

Huxley, J., & Kettlewel, H. (1965). *Charles Darwin and his world*. Thames & Hudson.

King, P. P., FitzRoy, R., & Darwin, C. R. (1939a). *Narrative of the surveying voyages of His Majesty's Ships Adventure and Beagle (1826–1836)* (Vol. 1). Henry Colburn.

King, P. P., FitzRoy, R., & Darwin, C. R. (1939b). *Narrative of the surveying voyages of His Majesty's Ships Adventure and Beagle (1826–1836)* (Vol. 2, with appendix). Henry Colburn.

King, P. P., FitzRoy, R., & Darwin, C. R. (1939c). *Narrative of the surveying voyages of His Majesty's Ships Adventure and Beagle (1826–1836)* (Vol. 3). Henry Colburn.

Palma, H. (2021). *Savage and civilized. Darwin, Fitz Roy and the Fuegians*. Biblos.

Penhos, M. (2018). *Paisajes con figuras. La invención de la Tierra del Fuego a bordo del Beagle (1826–1836)*. Ampersand.

Powell, J. (1993). *Britain's imperial century, 1815–1914: A study of empire and expansion*. Barnes & Noble.

Stone, I. (1981). *The origin: A biographical novel of Charles Darwin*. Doubleday.

Taladoire, E. (2017). *Cuando los indígenas descubrieron el viejo mundo (1493–1892)*. Fondo de Cultura Ecconómica.

Héctor A. Palma is full professor of Philosophy of Science at the Universidad Nacional de San Martín (Argentina). He has devoted himself to the history and philosophy of biology, particularly evolutionism, Darwinism, and its derivations. He currently directs a research project at the Laboratory of Research in Human Sciences (UNSAM-CONICET), entitled: "Studies on eugenics, gene editing and biological metaphors in social sciences and humanities."

Part II
Constructing a Theory

Chapter 5
Darwin's First Writings: From the *Beagle* Voyage to His Transmutation Notebooks (1837–1839) and Essay (1844)

Janet Browne

Abstract This chapter discusses the importance of Charles Darwin's writing and recording practices both for the development of his ideas about evolution by natural selection and in relation to nineteenth-century scientific practices. It is argued that Darwin's recording activities on the *Beagle* voyage contributed to British colonial appropriation of information. Darwin's successive notes and essays are discussed as a form of thought on paper.

Imagine you are Charles Darwin seated at your desk in the London house that you have rented since you returned from the *Beagle* voyage. You are 29 years old and feel grateful that you do not have to find a job. Your father has given you a substantial family inheritance and now you are an independently wealthy "gentleman of science," mingling with natural philosophers, scientists, and professional men in the clubs and learned societies of the day. You are extremely glad the voyage is over (those ocean waves). Yet it was the most transformative experience of your life. Ever since your return to England, you have been secretly pondering subversive questions about the origin of species and have already filled several notebooks with your thoughts and readings on the question. It is September 1838.

You pick up a book that has been loaned to you by your older brother Erasmus, "for amusement" you think. It is by Thomas Robert Malthus, about human population statistics, dating from the previous century but recently reissued in much longer format, packed with information about unrestrained population growth and the problems this brings to nations that cannot grow sufficient food to feed the masses (Malthus, 1798). You know this is a hot political topic because Malthus claims that it is the indigent poor who reproduce most prolifically. Malthus states that many of these impoverished people will die from starvation or illness and that this is a part of the natural order of things, what he calls "checks" designed by the Christian God to keep the overall numbers of people in Britain at a steady level.

J. Browne (✉)
Harvard University, Cambridge, MA, USA
e-mail: jbrowne@fas.harvard.edu

M. Elice Brzezinski Prestes (ed.), *Understanding Evolution in Darwin's "Origin"*, History, Philosophy and Theory of the Life Sciences 34,
https://doi.org/10.1007/978-3-031-40165-7_5

You turn the pages and what do you discover?

Darwin famously read Malthus's book in September 1838 and took from it the central concept of differential survival that lay at the heart of what would become his theory of evolution by natural selection. The moment was recorded in a notebook (labeled by Darwin as Notebook D) in an entry dated "28th." Too many individuals are born, he paraphrased from Malthus. There is a war in nature, a struggle for existence. In the fight to live, the worst adapted or weakest organisms tend to die first, leaving only the better forms, the healthiest or better adapted. These survivors would be the ones that generally had offspring. If such actions were repeated, over and over, organisms would become ever more appropriately suited to their conditions of existence. He called this process "natural selection" in analogy with the selection that agriculturists and animal breeders used on domestic beings to produce a wide variety of different breeds.

Darwin's actual words do not read like a great discovery. His thoughts were complex and hard for us to disentangle. Here, the extract has been shortened to help the sense:

> 28th [...] in nature production does not increase, whilst no check prevail, but the positive check of famine & consequently death [...] Population is increase at geometrical ratio in far shorter time than 25 years – yet until the one sentence of Malthus no one clearly perceived the great check amongst men [...] even one species of hawk decreasing in number must affect instantaneously all the rest. – The final cause of all this wedging, must be to sort out proper structure, & adapt it to changes. [...] One may say there is a force like a hundred thousand wedges trying [to] force every kind of adapted structure into the gaps in the oeconomy of nature, or rather forming gaps by thrusting out weaker ones. (Notebook D, pp. 134e–135e)[1]

The several layers of meaning embedded in these sentences have been well explored by generations of Darwin scholars (Kohn, 1980; Hodge & Kohn, 1985). The wedging that he talks about refers to competition for what we now call niches in the ecosystem: any slightly improved variant can force its weaker sibling out of its place in nature.

It is perhaps easier to understand the impact of this moment by referring to Darwin's autobiographical reflection, written at the end of his life:

> Being well prepared to appreciate the struggle for existence [...] it at once struck me that under these circumstances favourable variations would tend to be preserved, and unfavourable ones to be destroyed. Here, then, I had at last got a theory by which to work. (Barlow, 1958, p. 120)

We now know from extensive work by historians and sociologists of science that the process of discovery is frequently different from the language of retrospective reflection. Karl Popper described this as the distinction between the logic of discovery and the logic of justification. Darwin's autobiographical words are retrospective

[1] The "e" indicates that at some later point (probably when writing his Sketch of 1844) Darwin excised these pages. They were located only in the twentieth century and reinserted into sequence. All the notebooks and associated materials are transcribed and annotated in Barrett et al. (1987).

and teleological. He constructed a story that he wanted to believe, in which he had pursued numerous objective investigations until a theory was created that united all the observations and proved to be correct. In reality, Darwin proposed countless theories while he collected information: his information gathering was integral to his theorizing. And his theory of natural selection did not emerge fully formed like a modern oven-ready meal. He could not have arrived at the theory of evolution by natural selection without the experiences of the *Beagle* voyage to draw on, his university training, the careful thought and gathering of information since his return from the voyage, and his reading of Malthus in September 1838: all of these, and more, encouraged an interpretation of the natural world based on struggle and competition for survival. Malthus provided the key to the much larger problem he sought to resolve.

Many fine scholars have discussed the intellectual trajectory of Darwin's ideas both before and after reading Malthus (for example, Young, 1969; Bowler, 1984; Schweber, 1985). This chapter takes a more material view and focuses instead on the important acts of writing and recording. It treats Darwin as a recorder of nature and dwells on the mutual reinforcement of his private writings with those that he felt could be read by an audience. It moves from his time on the *Beagle* to the so-called "notebook" period, then to his pencil sketch of a complete theory made in 1842, to the fuller version written out in 1844 that is commonly called the "Essay of 1844."

The focus on writing has several justifications. First, the fact that Darwin wrote down most of his thoughts, ideas, and feelings is exceedingly helpful for historians, although of course he did not intend this. Second, the compilation of lists and catalogues was an essential part of nineteenth-century science, especially during expeditions, and Darwin's own writings show that, like his contemporaries, he believed science advanced primarily through the documentation and exchange of written views. He needed written material as an archive of information. Third, and mostly applicable to Darwin's time on the *Beagle*, the scientific records of a voyage can broadly be considered part of the colonizing process. Darwin was a white European male aiming to understand and collect materials from lands foreign to him, which can now be regarded as a form of intellectual bioprospecting. His writings – just as much as his collection of natural objects – appropriated local information, while his European eyes reinterpreted indigenous natural phenomena by repositioning them as "facts" in western thought. Writing and recording, we now know, can lay claim to natural phenomena in much the same way as removing plants, rocks, and animals from a region. Darwin's activities on the *Beagle* contributed to the construction of global natural history as an object of western inquiry. Lastly, the extensive written record that Darwin left behind not only provides an invaluable resource for understanding the development of the theory of evolution but also supplies rare insight into creative thought in science. Darwin's early writings can be seen as among the most highly significant documents available to historians for comprehending science's past.

So, it is relatively easy to think of Darwin as a writer as well as a thinker. A brief biographical look at his early life reveals that he regularly wrote things down. Important dates are that Darwin was born in 1809 and died in 1882. He traveled on the *Beagle* voyage with Captain Robert FitzRoy from 1831 to 1836. His major book, *On the Origin of Species by Means of Natural Selection, or the Preservation of Favoured Races in the Struggle for Life* was published in 1859. Another significant book was *The Descent of Man and Selection in Relation to Sex*, published in 1871. He researched and published many other books and articles, especially in geology and botany.

In his youth, he wrote many letters to his sisters and friends, a habit that continued during his university days, first at Edinburgh University and then at Cambridge University in England 1828–1831. In September 1831, he received an invitation to travel on the *Beagle*, a British surveying voyage commissioned to complete a survey of the coastline of Tierra del Fuego and subsequently to take a series of chronometric measurements around the globe for refining the British understanding of longitude. The voyage lasted from December 1831 to October 1836. This voyage had geographical and geopolitical aims. Such aims were fully integrated with scientific objectives because imperial endeavors of the past relied heavily on what we now call the field sciences, including the collection of a wide range of natural history materials and activities like observing, mapping, measuring, and recording. The *Beagle* voyage, like many others, included a variety of scientific undertakings. Captain FitzRoy had offered to take a naturalist with him to collect specimens and observe natural history particulars. That naturalist was Darwin. Historians now know a great deal about his time on board, his thoughts, his adventures, and his collecting activities from the wide range of written materials that he diligently composed.[2]

Darwin quickly found out that it was important to capture his voyage in words. Before the ship sailed, he visited experienced naturalists, travelers, and museum curators to find out what was most needed and how best to supply it. Everyone told him that the written record about objects was just as significant as any object itself: without written details, unnamed specimens would appear on a naturalist's table disconnected from their habitat, original colors, behavior, and any other ecological context. It was equally important to compile numbered lists of collected materials and to tag each object with the same number, for otherwise how would a naturalist be confident that the specimen was the same one described in the notes? These recommendations were obvious, but it was good that Darwin absorbed them at the start. Generations of naturalists, from the seventeenth century onward, compiled written records – and frequently drawings too – of their observations, collections, and dissections; and these notebooks offer many insights into the actual practice of respected scholars such as Reaumur, Spallanzani, or Humboldt.

Primed with advice from experienced naturalists, he began his voyage determined to record everything. At one level this was manifested in his field notebooks. Before he sailed, he purchased a quantity of small leather-bound notebooks that he could slip into his saddle bag or pocket while on expeditions. More than 20 of these survive

[2]Important online resources are the Darwin Correspondence Project (https://www.darwinproject.ac.uk), published in paper format as Burkhardt et al. (1985–present) and the website edited by John van Wyhe, Darwin Online (http://darwin-online.org.uk/).

in the Darwin archive. He used them to record his observations in the field – the equivalent of a laboratory notebook. He transferred information at regular intervals into larger ledgers, some of which comprised handwritten catalogues of specimens subdivided into animals, plants, geological, etc. For instance, he compiled a numbered list of the birds he collected, leaving sufficient space for remarks if he was so inclined. This very careful documentation of his collecting activities was necessary in order to make sense of his specimens when he returned. He envisaged donating much of the material to specialists or perhaps the British Museum for accurate identification and to enhance knowledge of the natural world. Some specimens he intended to keep in order to work on them himself. The written notes would be essential for describing living beings and geological items in his collections.

With the benefit of hindsight, we can see that such collecting and exploring depended not just on the ability to pick up and transport natural history specimens, although this was very important. Just as important was the ability to transfer information onto paper – to make a written record, or as Bruno Latour calls it, an "inscription" (Latour, 1988). Such inscriptions made the phenomena accessible to western inquiry. This was a form of appropriation that is only now being fully appreciated by scholars (Cañizares-Esguerra, 2006; Leitão & Sanchez, 2017). Moreover, historians have known for a long time that the imperial frame of mind during the nineteenth and twentieth centuries involved a passion for information presented in the form of documents, statistical charts, topographic measurements, paintings and drawings, and as time moved onward, in photographs. To imperial eyes, knowledge was not really knowledge until it was on paper. This again may sound obvious, but it is significant to note that the process of inscribing information on to paper for readers in the First World was an unmistakable way of excluding – or transforming – the knowledge of indigenous peoples.

The point is that Darwin not only made a large collection of animals, plants, and geological specimens during the *Beagle* voyage but that he also made extensive written notes about these objects as well as the landscapes, habitats, and peoples he encountered. He imitated the captain and composed a daily diary or "logbook" about his activities. He wrote long letters home to his family, university friends, and to his professor at Cambridge University, John Stevens Henslow, who had acquired for him the invitation to join the *Beagle* and was able, throughout the voyage, to answer scientific questions and receive the crates of material that Darwin shipped back.

Darwin's earliest speculations about the possibility of evolutionary change (transmutation) were probably made on the *Beagle*. Even so, he had no reason to become unorthodox in his thinking. As a young man, he admired the works of the English theologian William Paley who regarded the shape and form of living organisms as perfectly designed by the Creator (Brown, 1986; Fyfe, 1997). Each of the "perfect adaptations" or "contrivances" that Paley described in the natural world was understood by him as features specially created by God to fit each living being to its place in an overall divine plan. Paley's argument was that the world was so full of design there must be a designer in the same way as a watch necessarily required the existence of a watchmaker.

Yet Darwin knew of radical alternatives to this natural theological view. As a young man at university, he read and considered the works of transformists such as Jean Baptiste Lamarck, Buffon, and his own grandfather Erasmus Darwin (Hodge, 1985; Desmond & Moore, 1990). On board the *Beagle*, he encountered a long attack on Lamarck printed in the second volume of Charles Lyell's *Principles of Geology* (1830–1833) – and this attack explained transmutation and its dangers so fully that he may well have begun to ponder the issue. He started to notice instances where Paley's idea of perfect adaptation was plainly inadequate or wrong. In recording these observations, he occasionally speculated about the origin of living beings and adaptation to habitat. One especially revealing passage concerned the bird species he found on the Galápagos archipelago. After describing in his bird catalogue the multiple kinds of finch on the archipelago, he expressed the view that some seemed to be confined to individual islets: "If there is the slightest foundation for these remarks the zoology of Archipelagoes – will be well worth examining; for such facts would undermine the stability of Species" (Barlow, 1963, p. 262). It is likely that this was written after the ship's visit to the Galápagos in September 1835 and before his arrival back in the United Kingdom in October 1836.

By the time he returned from the *Beagle* voyage, Darwin's writing ambitions had greatly enlarged. He aimed to write a book about the geology of the voyage and, in a separate volume, publish illustrations of the most unusual organisms he had personally collected. Just before returning he also agreed to convert his daily journal into a descriptive natural historical account of the voyage that would be published alongside Captain FitzRoy's narrative of the voyage. All this written work shows that on board the ship, Darwin transformed himself into an author as well as a thinker.

On returning from the *Beagle* voyage, and for many years after, Darwin continued this habit of recording his thoughts. He used the remaining small leather-bound notebooks that he had employed on the voyage: indeed every excursion to a library or scientific meeting was treated like one of his previous natural history collecting trips. It is not certain exactly when he became convinced that species could have originated without divine agency. Early in 1837, he began recording a flurry of such ideas in his notebooks. He labeled the notebooks A to E, and then M and N, now known as the Transmutation Notebooks. Notebook A mostly concerns geology. M and N delve into religion, morals, and metaphysics. These notebooks were transcribed in their entirety, along with associated papers, and with all the deletions and insertions recorded, in a milestone publication in 1987 (Barrett et al., 1987).

From the moment of opening Notebook B, in March 1837, he expressed the belief that evolution had taken place, not just among the Galápagos Island birds but everything, including humans. Clearly, his intention was to record instances of possible transmutation and search for a mechanism that would drive the change. Page after page, he built theories that stretched as far as his imagination would take him. He thought about humans as much as animals and plants. He weighed up religious and moral questions and investigated metaphysics. He began an additional notebook to list the books he should read. Taken collectively, these notebooks reveal one of the most thorough research programs in the history of science, encompassing his notes on library work, quotations from conversations and letters, information about people to contact, extracts from periodicals, personal thoughts, questions, and reflections. Scholars are frequently dazzled by the intensity and range of Darwin's inquiry.

The notebooks also reveal that even before he devised the theory that ultimately appeared in *Origin of Species*, Darwin imagined important parallels between changes in domestic breeds of plant and animal and changes in wild species. That parallel lay at the heart of all his early notebook writings. It was not shaken by reading Malthus. In fact, Malthus's account of differential survival gave him the mechanism of "selection" to make this parallel real. It remained deeply significant for the rest of his life, that is, the analogy between artificial selection (choices made by breeders in the domesticated world) and natural selection (where nature does the selecting). From the start, he also understood that the history of life could be visualized as a branching tree. The well-known diagram of what looks like an evolutionary tree appears early in Darwin's Notebook B, written in 1837 (Fig. 5.1). It depicts a branching system of connections that implies continuity between species. It was drawn long before he had a final theory of natural selection in mind. It can perhaps best be thought of as a sketch illustrating the way extinction

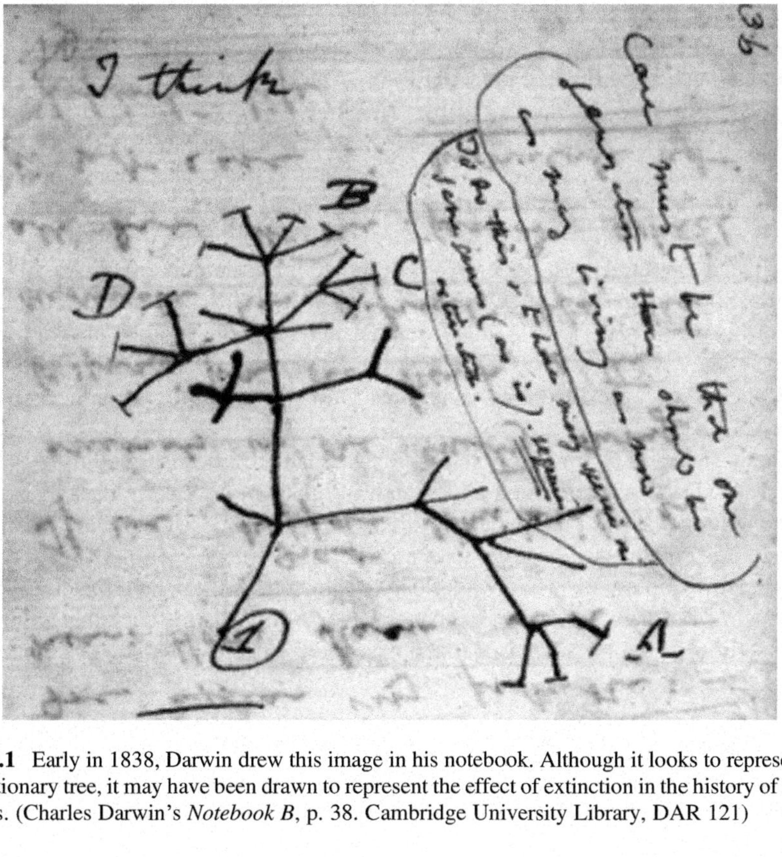

Fig. 5.1 Early in 1838, Darwin drew this image in his notebook. Although it looks to represent an evolutionary tree, it may have been drawn to represent the effect of extinction in the history of living beings. (Charles Darwin's *Notebook B*, p. 38. Cambridge University Library, DAR 121)

may have cut away at the tree of life to create divergences between the major groups as captured by classification schemes.

The Galápagos birds were classified in March 1837 by John Gould, a taxonomist from the Zoological Society of London, who also helped Darwin with his illustrated book *The Zoology of the Voyage of H.M.S. Beagle* (1839–1843). Gould stated that there were several species of finch, not simply a series of geographical varieties as Darwin had initially thought. Each species had a beak adapted to eat either insects, cactus, or seeds and some appeared to be confined to individual islets. Surprised, Darwin mulled this information over. If each island had its own birds, as Gould suggested, his speculations about the instability of species were truer than he thought. Could the similarities be explained if the finches had diversified from a common ancestor? (Sulloway, 1982a, b)

Darwin deliberately kept his notebook ideas about transmutation secret. He realized the need to be cautious. He told the geologist Charles Lyell, who had become a close friend, that he was filling "note-book after note-book [. . .] with facts which begin to group themselves clearly under sub-laws." Displaying extraordinary self-discipline, he worked intensively on these ideas in private.

And so, in September 1838, he read Malthus.

Four years later, he felt ready to write a short sketch of the theory that he had come to call "descent with modification." His notebooks provide the only record of the ways his thoughts were expanding and developing (Hodge, 2009).

He had also been busy in personal terms. During that 4-year period, he married his cousin Emma Wedgwood, and by 1842 they had two small children (William, b. December 1839; Anne, b. March 1841) and another on the way (Mary Eleanor, b. September 1842, d. October 1842). In September 1842, the family moved from London to a big house in Kent with extensive grounds. He had written and published three books. Physically, the overwork was taking its toll. Darwin was intermittently unwell, and the family consequently spent much of the summer of 1842 at his wife's parents' home in Staffordshire where he could be away from the polluted London air and (as his wife hoped) relax a little from this demanding publishing program. Along the way they visited Darwin's parental home in Shrewsbury where his father Dr. Robert Darwin resided. It is likely that Darwin wished for his father's medical advice.

There, successively in Shrewsbury and Maer Hall, Staffordshire, during May and June 1842, Darwin wrote out a first sketch of his theory.[3] It comprises 37 foolscap pages, written on poor-quality paper with a soft pencil and in many places is difficult to read. It is compressed in style, with much erasure and correction. It seems to be a hasty memorandum of what was clear to himself, rather than material for the convincing of others. Several of the pages have writings on the back that are marked for insertion in the text. The manuscript is accompanied by a single unnumbered

[3]Published by F. Darwin (1909); the manuscript of the 1842 sketch is in Cambridge University Library, Archives, call mark DAR 6. It is also transcribed in Darwin Online.

page, also written in pencil, headed "Maer, May 1842, useless." It also bears the words "This page was thought of as introduction."

The logical structure of this sketch roughly corresponds with the structure of his later "Essay of 1844" and in many ways too with *On the Origin of Species* (Hodge, 1977; Waters, 2009). The argument proceeds from an initial statement about the variability of domestic organisms followed by the claim that similar variations can also be seen in animals and plants in a state of nature. Darwin then presented his analogy: that a breeder's selection of variations in domestic organisms is analogous to selection in nature. He continued with a brief account giving the bones of his theory, that is, competition between variant organisms will result in differential survival. The better adapted variants are the ones most likely to survive and reproduce their kind, ultimately leading to gradual changes in form. This is followed by a discussion of the difficulties of the theory and a short section on instinct which was treated by him as a special case of difficulty. Later pages cover geology, geographical distribution, affinities and classification, unity of type and morphology, and abortive or rudimentary organs. He ended with a short recapitulation and conclusion.

This structure follows early Victorian guidelines about the way to present scientific theories as laid out by philosophers John Herschel and William Whewell (Hull, 1973). First, as Darwin conceived it, he gave the facts of nature, described without any theory, followed by the proposal of the theory. This was often called the Baconian method. The remainder of his text concerned itself with examples of how the theory can explain a range of phenomena, giving rise to what Whewell called a consilience of inductions, where the more aspects of nature that a theory can explain, the more likely it is to be true.

Darwin was unusual in including a discussion of the difficulties that he encountered. This would become a well-known characteristic of *On the Origin of Species* (Lustig, 2009). Yet Darwin's "difficulties" were not the ones that most people would think of – they were not metaphysical difficulties pertaining to the origin of humans, the emergence of life on Earth, or the removal of a divine creator. Darwin's difficulties were entirely biological: how could animal instincts and the mental qualities of organisms emerge and change; how did the self-evident sterility between species emerge; why did ants and bees have several different shapes of body; how could apparently perfect organs like the eye develop through gradual incremental changes? The big difficulties were deliberately put to one side.

Why did Darwin do this? The remarkable thing about this 1842 manuscript sketch was the absence of any reference to the origin of humankind and the creator. Probably, Darwin's talks with his wife and scientific colleagues, his orthodox Cambridge university training and extensive reading in contemporary science and philosophy warned him how radical his ideas were in relation to contemporary religious belief. He was highly motivated to avoid discussion of human origins and the divine creative force because his writing ran counter to everything then assumed about Adam and Eve and the Garden of Eden, even though few British people of the era were biblical literalists. He might additionally have decided that he needed to include a great deal more before he could argue convincingly about humankind's non-divine origins. Whatever the reason, he systematically drained the manuscript of human beings and theological questions, never to let them return until he published *The Descent of Man* in 1871. Even then he avoided any discussion of Christian doctrine.

Two years later, during the first part of 1844, Darwin expanded his sketch into a fully articulated essay.[4] It was completed on 5 July 1844. From several clues in his correspondence, it appears that he deliberately wrote this essay for an audience of scientific men. He did not envisage publishing it straight away for he was a cautious strategist, often uneasy about his proposals, and always attempting to gauge the kind of response that a theory of transmutation would generate. Moreover, given his intensive work on the topic, he was certain that he needed to do more research to make the theory fully hold water.

We know much of this because he wrote a note to his wife Emma, dated 5 July 1844 (Burkhardt et al., 1985–present, v. 3, pp. 43–45). It was in the form of a request. He asked her to ensure that the essay would be published in the event of his death and stipulated a sum of money to be bequeathed, together with his extensive library and portfolios of notes on species, to a scientific friend who would undertake to see the work through the press. Darwin listed the names of people to approach, all of whom were known to him as serious thinkers about species: the list included his close colleague Charles Lyell and his old university professor John Stevens Henslow. He altered this list several times afterward. Ten years later, in the autumn of 1854 when sorting out his notes in preparation for writing up his "big book" on species (never published in full, usually called *Natural Selection*), he decided that the botanist Joseph Dalton Hooker was by far the best man for the task and added a note on the cover to that effect. In his letter to Emma Darwin he said, "If, as I believe that my theory is true, & if it be accepted even by one competent judge, it will be a considerable step in science" (Burkhardt et al. (1985–present, v. 3, p. 43).

Was this a deliberate delay? Was he scared of publishing? Many people think so (Desmond & Moore, 1990; Van Wyhe, 2007; Buchanan & Bradley, 2017). And yet, in a sense, he was not in any hurry. Nowadays, in the light of all that is known about his personality and correspondence, it seems feasible to suggest that a strong commitment to scientific accuracy and a proper sense of scientific caution were at least as high in his mind as any fear of the consequences of publication. He did not feel ready for full-scale publication. The scope of his musings in the transmutation notebooks indicate the very wide range of investigations and topics that he thought were relevant. He had barely started to chip away at the surface.

One striking event, however, gave him reason to pause. Through the summer of 1844, he anxiously watched the controversy seething around an evolutionary book, *Vestiges of the Natural History of Creation*, published anonymously (Secord, 2000). We now know that the author was Robert Chambers, a talented journalist from Edinburgh. This book dramatically changed the texture of debate on evolution –

[4]Published by F. Darwin (1909). The original manuscript of the essay of 1844 is in Cambridge University Library, Archives, call mark DAR 7. The fair copy by an amanuensis is in DAR 113.

firing up the theologians, pushing secular ideas uncomfortably into Victorian drawing rooms, and inspiring violent criticism on the one hand and fascinated attention on the other. *Vestiges* became a popular publishing phenomenon. It raced across the English-speaking world in cheap editions and made a splash in translation in other countries. The unnamed author wrote fluently of the self-generated development of the living world from specks of animate matter right through to men and women. Although the scientific content was shaky, and the proposed mechanisms of change at times were laughable, its general evolutionary thrust was clear. It was not like Darwin's carefully structured essay with strictly limited boundaries. It took the development of the whole universe as its subject matter. It discussed the emergence of humankind from primates and pushed the divine creator into a backroom. It was a book, moreover, that tapped into the progressive, politically reformist aspirations of the age and called on the subterranean world of radical comparative anatomy and embryology, as well as on phrenological doctrines that the mind could be adjusted by willpower so that anyone could advance through the hierarchical social order of Britain or even overthrow it.

In the event, Darwin put the essay to one side, with his covering instructions, and ensured its survival by paying for it to be copied out. This latter version he called the "fair" copy. He energetically continued to work on all aspects of the theory, by now being able to conduct supporting experiments in his gardens and greenhouses and participate in an ever-widening correspondence with experts in various fields. In contrast to the usual image of Darwin as a recluse, and as an invalid, these letters show him as an established naturalist at the heart of British scientific society. He used letters as a primary research tool.

Writing the essay made him cautious. Paradoxically it also gave him confidence. Around now, he began to let his closest scientific friends into his secret. They did not seem outraged by Darwin's heterodox opinions. Only 2 months after their first exchange of letters, early in 1844, Darwin told Hooker that he was engaged in a "very presumptuous work" which had led him to the conviction that "species are not (it is like confessing a murder) immutable" (Burkhardt et al., 1985–present, v. 3, p. 2). He explained to another friend:

> I have continued steadily reading & collecting facts on variation of domestic animals & plants & on the question of what are species; I have a grand body of facts & I think I can draw some sound conclusions. The general conclusion at which I have slowly been driven from a directly opposite conviction is that species are mutable & that allied species are co-descendants of common stocks. I know how much I open myself, to reproach, for such a conclusion, but I have at least honestly & deliberately come to it. (Burkhardt et al., 1985–present, v. 3, pp. 67–68; letter to L. Jenyns, 12 October 1844)

Late in 1846, Hooker was invited to read the fair copy of the essay. He sent Darwin some notes about it in 1847 that are now unfortunately lost.

Darwin dropped a hint in public, too. In the second edition of his popular narrative about the natural history of the *Beagle* voyage, called *Journal of Researches into the Geology and Natural History of the Various Countries Visited by H.M.S. Beagle*, published in 1845 (Fig. 5.2), he tried a small verbal experiment to test the waters. In his description of the Galápagos finches, he wrote: "One might

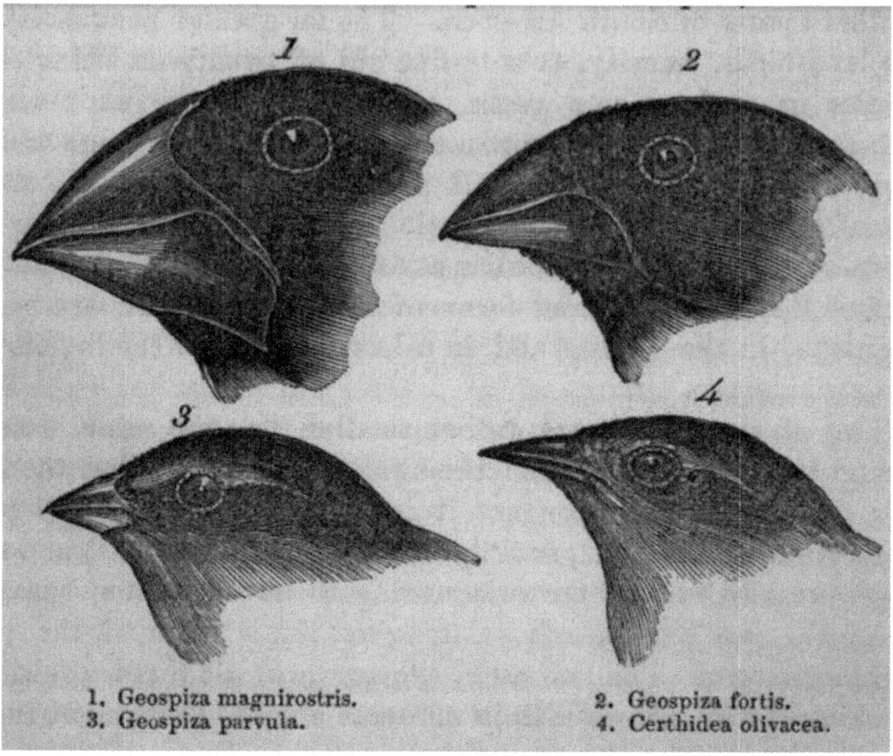

Fig. 5.2 Darwin's published image of four Galapagos finches, from the second edition of his *Journal of Researches*. In the text, he wrote: "One might really fancy that from an original paucity of birds in this archipelago, one species had been taken and modified for different ends". (Charles Darwin, *Journal of Researches* 2nd ed. 1845, p. 379)

really fancy that from an original paucity of birds in this archipelago, one species had been taken and modified for different ends" (Darwin, 1845).

Nevertheless, despite this new confidence, he did not publish the essay. Instead, he began a comprehensive study of barnacles starting in 1846. Eventually he devoted 8 years to their study, considering both the living and fossil forms and publishing a book about each. Scholars have asked: Was this, too, a deliberate delay in placing his evolutionary theory before the public? It might have been. Perhaps Hooker's reaction to the essay had been negative (Mannouris, 2011; Buchanan & Bradley, 2017). Perhaps the furor over the *Vestiges of the Natural History of Creation* set him back. Perhaps he was over-cautious. Equally likely, his biological interest in one anomalous barnacle collected in Chile during the *Beagle* voyage was high, and he had kept specimens in his possession ever since, hoping one day to have time to work on it. Now he found the time. In a move that sometimes perplexes historians he decided to explore the entire family of barnacles in order fully to understand this one species.

There were further ongoing reasons. During these years, he came to recognize that his theory needed modification. His anatomical work with barnacles was revealing that every individual organism in some way slightly varied from the norm – small variations occurring without any apparent causal factor. This contradicted his statement in the essay of 1844. There, he had stated that variability depended on the reproductive system being unsettled in some way, either by the process of domestication or by alterations in the environment. He was now seeing that variation in a living body needed no such external cause – it was a given. His study of barnacles was providing insight into evolution in the real world.

Additionally, at some point in the early 1850s, Darwin came to think that he needed to add what he called "the principle of divergence," the only major alteration that he made to his original "notebook" theory of natural selection (Kohn, 2009). In the *Origin of Species* – but not in his essay of 1844 – Darwin discussed this principle. It was an important innovation, for he needed it to explain how natural selection could produce the branches of the tree of life. He was sure that natural selection could explain the gradual shift of one species into another, each variant form becoming better adapted to conditions. But how to explain large steps like the divergence of reptiles into birds? Funnily enough, he had not realized that he needed such an explanation until quite late in the day. This additional concept concerned the economy of nature – an old metaphor from Linnaeus's time. The most successful variant, Darwin said, was one that could seize on unexploited places or roles in the natural economy and that this process would, over many generations, lead to divergence:

> I overlooked one problem of great importance [...] and I can remember the very spot in the road, whilst in my carriage, when to my joy the solution occurred to me; and this was long after I had come to Down. (Barlow, 1958, p. 120)

In 1854, after working for so many years on barnacles, he was at last ready to take up the challenge of writing his theory fully for publication. He did not go back to the essay of 1844 now that he had new things to add. This time around, he began a fresh manuscript expressly intending to publish, longer and more detailed than the essay, directed toward explaining his theory to expert scientific readers. He came to call it his "big book on species" (Stauffer, 1975). As soon as he had completed the last aspects of his barnacle studies, he sorted through his extensive piles of notes, read his transmutation notebooks again, cut some pages out of them that he wished to use, and began to write. He worked in his study in his home at Down House, sitting in the old leather chair (Fig. 5.3). This big book was never published although it is now available in print form, called *Natural Selection*. He poured everything he now knew into it, while retaining the original structure of the argument as first mapped out in his sketch of 1842 and expanded essay of 1844. By 1858, he was more than two thirds through this new volume.

But the Essay of 1844 remained important in other ways. When Alfred Russel Wallace sent Darwin a letter in June 1858 enclosing a description of his own theory of evolution by natural selection, the essay became Darwin's primary evidence that he had been working on a similar theory of evolution for more than a decade.

Fig. 5.3 Early twentieth-century photograph of Darwin's study in Down House, Kent, where he wrote and worked. The house is now a museum operated by English Heritage. (Wellcome Library London)

Wallace's intervention has caught the imagination of historians, who see this simultaneous discovery as far more than a mere coincidence (Berry, 2002, 2013; Beccaloni, 2008). For most scholars this remarkable event indicates that the intellectual and social context of Victorian Britain was so pervasive that two people, who barely knew each other, could both engage in natural history collecting expeditions that enlarged their understanding of the abundance of life, could both read the same scientific texts, and both puzzle over the variety and interrelations of species to the point where they came up with the same theory. The simultaneity suggests that colonial, industrial, and mercantile Britain provided each man with deeply powerful metaphors relating to competition, selection, survival, and change.

The date that Wallace's manuscript arrived at Darwin's home will never be known for sure. The letter and enclosure have never been found. But on 18 June 1858, Darwin wrote a letter to Lyell to express his shock at being well and truly forestalled. "I never saw a more striking coincidence... if Wallace had my MS sketch written out in 1842 he could not have made a better short abstract!" (Burkhardt et al., 1985–present, v. 7, p. 107; Costa, 2014).

The full story of the joint announcement of Darwin's and Wallace's theories at the Linnean Society of London on 1 July 1858 is well-known and not recounted here (it is covered in Browne, 2003). It is, however, worth mentioning that Darwin's well-established habit of writing everything down suddenly became critical for establishing his priority over Wallace. Darwin sent the fair copy of his Essay of 1844 to Joseph Hooker to establish that Hooker had read it some 10 years beforehand. Over the space of only a few days, it was decided that Hooker and Lyell would present extracts from Darwin's manuscripts, alongside Wallace's paper, at the Linnean Society, so that Darwin and Wallace would share the discovery. Short passages from Darwin's Essay of 1844 were copied out for this purpose. The other element of Darwin's contribution was a draft of a letter he had written to his American friend the botanist Asa Gray in 1857 (Burkhardt et al., 1985–present, v. 6, pp. 445–450). These bits and pieces were not fully coherent. Wallace's writing, by contrast, sparkled with continuity and clarity of thought. So, the final, and ultimately most strategic, function of Darwin's written Essay of 1844 was fortuitously to prove that he had been writing about evolution by natural selection for more than a decade.

After the announcement at the Linnean Society, Darwin began to write *On the Origin of Species*, compressing and supplementing the big manuscript that was already partly underway. He completed the text in March 1859. It was published by John Murray in London in November 1859.

In conclusion, what can this discussion of Darwin's early writings tell us? Primarily, it shows Darwin's thoughts developing through the writing process. To write was also to think. Furthermore, and in the larger historical sense, it is useful to acknowledge that the practice of recording can be just as significant a part of scientific activity as thinking, collecting, and experimenting. This chapter shows that without a record, there can be no science. It also helps us see that science can only move forward if its practitioners communicate their views, either through letters, conversation, reading papers at learned societies, or in printed material. In Darwin's day, it was already felt that print publication was the primary vehicle for placing new views into the community. Darwin did not do so until shocked into action by Wallace's communication. Hence, the story of his writings from notebooks to published volume was long and convoluted. We see a remarkable mind at work, growing in knowledge and maturity. Darwin thought as he wrote, at first accompanying a colonial expedition, and then buffeted by contemporary doubts and setbacks, always deeply conscious of the meaning and possible reception of his views.

References

Barlow, N. (Ed.). (1958). *The autobiography of Charles Darwin, 1809–1882, with original omissions restored*. Collins.

Barlow, N. (Ed.). (1963). Darwin's ornithological notes. *Bulletin of the British Museum of Natural History*, 203–278.

Barrett, P. H., Gautrey, P. J., Herbert, S., Kohn, D., & Smith, S. (Eds.). (1987). *Charles Darwin's notebooks, 1836–1844: Geology, transmutation of species, metaphysical inquiries*. British Museum (Natural History) and Cambridge University Press.

Beccaloni, G. (2008). *Natural selection and beyond: The intellectual legacy of Alfred Russel Wallace*. Oxford University Press.

Berry, A. (2002). *Infinite tropics: An Alfred Russel Wallace anthology*. Verso.
Berry, A. (2013). Alfred Russel Wallace: Evolution's red-hot radical. *Nature, 496*, 162–164.
Bowler, P. J. (1984). *Evolution: The history of an idea*. University of California Press.
Brown, F. B. (1986). The evolution of Darwin's theism. *Journal of the History of Biology, 19*, 1–45.
Browne, J. (2003). *The power of place: Vol. 2 of a biography of Charles Darwin*. Knopf.
Buchanan, R. D., & Bradley, J. (2017). "Darwin's delay": A reassessment of the evidence. *Isis, 108*, 529–552.
Burkhardt, F. H., et al. (Eds.). (1985–present). *The correspondence of Charles Darwin* (30 vols). Cambridge University Press.
Cañizares-Esguerra, J. (2006). How derivative was Humboldt? Microcosmic narratives in early modern Spanish America and the (other) origins of Humboldt's ecological sensibilities. In J. Cañizares-Esguerra (Ed.), *Nature, empire, and nation: Explorations of the history of science in the Iberian world*. Stanford University Press.
Costa, J. T. (2014). *Wallace, Darwin, and the origin of species*. Harvard University Press.
Darwin, C. R. (1845). *Journal of researches into the natural history and geology of the countries visited during the voyage of H.M.S.* Beagle *round the world under the command of Capt. FitzRoy* (2nd ed). John Murray.
Darwin, F. (Ed.). (1909). *The foundations of the "Origin of Species": two essays written in 1842 and 1844 by Charles Darwin*. Cambridge University Press.
Desmond, A. J., & Moore, J. R. (1990). *Darwin*. Penguin.
Fyfe, A. (1997). The reception of William Paley's *Natural Theology* in the University of Cambridge. *British Journal for the History of Science, 30*, 321–335.
Hodge, M. J. S. (1977). The structure and strategy of Darwin's "long argument". *British Journal for the History of Science, 10*, 237–245.
Hodge, M. J. S. (1985). Darwin as a lifelong generation theorist. In D. Kohn (Ed.), *The Darwinian heritage* (pp. 207–244). Princeton University Press in association with Nova Pacifica.
Hodge, M. J. S. (2009). The notebook programmes and projects of Darwin's London years. In J. Hodge & G. Raddick (Eds.), *The Cambridge companion to Darwin* (2nd ed., pp. 44–72). Cambridge University Press.
Hodge, M. J. S., & Kohn, D. (1985). The immediate origins of natural selection. In D. Kohn (Ed.), *The Darwinian heritage* (pp. 185–206). Princeton University Press in association with Nova Pacifica.
Hull, D. (1973). Introduction. In D. Hull (Ed.), *Darwin and his critics: The reception of Darwin's theory of evolution by the scientific community*. Chicago University Press.
Kohn, D. (1980). Theories to work by: Rejected theories, reproduction, and Darwin's path to natural selection. *Studies in the History of Biology, 4*, 67–170.
Kohn, D. (2009). Darwin's keystone: The principle of divergence. In M. Ruse & R. Richards (Eds.), *The Cambridge companion to the origin of species* (2nd ed., pp. 87–108). Cambridge University Press.
Latour, B. (1988). *Science in action: How to follow scientists and engineers through society*. Harvard University Press.
Leitao, H., & Sanchez, A. (2017). Too much to tell: Narrative styles of the first descriptions of the natural world of the Indies. *History of Science, 55*, 167–186.
Lustig, A. J. (2009). Darwin's difficulties. In M. Ruse & R. Richards (Eds.), *The Cambridge companion to the origin of species* (2nd ed., pp. 109–128). Cambridge University Press.
Malthus, T. R. (1798/1970). *An essay on the principle of population* (A. Flew, Ed.). Penguin Books.
Mannouris, C. (2011). Darwin's "beloved barnacles": Tough lessons in variation. *History and Philosophy of the Life Sciences, 33*, 51–70.
Schweber, S. S. (1985). The wider British context in Darwin's theorizing. In D. Kohn (Ed.), *The Darwinian heritage* (pp. 35–69). Princeton University Press in association with Nova Pacifica.
Secord, J. A. (2000). *Victorian sensation: The extraordinary publication, reception, and secret authorship of Vestiges of the Natural History of Creation*. University of Chicago Press.

Stauffer, R. C. (Ed.). (1975). *Charles Darwin's Natural Selection, being the second part of his big species book written from 1856 to 1858*. Cambridge University Press.
Sulloway, F. J. (1982a). Darwin and his finches: The evolution of a legend. *Journal of the History of Biology, 15*, 1–53.
Sulloway, F. J. (1982b). Darwin's conversion: The *Beagle* voyage and its aftermath. *Journal of the History of Biology, 15*, 325–396.
Van Wyhe, J. (2007). Mind the gap: Did Darwin avoid publishing his theory for many years? *Notes and Records of the Royal Society, 61*, 177–205.
Waters, C. K. (2009). The arguments in the *Origin of Species*. In J. Hodge & G. Raddick (Eds.), *The Cambridge companion to Darwin* (2nd ed., pp. 120–143). Cambridge University Press.
Young, R. M. (1969). Malthus and the evolutionists: The common context of biological and social theory. *Past & Present, 43*, 109–145.

Janet Browne's biography of Darwin explored the ways in which scientific knowledge was created, distributed, and accepted, moving from private to public. She is currently exploring the history of Darwin's impact on popular culture from the time of his death to today.

Chapter 6
The Development of Darwin's Theory: From Natural Theology to Natural Selection

Kostas Kampourakis

Abstract It is often said that Darwin's study of nature drove him to atheism. Whereas this might be, in principle, possible, it does not seem to have actually been the case for him. Both in his autobiography, which was not intended to be published, and in his personal correspondence, Darwin consistently described himself as an agnostic. It is true that he underwent several fluctuations of belief during his life, but in the end, he never explicitly rejected the existence of God. What is even more important is that the main shift he underwent during his life was conceptual, not emotional or religious. That was his shift from natural theology and a view of adaptation as perfect to natural selection and a view of adaptation as relative to the environment, something historian Dov Ospovat pointed out 40 years ago. In this chapter, I present the evidence from Darwin's writings about his agnosticism and about his conceptual shift from the perfect adaptation of natural theology to the relative adaptation of natural selection.

6.1 Introduction

A common argument among anti-evolutionists, at least in my own experience, is that evolution leads to atheism and that Darwin was an atheist. For instance, consider the following excerpt from an article on the website of Discovery Institute titled: "What Were Darwin's Religious Views?"

> But what about Charles? His views may have changed, but for at least the second half of his life, he was—for all practical purposes—an atheist. Anyone doubting that must read the detailed accounts of his life in Adrian Desmond and James Moore, Darwin and Janet Browne's two-volume, Charles Darwin. Darwin did refer to himself as an "agnostic"rather than an "atheist." However, his preference for the term "agnostic"seems to have been dictated primarily by his worries about offending people unnecessarily. Sorry, folks. There's no deathbed conversion (Wiker, 2009).

K. Kampourakis (✉)
Section of Biology and IUFE, University of Geneva, Geneva, Switzerland
e-mail: Kostas.Kampourakis@unige.ch

M. Elice Brzezinski Prestes (ed.), *Understanding Evolution in Darwin's "Origin"*, History, Philosophy and Theory of the Life Sciences 34,
https://doi.org/10.1007/978-3-031-40165-7_6

In this chapter, I aim to address the conception – especially popular among anti-evolutionists – that Charles Darwin's study of nature and his theory about evolution (called transmutation, among other names, at the time) made him become an atheist. Based on the work of various historians, especially the late Dov Ospovat, I show that the shift he underwent was not from theism to atheism but rather from natural theology to natural selection. This entails that Darwin's main shift was conceptual and had to do with his understanding of nature, not emotional, and having to do with his worldviews, especially his religious beliefs.

I should make clear that I'm not arguing that evolution cannot drive one to become an atheist; indeed, it can. But this strongly depends on the inferences one makes from evolutionary theory and therefore is independent of the theory itself. Evolutionary theory is not inherently atheistic; one should understand the theory and then decide what one believes about religion (Kampourakis, 2020). I am arguing then that this was not the case for Charles Darwin. How can we know? Because currently, we have available not only every book or manuscript that he has ever written on the so-called Darwin online website (http://darwin-online.org.uk/) but also his correspondence on the Darwin Correspondence Project website (https://www.darwinproject.ac.uk). Therefore, it is possible to follow his thinking and the development of his theory over the years by reading what he was writing in each particular case. Let us then see what happened.

6.2 A Note on Conceptual Change

Our knowledge of the world takes the form of concepts that are mental representations. Scientific concepts provide systematic representations through which explanations of and predictions about phenomena are possible (Nersessian, 2008, p. 186). Concepts should be distinguished from conceptions, the latter being the different meanings of or meanings associated with particular concepts. From our early childhood, we experientially formulate conceptions of the world described as preconceptions. These can be either correct or wrong. What is important is that new knowledge (e.g., through schooling) does not guarantee a correct understanding of concepts if the preconceptions are not properly restructured. This is what conceptual change is about: the change of our conceptions with development and learning. For conceptual change to occur, existing conceptions must be properly addressed so that individuals understand that they are wrong or explanatorily insufficient. Ideally, a conceptual change process's outcome is replacing preconceptions with accurate concepts (see the various chapters in Vosniadou (2013)). Consequently, conceptual change can be defined as the change of conceptions in the wider sense (including the change in the meaning of concepts or the change in the relationship among concepts within an explanatory scheme or model) as a result of conceptual conflict, i.e., the realization that prior conceptions are wrong or explanatorily insufficient (see Kampourakis, 2020).

In the philosophy of science, conceptual change is usually related to the work of Thomas Kuhn. In this view, conceptual change is characterized as a paradigm shift, during which the new paradigm emerges at once in the mind of someone who is not committed to the old paradigm. Such a person is able to see that the old paradigm no longer works and so is able to conceive of a new one to replace it. The resulting transition to a new paradigm is a scientific revolution. Kuhn thought that a paradigm shift was a transition between incommensurable and competing paradigms, which could not be made gradually but rather occurred at once as a gestalt switch (Kuhn, 1962/1996, p. 122). However, this does not seem to have been the case for evolutionary theory. For instance, the influences on Darwin's theory were so many, so deep, so prolonged, and so various that no single transition can be identified as a shift that replaced a pre-Darwinian with a Darwinian paradigm (Hodge & Radick, 2009).

However, the conceptual changes that historical figures themselves underwent are of interest. A careful study of the cognitive processes of past scientists shows that novel concepts do not emerge all at once and fully developed in the minds of scientists. Rather, they are the products of lengthy cognitive processes under the influence of a combination of conditions (Nersessian, 2008, p. 5). Indeed, Darwin underwent shifts due to conceptual conflicts. When he realized that his conceptions could not sufficiently account for the observed phenomena, he replaced them with new ones. Darwin developed his theory for 20 years and underwent two major shifts from his initial views. The first shift was that from special creation to transmutation, which was completed around March 1837. The second and more prolonged shift was from perfect adaptation to relative adaptation, which was completed around March 1857.

6.3 Beginnings

Charles Darwin was neither the first evolutionist nor the first to propose an evolutionary theory. Many naturalists had developed a variety of evolutionary views before Charles Darwin in the context of debates about the stability of species (Corsi, 2005). One interesting figure was Erasmus Darwin, the grandfather of Charles. However, it was Jean Lamarck who developed the first evolutionary synthesis, in which, among other things, he proposed two laws for the transmutation of species. The first one was that the continuous use of particular organs strengthens them and makes them better for a role, whereas permanent disuse may weaken and deteriorate them. The second law was about the inheritance of acquired characters, for which Lamarck is most famous. Accordingly, the changes acquired during life because of extended use or disuse can then be transmitted to the next generation. Here, I should note that a common contrast is between the inheritance of acquired characters in Lamarck's and natural selection in Darwin's theory, even though Darwin also accepted that the inheritance of acquired characters was possible (Burkhardt, 1995). Lamarck's views received harsh criticism from some naturalists. One of them was Charles Lyell, a prominent geologist. He had a major influence on

Darwin and criticized Lamarck for developing a speculative theory and noted that he had not been able to establish properly with empirical evidence what he was claiming (Lyell, 1832).

Charles became familiar with Lamarck's theory during his time at Edinburgh. Charles had been sent there to study medicine, along with his brother Erasmus, perhaps to follow the steps of their father Robert and their grandfather Erasmus, who were successful physicians. While at Edinburgh, Charles studied natural history and worked with anatomist Robert Grant, probably the only teacher in Britain sympathetic to Lamarckism in the three decades before 1859 (Desmond, 1984). But as Charles did not really want to become a physician, especially since he despised the brutality of surgery as it was done at the time, he was sent by his father to study at Cambridge, where he would get a qualification that would make him something like a clergyman. But what Charles studied there with interest was natural history, with geologist Adam Sedgwick and botanist John Henslow. Charles learned to study rock formations during his geological tour with Sedgwick in the summer of 1831. Henslow also taught him to collect and label plants and instilled in him the idea of botanical endemism. Both of these were crucial influences on Darwin's later work, especially during his famous Beagle voyage that was arranged by Henslow (Kohn et al., 2005; Roberts, 2001).

While studying at Cambridge, Charles became familiar with the writings of William Paley, including the argument of divine design in nature. Paley had written:

> [. . .] for, in the watch which we are examining, are seen contrivance, design; an end, a purpose; means for the end, adaptation to the purpose. And the question, which irresistibly presses upon our thoughts, is, Whence this contrivance and design? The thing required, is, the intending mind, the adapted hand, the intelligence by which that hand was directed.
>
> This mechanism being observed [. . .] the inference, we think, is inevitable, that the watch must have had a maker - that there must have existed, at some time, and at some place or other, an artificer or artificers who formed it for the purpose which we find it actually to answer, who comprehended its construction, and designed its use (Paley, 2006/1802).

Paley compared organisms to instruments with mechanisms, such as a watch, and argued that whenever we see divine design and contrivance (this is another word for adaptation) therein, and ask where this comes from, we're trying to infer who was the intelligent person who designed that for a purpose. Therefore, Paley suggested, we can make similar inferences for organisms. Likewise, whenever we see purpose and design in a structure in nature, we must infer that there must have been an intelligent designer behind this who designed the structure as we see it. And if humans are able to construct such complex instruments such as a watch or a telescope, we can imagine that the agent who designed the eye, which is much more complex than the telescope, must be much more competent than humans, and this can only be God. This is a view that Charles initially accepted with pleasure, and his lifelong fascination with the study of adaptation was due to Paley's influence on him (Brooke, 1985).

As I mentioned, thanks to Henslow, Charles had the opportunity to travel aboard the HMS Beagle and made a voyage around the world. While aboard, he read Lyell's principles of geology and was influenced by his theory of uniformitarianism. This was the idea that the natural processes operating in the past are the same as those that can be observed operating today; by operating for a long time, such processes can gradually bring about the changes that we observe. An alternative view was catastrophism, according to which it was possible to explain the changes on the earth's surface only based on big catastrophes. However, perhaps the most famous incident of the Beagle voyage was the visit that Charles made to the Galápagos islands. There, he collected plants that he had intended to send back to Henslow, but also birds. However, it was not until after his return to England that he realized that the Galápagos birds that he had collected belonged to different species, thanks to ornithologist John Gould (Sulloway, 1984).

After that, Charles began to put together the pieces of information he had collected and eventually realized that Paley's views might not be adequate to explain what he had observed in nature. Having observed endemic species, he realized that the organisms in the Galápagos islands were different from the respective species in similar environments elsewhere. This could not be explained by the argument from design; why would the divine designer design different species for living under the same conditions? Thus, he started writing down his thoughts about transmutation, and in 1837 he came up with a drawing resembling a phylogenetic tree. However, his thoughts about transmutation allowed a place for divine creation:

> Astronomers might formerly have said that God ordered each planet to move in its particular destiny. — In same manner God orders each animal created with certain form in certain country, but how much more simple & sublime powers let attraction act according to certain law such are inevitable consequen[ces]. Let animals be created, then by the fixed laws of generation, such will be their successors (Darwin, 1837).

In other words, Charles wrote that God may have created species, but then he may have let them evolve through natural processes. This is a view that is often described as deism: God created species, and then natural processes have been operating for years, guiding their evolution. Whereas Charles began thinking about transmutation, he did so under a strict sense of perfect adaptation. In this view, there are places in the environment to which organisms can adapt, and this adaptation is perfect.

6.4 Becoming an Evolutionist

On September 1838, Charles read the *Essay on the Principle of Population* by Malthus, describing the idea of struggle for existence in human societies. The main idea was that whereas the human population grows exponentially, this is not the case for resources. Therefore, after a point, there will be a struggle for existence among humans because the resources available will not be sufficient to support them all. In his autobiography, he wrote that he happened to read that for amusement, but it seems that he had heard the ideas of Malthus before as he was widely read at the time (Schweber, 1980). After reading Malthus, Charles realized that there were actually two types of struggle: interspecific struggle, that is, struggle between different species, which Malthus also recognized; but Charles also noted that there also existed intraspecific struggle, that is, struggle within species (Bowler, 1976; Vorzimmer, 1969). Around the same time, Charles became interested in breeding and artificial selection. The pamphlets written by animal breeders, such as John Sebright, were explicit about the nature and power of artificial selection. The work of breeders was widely known, but most of his contemporary men of science did not pay any attention to it. So what Charles did was to draw an analogy between artificial and natural selection. His thinking was that if breeders could arrive at their wished results by consciously selecting those varieties that they wanted to bring together, it could be possible for nature to do the same. Charles even joined several pigeon breeding clubs to see for himself how far selective breeding could go in producing new varieties (Evans, 1984; Ruse, 1975a).

Darwin's key insight was combining artificial selection with Malthus and developing the analogy between artificial selection and natural selection. Here is how it works. Imagine we have a population consisting of two kinds of individuals, say green and brown. During artificial selection, a conscious human selector can decide to let only some of these individuals breed but not the others, say the brown ones. As a result, the subsequent generations will have more and more brown individuals, and eventually, there might be a point when all individuals are brown. This is the outcome of the conscious selection of brown color made by a human selector. The question then for Darwin was, how can the same process take place in nature if there is no conscious selection? Natural selection cannot rely on an intelligent, conscious agent that makes a selection. Malthus gave the solution to this problem. Charles realized that competition among species could play the same role in natural selection as the role of the human selector in artificial selection. So, Malthus was critical because he provided Charles with a process that could explain how natural selection takes place in nature (Kohn, 2009).

There was nevertheless another reason for using Malthus. Charles wanted his theory to align well with the philosophy of science of his time. Its two prominent representatives were John Herschel and William Whewell. They believed that the aim of science was to find the laws of nature and then to identify the true causes (*vera causae*) that guided these laws. Malthus' theory provided Charles with the quantitative law his theory required (Ruse, 1975b, 2000). In order to show that natural selection was a true cause, Charles had to establish the existence of a cause by studying it in action. To achieve this, he was based on the analogy of artificial selection. Then Charles had to show that the cause is competent to produce the phenomena to be explained. Charles relied on a series of thought experiments to show the competency of natural selection in producing new species, even if it had never been observed to do so. Finally, Charles had to show that the cause is indeed responsible for these phenomena. The responsibility of natural selection was based on the fact that it seemed more probable than any other theory in explaining several kinds of facts about species (e.g., adaptations, geographical distribution) (Hull, 2009; Lennox, 2005). The latter was actually something that Charles never actually managed to do. He did not show that natural selection was actually responsible for speciation, which was one of the major criticisms he received when the *Origin of Species* was published.

By 1844, Charles had developed a sketch of his theory. He shared it with his wife Emma and asked her in the event of his sudden death to see it to publication. He thought his theory was important but was not yet ready to publish it. At around the same time, he sent an abstract of this sketch of his theory to Joseph Hooker, a botanist and a close colleague. He also explained that his observations had led him to conclude that species are not immutable. There were at least two reasons for not publishing his theory. One was the publication in the same year of the book *Vestiges of the Natural History of Creation*, which was published anonymously by Robert Chambers. That book caused an evolutionary scandal because it brought to the fore issues that were not widely discussed. Charles was concerned that publishing at that time, his theory would be compared to the *Vestiges* (Secord, 2000). At the same time, Charles was aware of the criticism of Lamarck's theory by Lyell and others about its speculative basis, so he wanted to establish his own theory as well as possible.

I must note here that the theory that Charles had developed by 1844 was not the theory we read in the *Origin of Species*. At that time, Darwin had just abandoned the idea of perfect adaptation (that there is only one best possible form for any given set of conditions) for a quite similar view, limited perfection, which nevertheless allowed the possibility of alternative forms and rudimentary organs. At that time, Darwin also thought there was little available variation in nature as perfectly adapted forms did not vary. Consequently, a change in external conditions was required for new variation to occur. Two years later, in 1846, Charles began his study of the barnacles, which took him 8 whole years and eventually made him realize that the variation available in nature was much more than he had initially thought (Ospovat, 1981). During that period, in 1851, his beloved daughter Annie died at the age of 10 years old. This really caused Charles great sorrow and pain and perhaps influenced his religious views (Brooke, 2009; Keynes, 2001; Spencer, 2009).

Returning to conceptual matters, the most important shift we can find during the period that Charles studied barnacles was toward a view that allowed for more and more variation and less and less perfection. Then around 1854–1855, we see a new idea in his writings, the principle of divergence. This was eventually described in the *Origin of Species* as:

> Here, then, we see in man's productions the action of what may be called the principle of divergence, causing differences, at first barely appreciable, steadily to increase, and the breeds to diverge in character both from each other and from their common parent. [. . .] But how, it may be asked, can any analogous principle apply in nature? I believe it can and does apply most efficiently, from the simple circumstance that the more diversified the descendants from any one species become in structure, constitution, and habits, by so much will they be better enabled to seize on many and widely diversified places in the polity of nature, and so be enabled to increase in number (Darwin, 1859, p.112).

This was based on the idea of the division of labor, which in turn explained how natural selection could produce the branches of the tree of life. It is unclear whether Adam Smith or Henri Milne-Edwards directly influenced Charles. However, adopting this idea brought about a critical modification to his theory. The key element here was that it was possible for some individuals, who represent a subset of variation within a population, to occupy a previously unoccupied niche and evolve therein, thus further differentiating from the initial population. According to the principle of divergence, this is why different varieties of a species may initially emerge, and these may eventually evolve into distinct species (Kohn, 2009).

By 1857, we can also see a shift from perfect to relative adaptation, which is evident in parts of the book *Natural Selection*. This was a multivolume book that Darwin had begun to write, and in the manuscript, one documents his shift toward the idea that adaptation depends on the environment, and it cannot be relative in any strict sense (Ospovat, 1981). For instance:

> Though climatal conditions may have no great influence on organisation or visible structure, yet it is notorious that the great majority of organic beings are adapted, within moderately narrow limits, to the climate <of the regions> which they inhabit. When, therefore, a Naturalist meets an animal with a very wide range, for instance the Puma in the reeking hot forests of Central America, on the dry deserts of Patagonia, in the damp cold woods of Tierra del Fuego & up to the limits of eternal snow on the Cordillera, he is much surprised; for he is accustomed to meet for instance, one species confined to the Tropics, another to the temperate & another to the cold regions; his surprise is, also, increased, from falsely attributing (as I believe) far too much weight/11/to the relations between climate & visible structure; climatal conditions are manifest; but the more important conditions determining each creature's power of getting food & escaping dangers are obscure in the highest degree. Nor must we overrate the degree of adaptation in the constitution of each living being to the climate of its own restricted home: when a new plant is introduced from a foreign land, until actual trial we cannot closely tell what range of climate it will endure. Even plants confined to certain islands, & which have never ranged, as far as we know beyond the narrow confines of their home, are found to endure very different climates: look at the Snowberry tree (*Chiococca racemosa*) how difficult to eradicate from our shrubberies, who would have ever supposed that it had been naturally confined to the West Indian islands? < Those who think each species created, as we now see it, will. Must we say that such island plants were created for the prospective chance of the island becoming joined to the mainland & then the plants in question spreading? (Stauffer, 1975).

While working on his *Natural Selection* book, Charles received a letter from Alfred Russel Wallace, in which he read a theory that was quite similar to his own. This made him seriously concerned that he would lose priority, and he famously wrote Lyell the following letter:

> My dear Lyell
>
> Some year or so ago, you recommended me to read a paper by Wallace in the Annals, which had interested you & as I was writing to him, I knew this would please him much, so I told him. He has to day sent me the enclosed & asked me to forward it to you. It seems to me well worth reading. Your words have come true with a vengeance that I shd. be forestalled. You said this when I explained to you here very briefly my views of "Natural Selection" depending on the Struggle for existence.— I never saw a more striking coincidence. If Wallace had my M.S. sketch written out in 1842 he could not have made a better short abstract! Even his terms now stand as Heads of my Chapters.
>
> Please return me the M.S. which he does not say he wishes me to publish; but I shall of course at once write & offer to send to any Journal. So all my originality, whatever it may amount to, will be smashed. Though my Book, if it will ever have any value, will not be deteriorated; as all the labour consists in the application of the theory.

> I hope you will approve of Wallace's sketch, that I may tell him what you say.
> My dear Lyell | Yours most truly | C. Darwin (Darwin Correspondence Project, letter no. 2285, n.d.-a).

Eventually, Charles' priority in conceiving natural selection was certified by an abstract of his theory that he had sent to the American botanist Asa Gray as early as 5 September 1857. He wrote:

> As you seem interested in subject, & as it is an immense advantage to me to write to you & to hear ever so briefly, what you think, I will enclose (copied so as to save you trouble in reading) the briefest abstract of my notions on the means by which nature makes her species (Darwin Correspondence Project, letter no. 2136, n.d.-b).

Darwin then rushed to write an abstract of his theory, which was published as *The Origin of Species* on 22 November 1859. There he described the main process, natural selection, as:

> Owing to this struggle for life, any variation, however slight and from whatever cause proceeding, if it be in any degree profitable to an individual of any species, in its infinitely complex relations to other organic beings and to external nature, will tend to the preservation of that individual, and will generally be inherited by its offspring. The offspring, also, will thus have a better chance of surviving, for, of the many individuals of any species which are periodically born, but a small number can survive. I have called this principle, by which each slight variation, if useful, is preserved, by the term of Natural Selection, in order to mark its relation to man's power of selection. We have seen that man by selection can certainly produce great results, and can adapt organic beings to his own uses, through the accumulation of slight but useful variations, given to him by the hand of Nature (Darwin, 1859, p. 61).

6.5 Conclusion

Following Darwin's writings between 1839, when he first fully developed the idea of natural selection in his notebooks, and 1859, when the *Origin of Species* was published, we can see a shift from perfect adaptation in a strict sense through limited perfection to relative adaptation. This is a conceptual shift, first documented by historian Dov Ospovat, which has nothing to do with religious views. This entails that Charles did not publish his theory for 20 years only because he was afraid of the reactions it would cause. Perhaps he was, but what is more important is that the theory was not fully developed right from the start in 1839, not even in 1844 when he wrote and shared the first draft of it. Therefore, the main change he underwent was from natural theology and Paley's ideas of perfect adaptation that had influenced him, to natural selection, not from theism to atheism.

In his private correspondence and autobiography, none of which were intended for publication, Darwin expressed his views about religion – quite clearly, in my view. For instance, in a letter to John Fordyce, written on 7 May 1879, Charles wrote:

Dear Sir

It seems to me absurd to doubt that a man may be an ardent Theist & an evolutionist.– You are right about Kingsley. Asa Gray, the eminent botanist, is another case in point– What my own views may be is a question of no consequence to any one except myself.– But as you ask, I may state that my judgment often fluctuates. Moreover whether a man deserves to be called a theist depends on the definition of the term: which is much too large a subject for a note. In my most extreme fluctuations I have never been an atheist in the sense of denying the existence of a God.– I think that generally (& more and more so as I grow older) but not always, that an agnostic would be the most correct description of my state of mind.

Dear Sir | Yours faithfully | Ch. Darwin (Darwin Correspondence Project, letter no. 12041, , n.d.-c).

Furthermore, in his autobiography, Charles wrote:

I cannot pretend to throw the least light on such abstruse problems. The mystery of the beginning of all things is insoluble by us; and I for one must be content to remain an Agnostic.

I feel most deeply that the whole subject is too profound for the human intellect. A dog might as well speculate on the mind of Newton. Let each man hope and believe what he can.

The old argument of design in nature, as given by Paley, which formerly seemed to me so conclusive, fails, now that the law of natural selection has been discovered. We can no longer argue that, for instance, the beautiful hinge of a bivalve shell must have been made by an intelligent being, like the hinge of a door by man. There seems to be no more design in the variability of organic beings and in the action of natural selection, than in the course which the wind blows. Everything in nature is the result of fixed laws (Barlow, 1958/2005, p. 87).

I have nothing else to add.

References

Barlow, N. (Ed.). (1958/2005). *The autobiography of Charles Darwin 1809–1882*. W.W. Norton.

Bowler, P. J. (1976). Malthus, Darwin and the concept of struggle. *Journal of the History of Ideas, 37*(4), 631–650.

Brooke, J. (1985). The relations between Darwin's science and his religion. In J. Durant (Ed.), *Darwinism and divinity* (pp. 40–75). Blackwell.

Brooke, J. H. (2009). Darwin and Victorian Christianity. In J. Hodge & G. Radick (Eds.), *The Cambridge companion to Darwin* (pp. 197–218). Cambridge University Press.

Burkhardt, R. W. (1995). *The Spirit of system: Lamarck and evolutionary biology*. Harvard University Press.

Corsi, P. (2005). Before Darwin: Transformist concepts in European natural history. *Journal of the History of Biology, 38*, 67–83.

Darwin Correspondence Project. (n.d.-a). Letter no. 2285, https://www.darwinproject.ac.uk/letter/DCP-LETT-2285.xml

Darwin Correspondence Project. (n.d.-b). Letter no. 2136, https://www.darwinproject.ac.uk/letter/DCP-LETT-2136.xml

Darwin Correspondence Project. (n.d.-c). Letter no. 12041, https://www.darwinproject.ac.uk/letter/DCP-LETT-12041.xml

Darwin, C. R. Notebook B: [Transmutation of species (1837–1838)]. CUL-DAR121.– Transcribed by Kees Rookmaaker. (Darwin Online, http://darwin-online.org.uk/).

Darwin, C. R. (1859). On the origin of species by means of natural selection, or the preservation of favoured races in the struggle for life. In J. van Wyhe (Ed.), *The complete work of Charles*

Darwin online. John Murray. http://darwin-online.org.uk/content/frameset?itemID=F373&viewtype=text&pageseq=1

Desmond, A. (1984). Robert E. Grant: The social predicament of a pre-Darwinian transmutationist. *Journal of the History of Biology, 17*(2), 189–223.

Evans, L. T. (1984). Darwin's use of the analogy between artificial and natural selection. *Journal of the History of Biology, 17*(1), 113–140.

Hodge, J., & Radick, G. (2009). Darwin's theories in the intellectual long run. In J. Hodge & G. Radick (Eds.), *The Cambridge companion to Darwin* (pp. 277–301). Cambridge University Press.

Hull, D. L. (2009). Darwin's science and Victorian philosophy of science. In J. Hodge & G. Radick (Eds.), *The Cambridge companion to Darwin* (pp. 173–196). Cambridge University Press.

Kampourakis, K. (2020). *Understanding evolution*. Cambridge University Press.

Keynes, R. (2001). *Darwin, his daughter and human evolution*. Riverhead Books.

Kohn, D. (2009). Darwin's keystone: The principle of divergence. In M. Ruse & R. J. Richards (Eds.), *The Cambridge companion to the "origin of species"* (pp. 87–108). Cambridge University Press.

Kohn, D., Murrell, G., Parker, J., & Whitehorn, M. (2005). What Henslow taught Darwin. *Nature, 436*(7051), 643–645.

Kuhn, T. S. (1962/1996). *The structure of scientific revolutions*. University of Chicago Press.

Lennox, J. G. (2005). Darwin's methodological evolution. *Journal of the History of Biology, 38*(1), 85–99.

Lyell, C. (1832). *Principles of geology*. John Murray.

Nersessian, N. J. (2008). *Creating scientific concepts*. MIT Press.

Ospovat, D. (1981). *The development of Darwin's theory: Natural history, natural theology and natural selection, 1838–1859*. Cambridge University Press.

Paley, W. (2006/1802). *Natural theology or evidence of the existence and attributes of the deity, collected from the appearances of nature*. Oxford University Press.

Roberts, M. (2001). Just before the beagle: Charles Darwin's geological fieldwork in Wales, summer 1831. *Endeavour, 25*(1), 33–37.

Ruse, M. (1975a). Charles Darwin and artificial selection. *Journal of the History of Ideas, 36*(2), 339–350.

Ruse, M. (1975b). Darwin's debt to philosophy: An examination of the influence of the philosophical ideas of John F.W. Herschel and William Whewell on the development of Charles Darwin's theory of evolution. *Studies in the History and Philosophy of Science, 6*, 159–181.

Ruse, M. (2000). Darwin and the philosophers: Epistemological factors in the development and reception of the origin of species. In R. Creath & J. Maienschein (Eds.), *Biology and epistemology* (pp. 3–26). Cambridge University Press.

Schweber, S. S. (1980). Darwin and the political economists: Divergence of character. *Journal of the History of Biology, 13*(2), 195–289.

Secord, J. A. (2000). *Victorian sensation: The extraordinary publication, reception, and secret authorship of vestiges on natural history of creation*. The University of Chicago Press.

Spencer, N. (2009). *Darwin and god*. SPCK.

Stauffer, R. C. (Ed.). (1975). *Charles Darwin's natural selection; being the second part of his big species book written from 1856 to 1858* (pp. 285–286). Cambridge University Press. Available at: http://darwin-online.org.uk/content/frameset?itemID=F1583&viewtype=text&pageseq=1

Sulloway, F. J. (1984). Darwin and the Galapagos. *Biological Journal of the Linnean Society, 21*(1–2), 29–59.

Vorzimmer, P. (1969). Darwin, Malthus and the theory of natural selection. *Journal of the History of Ideas, 30*(4), 527–542.

Vosniadou, S. (2013). *International handbook of research on conceptual change*. Routledge.

Wiker, B. (2009). What were Darwin's religious views? *Discovery Institute*. https://www.discovery.org/a/9501/

Kostas Kampourakis, teaches biology and science education courses at the University of Geneva. He is the author and editor of several books about evolution, genetics, philosophy, and the history of science. He is also the editor of the book series *Understanding Life*, published by Cambridge University Press. In the past, he was Editor-in-chief of the Springer journal *Science & Education*, co-editor of the Springer book series *Contributions from Biology Education Research*, and editor of the Springer book series *Science: Philosophy, History and Education.*

Chapter 7
"Great as Immensity, Deep as Eternity": What Could the *Grandeur* of Life Say About God's Existence, According to Darwin?

João F. N. B. Cortese

> *For so I created them free and free they must remain.*
> *John Milton,* Paradise Lost.

Abstract To what extent Charles Darwin's theory of natural selection is compatible with a religious view remains an open question for many, despite being debated in different contexts. The religious reception of the *Origin of Species* had a lot of impact on this, Darwin himself making additions in the subsequent editions of the book. Besides, in his writings, Darwin makes statements that one could well qualify as arguments regarding the reasonability of the existence of God – arguments that, being primarily skeptical, can be related to theological arguments. Darwin's described his position as an "agnostic," not saying that his theory would imply the inexistence of God. As far as natural theology is concerned, it is valuable to identify points he had to deal with as he departed from it throughout his life. This chapter discusses Darwin's assertions with theological implications, addressing passages of the *Origin* but also from his autobiography and some of his letters. Moreover, his responses to the *Vestiges of the Natural History of Creation* are considered, as well as his exchange with Asa Gray and some of his correspondence with other authors.

7.1 What Does the Diversity of Life Tell Us About God?

Let us begin by the end. The first edition of Darwin's *On the Origin of Species* finished with the following passage:

> There is grandeur in this view of life, with its several powers, having been originally breathed into a few forms or into one; and that, whilst this planet has gone cycling on

J. F. N. B. Cortese (✉)
Institute of Biosciences, University of São Paulo, São Paulo, Brazil

SPHERE Laboratory, UMR 7219, CNRS / Université Paris Cité, Paris, France
e-mail: joaocortese@gmail.com

M. Elice Brzezinski Prestes (ed.), *Understanding Evolution in Darwin's "Origin"*, History, Philosophy and Theory of the Life Sciences 34,
https://doi.org/10.1007/978-3-031-40165-7_7

> according to the fixed law of gravity, from so simple a beginning endless forms most beautiful and most wonderful have been, and are being, evolved (Darwin, 1859, p. 490).[1]

The second edition, from 1860, as well as the following editions, saw a change, as one could read the same passage with a small addition, life "having been originally breathed *by the Creator* into a few forms or into one" (Darwin, 1860b).

One can argue why Darwin added this passage the following year – perhaps wanting to show the public that his theory wouldn't be incompatible with the existence of God. We will come back to Darwin's personal beliefs. For now, this passage motivates the question: would Darwin's views be compatible with a religious view of the world? In which sense?

To answer those questions, we must address the complex relations between science and religion, discussed here in the case of Christianity. Darwin, we should remember, was raised in an Anglican Christian society, and part of his education was at Cambridge, where he arrived with the intention of becoming a clergyman. So, for evaluating his beliefs, we cannot ignore this initial background, and we should perceive it through the relevant particularities of such reformed Anglicans.[2]

Considering the Anglican Christian faith in which he was raised, we must first differ between "revealed" and "natural" theology. The first relates to aspects of the Christian faith expressed in phrases such as "the Bible as the Word of God" and "Jesus Christ as the Son of God" (Ruse, 1979, p. 64). The knowledge of such claims comes from special, divine revelation. The second defines a "knowledge of God derived through reason and the senses: the evidence of divinity directly discernible in the world."[3] Revelation would be supernatural – God reveals to men what they could not, by their own reason or any "natural" evidence, achieve to understand – one example being the Trinity: the fact that Father, Son, and Holy Spirit are three persons and one God. On its side, natural theology deals with how to evaluate the existence of God departing from the senses or reason.[4]

---

[1] Darwin's works are consulted in the Darwin Online project (Van Wyhe, 2002).

[2] Roman Catholic and Protestant theologies differ in important aspects, including the possibility of a rational demonstration of God (even if generalizing among thinkers is not an easy task here). In any case, we should keep in mind Darwin's Anglican background, recognizing also that several peculiarities could be considered here. One could think, for instance, on the importance of unitarianism (which denies the Trinity) at the time, including for Emma Darwin's religious beliefs (see "What did Darwin believe?" https://www.darwinproject.ac.uk/commentary/religion/what-did-darwin-believe; consulted on February 10, 2022). On Darwin's relationship to religion, see Ruse (2009), Brooke (2009) and Interdisciplinary Encyclopedia of Religion and Science (2022); more specifically on his relationship to natural theology, see Ospovat (1981). For a good biography of Darwin, see Browne (2006).

[3] Ruse (1979, p. 64). Ruse adds that it is not clear if "these two aspects of religion can or should be kept separated."

[4] "In contemporary philosophy [. . .] both 'natural religion' and 'natural theology' typically refer to the project of using all of the cognitive faculties that are 'natural' to human beings – reason, sense-perception, introspection – to investigate religious or theological matters" (Chignell & Pereboom, 2020).

So, regarding natural theology and evolution, a relevant question is what the plurality of species says about the existence of God. Do existing forms of life indicate anything about God's act of Creation?

Or, in the opposite direction, one can ask *how to explain the diversity of life*, a question for which one can find a kind of overlap between the attempts made in the domains of science and natural theology (at least in Darwin's time).

We need to define some other critical terms before proceeding further. Natural theology generally relies on two main arguments: cosmological and teleological.[5]

A "cosmological argument" stands for "a general pattern of argumentation (*logos*) that makes an inference from particular alleged facts about the universe (*cosmos*) to the existence of a unique being, generally identified with or referred to as God" (Reichenbach, 2021). That is to say, looking for a feature of the universe or life – say, its complexity – one would infer that God exists as its Creator, the first cause of natural phenomena.

"Teleological" or design arguments in natural theology "claim that the natural world displays some sort of purposive or end-directed design, and that this licenses the conclusion that the natural world has a very powerful and intelligent designer" (Chignell & Pereboom, 2020). Here, in natural theology (but not in all teleological thinking), one can see the tradition of searching for a final cause related to a designer.[6]

These general categories will help to analyse see how Darwin dealt with these problems in his time. In what follows, we will first consider Darwin's comments about a previous work that defended the transformation of species, explaining it by arguments that also appear in natural theology: the *Vestiges of the Natural History of Creation*. Then we will turn to cosmological arguments identified in his autobiographical account and design arguments in letters exchanged with Asa Gray. Finally, coming to conclusions, we will consider some of Darwin's declarations on the relation between his religious position and his theory in other correspondences.

---

[5] Chignell and Pereboom (2020) consider "arguments from religious experience" to be a *third* kind of a posteriori arguments in natural theology (side by side with the cosmological argument and with teleological or design arguments). It refers to a special "sense of divinity" included, for some authors, in "our ordinary human cognitive faculties." For others, it is caused by something other than that, not belonging, therefore, to natural theology.

[6] Refusing natural theology arguments in biology is not necessarily the same as refusing teleology. Indeed, one can talk on the one hand of a teleology associated with natural theology and on the other hand of a "teleological character of many explanations in the biological sciences" (Lennox, 2013, p. 157). Theodor H. Huxley wrote that "Teleology, as commonly understood, had received its deathblow at Mr. Darwin's hands" (Huxley, 1896, p. 82); but he also wrote that "there is a wider Teleology, that is not touched by the doctrine of Evolution, but is actually based upon the fundamental proposition of Evolution" (p. 86). In this sense, one could say, reconciling Huxley's comments, "that the teleology associated with natural theology had received its deathblow but that a selection-based teleology survives" (Lennox, 2013, p. 157).

7.2 *Mr. Vestiges'* Account of Creation

In 1844, the book *Vestiges of the Natural History of Creation* was published. His author was anonymous, called by some critics "Mr. Vestiges." Only in 1884, 13 years after the death of Robert Chambers (1802–1871), a Scottish journalist and editor, it was publicly revealed that he was its author (Secord, 2000, p. 22). Before that, however, such authorship was informally known to almost everyone (Secord, 2000, p. 434), but not before many months of speculation. Darwin's name appeared in the long list of likely authors, which included, beyond celebrated naturalists, "reformers and reactionaries, women and men, aristocrats and working-class socialists, novelists" (p. 433). Although the book was considered significantly different from what a clergyman might write (p. 20), this was precisely the assumption recorded by Darwin, more than a decade later, in the *Origin*. For in the "Historical Sketch," included from the third edition (Darwin, 1861b) of the book, Darwin explicitly refers to the author of the *Vestiges of Creation* as "the natural theologian."

Darwin points out that the tenth edition of the *Vestiges* contained considerable improvements. However, with clear critical intent, he quotes a not-so-short passage in which the "anonymous author" summarizes his explanation of the two impulses of change: living beings undergo adaptations by sudden leaps and vital forces. Then, advocating a gradual process of change (and without mention of that cause other than physics), Darwin concludes that those two supposed impulses explained nothing "in a scientific sense."[7]

The same opinion appeared in a short paragraph of the Introduction of the *Origin's* first and second editions – before being suppressed to the detriment of the Historical Sketch of the third edition:

> The author of the 'Vestiges of Creation' would, I presume, say that after a certain unknown number of generations, some bird had given birth to a woodpecker [. . .] and that these had been produced perfect as we now see them; but this assumption seems to me to be no explanation, for it leaves the case of the coadaptations of organic beings to each other and to their physical conditions of life, untouched and unexplained (Darwin, 1859, pp. 3–4).

In contrast, the most extended commentary in the Historical Sketch ends with what appears to be the actual intent or justification for dealing with such a "natural theologian": his incisive argument that species are not "immutable production":

> In my opinion it has done excellent service in calling in this country attention to the subject, in removing prejudice, and in thus preparing the ground for the reception of analogous views.[8] (Darwin, 1861b, p. xvi).

---

[7]Darwin wrote in 1845 to Hooker about the *Vestiges*: "his geology strikes me as bad and his zoology far worse" (Darwin, 1845).

[8]The *Vestiges* being now mentioned in the Historical Sketch (including the woodpecker example), the paragraph in the Introduction criticizing the *Vestiges* that appeared in the first edition disappeared in the third.

Indeed, one could read on the *Vestiges*:

> – In what way was the creation of animated beings effected? The ordinary notion may, I think, be not unjustly described as this, – that the Almighty author produced the progenitors of all existing species by some sort of personal or immediate exertion. But how does this notion comport with what we have seen of the gradual advance of species, from the humblest to the highest? (Chambers, 1844, pp. 162–163).

Accepting an action by God, the *Vestiges* admitted the transformation of species.[9] That is to say, although not constituting a sufficient explanation for Darwin (and having scientific imprecisions), the *Vestiges* removed prejudice (claiming that species could change)! Not bad as an opinion held by a "natural theologian," which shows that Darwin gave some importance to the works of natural theology.

7.3 Darwin's Religious Views: His Autobiography

Darwin declares in his autobiography that after his voyage on the Beagle, he "gradually came to disbelieve in Christianity as a divine revelation."[10] One of the reasons that made it difficult for him to see "how anyone ought to wish Christianity to be true" was the fact that those who did not believe would be "everlastingly punished," a case in which Darwin declared to recognize his father, brother, and almost all his best friends. The Old Testament's presenting God as a "revengeful tyrant," in Darwin's words, made him suspicious of the truth of the Scriptures (Darwin, 1958, p. 86, p. 87, and p. 85).

Darwin says that at a slow rate, but in a definite way, he reached his disbelief in Christianity. This did not prevent him from admiring the message of Christianity, but still, he thought this would be difficult to find in a coherent reading of the Bible:

> Beautiful as is the morality of the New Testament, it can hardly be denied that its perfection depends in part on the interpretation which we now put on metaphors and allegories (Darwin, 1958, p. 86).[11]

[9] Not only the *Vestiges* admitted the transformation of species, but it also argued against a simplistic view on the creative action of God: "how can we suppose that the august Being who brought all these countless worlds into form by the simple establishment of a natural principle flowing from his mind, was to interfere personally and specially on every occasion when a new shell-fish or reptile was to be ushered into existence on one of these worlds? Surely this idea is too ridiculous to be for a moment entertained." (Chambers, 1844, p. 164)

[10] Darwin's autobiography was published posthumously, and only in 1958 the edition "with the original omissions restored" was published by his grand-daughter Nora Barlow (Darwin, 1958). We follow this edition here.

[11] It is interesting to see that the reason that brings Darwin in tension with the Bible is one regarding how to interpret it (in literal or in a figurative way) – a problem related to what brought so much incomprehension regarding Galileo's position. In the Catholic context, Pope Leo XIII, in his encyclical *Providentissimus Deus* (1893), and much before Augustine of Hippo (354–430), in his *De Genesi ad Litteram*, made the point that one should not "take literally the cosmology of Genesis

More philosophically, if we may, Darwin says that he "did not think much about the existence of a personal God until a considerably later period" of his life, and he presents his conclusions on it:

> The old argument of design in nature, as given by Paley, which formerly seemed to me so conclusive, fails, now that the law of natural selection has been discovered. We can no longer argue that, for instance, the beautiful hinge of a bivalve shell must have been made by an intelligent being, like the hinge of a door by man. There seems to be no more design in the variability of organic beings and in the action of natural selection, than in the course which the wind blows. Everything in nature is the result of fixed laws (Darwin, 1958, p. 87).

Darwin is explicit about the fact that William Paley's design argument fails regarding the existence of a personal God.[12] Impressive findings in nature, such as the hinge of a bivalve shell, would not indicate an intentional design behind it once the action of natural selection has been revealed, such that "everything in nature is the result of fixed laws." One can note that Darwin doesn't touch here the question of whom or what established these "laws," and his use of this term also appears in the letters we analyze below.

Paley's famous analogy compared an organism with a watch and inferred that an artificer should have designed organisms according to a purpose[13] as the watch has its watchmaker. In 1868, nine years after the *Origin*'s first edition, Darwin discussed the subject at the end of his book, *The variation of animals and plants under domestication* (Darwin, 1868). There he exposes an argument – we will come back to it below – that later he would consider, in his autobiography, as never answered. In what follows in his autobiography, however, Darwin writes:

> But passing over the endless beautiful adaptations which we everywhere meet with, it may be asked how can the generally beneficent arrangement of the world be accounted for? (Darwin, 1958, p. 88).

It is legitimate to interpret these words as saying that not everything seemed well explained to Darwin: that "adaptations" claim for more explanations and that Darwin remained with a *question* about the possibility of understanding a sufficient reason for the world.

One could say that Darwin is struggling with the plausibility of the cosmological argument. That passage of his autobiography shows that he is discussing a central question to theology – the very existence of God – as he talks not just about the history of his personal beliefs but also about *reasons* for believing or not in God.

Besides denying Paley's argument of design, he considers the question regarding the existence of suffering in the world and even concerning the "argument from

because drawing scientifically precise pictures of nature is not the Bible's concern" (Haught, 2013, p. 487).

[12] This does not prevent Darwin from recognizing, also in his autobiography, the importance Paley had on his formation, saying that, together with Euclid, the study of their works "was the only part of the Academical Course which, as I then felt and as I still believe, was of the least use to me in the education of my mind" during his formation in Cambridge (Darwin, 1958, p. 59).

[13] Paley (1802/2006). See Paley's quote in Kampourakis' contribution to this volume.

religious experience." Concerning the last, Darwin remembers in his autobiography that he wrote in his journal, when he was in the middle of a Brazilian forest, that "it is not possible to give an adequate idea of the higher feelings of wonder, admiration, and devotion which fill and elevate the mind" (Darwin, 1958, p. 91), but he also says in the autobiography that this "sense of sublimity," despite being hardly explainable, "can hardly be advanced as an argument for the existence of God" (Darwin, 1958, p. 92).

Thus, Darwin is arguing not only if God exists but which arguments could be considered to decide if God exists – a proper theological debate.[14]

Darwin then returns to a kind of cosmological argument as something that touches him most strongly:

> Another source of conviction in the existence of God, connected with the reason and not with the feelings, impresses me as having much more weight. This follows from the extreme difficulty or rather impossibility of conceiving this immense and wonderful universe, including man with his capacity of looking far backwards and far into futurity, as the result of blind chance or necessity. When thus reflecting I feel compelled to look to a First Cause having an intelligent mind in some degree analogous to that of man; and I deserve to be called a Theist (Darwin, 1958, p. 93).

But Darwin says he doubts if one can trust such conclusions drawn by the "mind of man, which has, as I fully believe, been developed from a mind as low as that possessed by the lowest animal" (Darwin, 1958, p. 93). He then comes to a definition of his own position:

> I cannot pretend to throw the least light on such abstruse problems. The mystery of the beginning of all things is insoluble by us; and I for one must be content to remain an Agnostic. (Darwin, 1958, p. 94).

Darwin, the "skeptical naturalist," discussing which arguments concerning the existence of God are more or less reasonable, arrives at the position of *not knowing* about the final answer to the subject – it is worthy to note that the term "agnostic" was claimed to have been coined by Thomas Huxley.[15] We will come back to this term to designate Darwin's position. But, for the time being, let us see how Darwin reacted to some "design" arguments.

---

[14] In the last chapter of *The Descent of man*, Darwin also presents an argument regarding a problematic argument concerning the existence of God: "I am aware that the assumed instinctive belief in God has been used by many persons as an argument for His existence. But this is a rash argument, as we should thus be compelled to believe in the existence of many cruel and malignant spirits, only a little more powerful than man; for the belief in them is far more general than in a beneficent Deity" (Darwin, 1871, p. 395).

[15] The term would have been coined in 1869, in one of the meetings of the *Metaphysical Society*. See Huxley (1899/1904).

7.4 A Direct Discussion with a Proponent of Design: Asa Gray

Asa Gray (1810–1888) was an American Presbyterian botanist who taught natural history at Harvard University. After a brief encounter with Darwin in 1838, he eventually became a supporter of the British naturalist in the United States. Darwin and Gray exchanged about 300 hundred letters[16] between 1854 and 1881.

One of the exciting aspects of this correspondence concerns the issue of design in nature precisely. On the one hand, Darwin and Gray would agree on the importance of adaptations and "contrivances," seeing a new kind of teleology appearing in evolution (different from the teleology of natural theology); on the other hand, their agreement was not so simple, since the idea of design seemed to bring them apart (Lennox, 2013).

On 5 June 1861, Darwin wrote to Gray that, studying domestic variations, he came to "differ more" from his correspondent, as he saw "what an enormous field of undesigned variability there is ready for natural selection to appropriate for any purpose useful to each creature" (Darwin, 1861a).

A difference in their position was indeed also recognized by Gray (Lennox, 2013), who wrote in a letter in 1863:

> Of course we believers in *real design*, make the most of your frank and natural terms, "contrivance, purpose", &c — and pooh-pooh your endeavors to resolve such contrivances into necessary results of certain physical processes! (Gray, 1863).

Darwin indeed had used the term "contrivance," even at the title of one of his works: *On the Various Contrivances by Which British and Foreign Orchids Are Fertilised by Insects, and on the Good Effects of Intercrossing*, from Darwin, 1862 – a work well appreciated by Gray. But the difficult question is: what should one understand by "contrivance" here?[17]

In the conclusion of *The variation of animals and plants under domestication* (Darwin, 1868), Darwin wrote that on the one hand fragments of rock falling from a cliff are perhaps not "accidental," "for the shape of each depends on a long sequence of events, all obeying natural laws," but at another hand, regarding their use, as in the case of a man building his house, "their shape may be strictly said to be accidental" – making an important distinction between two senses of "accidental":

[16] For Darwin and Gray's exchange of letters, see https://www.darwinproject.ac.uk/asa-gray. We follow here the analysis made by Lennox (2013).

[17] In a review of this work, G. D. Campbell wrote: "It is curious to observe the language which this most advanced disciple of pure naturalism instinctively uses when he has to describe the complicated structure of this curious order of plants" (Campbell 1862, pp. 292–293; *apud* https://www.darwinproject.ac.uk/letter/?docId=letters/DCP-LETT-4056.xml). In Darwin's 1942 sketch for the *Origin*, one can read "We must look at every complicated mechanism and instinct, as the summary of a long history, ⟨as the summing up⟩ of useful contrivances, much like a work of art" (Darwin, 1909, p. 51).

> And here we are led to face a great difficulty, in alluding to which I am aware that I am travelling beyond my proper province. An omniscient Creator must have foreseen every consequence which results from the laws imposed by Him. But can it be reasonably maintained that the Creator intentionally ordered, if we use the words in any ordinary sense, that certain fragments of rock should assume certain shapes so that the builder might erect his edifice? (Darwin, 1868, vol. 2, p. 431).

As Darwin insinuates, one should not think that such a pre-established order by the Creator could be recognized in this case. Thus, regarding the problem of how would the innumerable variations of domestic animals be so ordered for the sake of the breeder, Darwin writes that we should not think that

> [. . .] variations, alike in nature and the result of the same general laws, which have been the groundwork through natural selection of the formation of the most perfectly adapted animals in the world, man included, were intentionally and specially guided. However much we may wish it, we can hardly follow Professor Asa Gray in his belief "that variation has been led along certain beneficial lines," like a stream "along definite and useful lines of irrigation" (Darwin, 1868, vol. 2, p. 432).

Darwin was thus led, despite his agreement with Asa Gray on the importance of adaptation, to recognize differences from his correspondent regarding the design that would act on variations.[18] Besides, Darwin realized that the problem of considering together variations in forms of life and the omnipotence and omniscience of God was such a problematic issue that we are thus "brought face to face with a difficulty as insoluble as is that of free will and predestination" (Darwin, 1868, pp. 431–432).[19]

In an 1860 letter to Gray, Darwin would acknowledge his limitations regarding how one could make sense of the "wonderful universe" just as a "result of brute force":

> Down Bromley Kent.
>
> May 22.
>
> [...] With respect to the theological view of the question; this is always painful to me.— I am bewildered.— I had no intention to write atheistically. But I own that I cannot see, as plainly as others do, & as I sh[d] wish to do, evidence of design & beneficence on all sides of us. There seems to me too much misery in the world. I cannot persuade myself that a beneficent & omnipotent God would have designedly created the Ichneumonidæ with the express intention of their feeding within the living bodies of caterpillars, or that a cat should play with mice. Not believing this, I see no necessity in the belief that the eye was expressly designed. On the other hand I cannot anyhow be contented to view this wonderful universe

---

[18] Regarding the analogy from the rock fragments, Lennox (2013, p. 155) writes: "the lengthy correspondence with Gray has helped [Darwin] to differentiate two notions of chance that are not clearly distinguished in the *Origin*." Lennox also highlights the fact that the rock fragments were not *designed* by the builder for their roles in the building, but *selected* by him.

[19] In a letter responding to M. E. Boole, the widow of the mathematician and logician George Boole (who had passed away in 1864), Darwin wrote in 14 December 1866 that "it has always appeared to me more satisfactory to look at the immense amount of pain & suffering in this world, as the inevitable result of the natural sequence of events, i.e. general laws, rather than from the direct intervention of God though I am aware this is not logical with reference to an omniscient Deity — Your last question seems to resolve itself into the problem of Free Will & Necessity which has been found by most persons insoluble." (Darwin, 1866)

> & especially the nature of man, & to conclude that everything is the result of brute force. I am inclined to look at everything as resulting from designed laws, with the details, whether good or bad, left to the working out of what we may call chance. Not that this notion *at all* satisfies me. I feel most deeply that the whole subject is too profound for the human intellect. A dog might as well speculate on the mind of Newton.— Let each man hope & believe what he can.—.
>
> Certainly I agree with you that my views are not at all necessarily atheistical. The lightning kills a man, whether a good one or bad one, owing to the excessively complex action of natural laws,—a child (who may turn out an idiot) is born by action of even more complex laws,—and I can see no reason, why a man, or other animal, may not have been aboriginally produced by other laws; & that all these laws may have been expressly designed by an omniscient Creator, who foresaw every future event & consequence. But the more I think the more bewildered I become; as indeed I have probably shown by this letter.
>
> Most deeply do I feel your generous kindness & interest.—.
>
> Yours sincerely & cordially | Charles Darwin (Darwin, 1860a).

Darwin still confessed to himself "bewildered" by these questions.

In any case, an important role was given to the "laws [. . .] designed by an omniscient Creator." The first edition of the *Origin* brought two significant epigraphs by Bacon and Whewell, showing an effort by Darwin to expose the possibility of conjugating his theory with laws of nature.[20] Darwin would perhaps then share a view acceptable by Gray that "biological adaptation was achieved by divinely instituted laws of nature" (Lennox, 2013, p. 155). But as we saw, the status of variation is not the same for Gray and Darwin, such that Gray's "design" could not be seen in the whole of Darwin's theory.

This did not prevent Darwin from still writing in 1860, on the second edition of the *Origin*:

> I see no good reason why the views given in this volume should shock the religious feelings of any one. A celebrated author[21] and divine has written to me that "he has gradually learnt to see that it is just as noble a conception of the Deity to believe that He created a few original forms capable of self-development into other and needful forms, as to believe that He required a fresh act of creation to supply the voids caused by the action of His laws"[22] (Darwin, 1860b).

[20]On the third edition, Darwin added to these epigraphs, one by Samuel Butler, that would have been suggested by Asa Gray (Moore, 1981, p. 324). I thank Nelio Bizzo for this reference. On the aforementioned epigraphs, see Brooke (2009, p. 265).

[21]The author in question is Charles Kingsley, in his letter to Darwin from 18 November 1859 (Kingsley, 1859).

[22]Darwin wrote at the first edition of the *Origin*: "I should infer from analogy that probably all the organic beings which have ever lived on this earth have descended from some one primordial form, into which life was first breathed" (Darwin, 1859, p. 484). However, as we said in the beginning of our text, in the *Origin*'s second edition, he added "by the Creator" at the end of the last phrase, as well as mentioned "the Creator" at the very last sentence of book. Not only Darwin changed the phrase, but he wrote to Hooker in 1863: "I have long regretted that I truckled to public opinion & used Pentateuchal term of creation, by which I really meant 'appeared' by some wholly unknown process.— It is mere rubbish thinking, at present, of origin of life; one might as well think of origin of matter" (Darwin, 1863). Darwin eliminated the reference to the Creator at the first mentioned phrase in the subsequent editions of the *Origin* (from the third to the sixth). However, despite this

Darwin would then write that his theory was compatible with divine laws.[23] But to see precisely how this could be was a matter that made him "bewildered."[24]

7.5 What Darwin Said to Others: The Importance of His Letters on the Subject

As we have seen, from the point of view of Darwin's writings there is not a necessary incompatibility between evolutionism and religion – at least from the point of view of combining evolution and the laws of Creation according to certain Christian views.[25] We can read as early as 1868, in a letter by John Henry Newman (canonized as a saint by the Catholic Church in 2019):

> It does not seem to me to follow that Creation is denied because the Creator, millions of years ago, gave laws to matter. He first created matter and then he created laws for it — laws which should *construct* it into its present wonderful beauty, and accurate adjustment and harmony of parts *gradually*. [...] Mr. Darwin's theory *need* not then to be atheistical, be it true or not; it may simply be suggesting a larger idea of Divine Prescience and Skill. [...] I do not [see] that 'the accidental evolution of organic beings' is inconsistent with divine design — It is accidental to *us*, not to *God* (Newman, 1868/1973, pp. 77–78).

We can classify as "theistic evolution" the position that reconciles creation and evolutionary process," which we can find inclusively – but as an exception in his position – in Darwin's writings.[26] But what about Darwin's general position? We already recognized the "accidental" variations on the forms of life as a possible obstacle to such "theistic evolution," in a sense just mentioned. But what did he

---

affirmation and this attitude, he never eliminated the last sentence of the book, which remained for all editions, such that a reference to the "Creator" endured (Numbers, 2013, p. 476).

[23] In the 1842 sketch to the *Origin*, Darwin wrote: "It accords better with [our modesty] the lowness of our faculties to suppose each must require the fiat of a creator, but in the same proportion the existence of such laws should exalt our notion of the power of the omniscient Creator" (Darwin, 1909, p. 52).

[24] We should note that considering the laws of nature as established by God allows to consider that God has an *action* on Creation that is not a direct *intervention* at each moment. With regard to the importance of the laws of creation in the interpretation of the changes of forms of life, one should also consider Charles Babbage and his importance to Darwin (see, for instance, Pimentel (2020) and Snyder (2011)). On Darwin's view of "Laws impressed on matter by the Creator", see also Brooke (2009).

[25] The present chapter does not address the issue of which *ethics* is implied by a Darwinian view of evolution and if it is compatible with a religious one.

[26] For a passage by Darwin himself that could be considered under the classification of "theistic evolution," see the quotation of Notebook B in Kampourakis' contribution to this volume. If one adopts the distinction between "theism (an immanent God)" and "deism (God as unmoved mover)" (Ruse, 2009, p. 372), a possibility is to consider Darwin a "deist" in some of his positions: according to Brooke (2009, p. 264). "in 1859 he [Darwin] was a deist – one who had rejected revelation as a source of knowledge but who was unwilling to regard the laws of nature as either self-explanatory or accidental".

conclude on the matter or God's place in a universe of evolving forms of life? Regarding this aspect, it is interesting to analyze Darwin's correspondence.

In a letter to the clergyman John Fordyce in 1879, Darwin wrote that it seemed "absurd to doubt that a man may be an ardent Theist & an evolutionist," giving Asa Gray as an example. Regarding Darwin himself, he wrote: "I may state that my judgment often fluctuates" (Darwin, 1879).

It is thus crucial to restate that Darwin acknowledges that his position oscillated. However, he stated, "in my most extreme fluctuations I have never been an atheist in the sense of denying the existence of a God" (Darwin, 1879). "Not knowing" or "agnostic," writes Darwin, seemed more and more the correct term to describe his state of mind.[27]

In a letter from 27 March 1873, N. D. Doedes, a student at the University of Utrecht, asked "on what grounds" Darwin believed in the existence of God, confessing that he could no longer believe, having left the study of theology. Doedes wrote:

> I suppose, you are Deïst (I mean, believing in a God who has created the universe with unvariable laws, and who now does not mind it anymore); else I cannot conceive how you do combine your faith with your knowledge. Is your chief ground for your belief in God perhaps *this*, that you think a first cause, a Creator, needed for the universe? (Doedes, 1873a).

On 2 April, Darwin replied:

> It is impossible to answer your question briefly; and I am not sure that I could do so, even if I wrote at some length. But I may say that the impossibility of conceiving that this grand and wondrous universe, with our conscious selves, arose through chance, seems to me the chief argument for the existence of God; but whether this is an argument of real value, I have never been able to decide. I am aware that if we admit a first cause, the mind still craves to know whence it came and how it arose. Nor can I overlook the difficulty from the immense amount of suffering through the world. I am, also, induced to defer to a certain extent to the judgment of the many able men who have fully believed in God; but here again I see how poor an argument this is. The safest conclusion seems to be that the whole subject is beyond the scope of man's intellect; but man can do his duty (Darwin, 1873).

Again, Darwin confessed himself bewildered by the "grand and wondrous universe," seeing there an argument for the existence of God, but still not conclusive.

On 4 April, Doedes wrote back, recognizing that the universe

> [. . .] *is* most wondrous and grand ("great as immensity, deep as eternity", by the beautiful expression of Mr. Carlyle)[28] (Doedes, 1873b).

But Doedes said he couldn't conceive how that could be if the universe "did not 'arise,' in the sense of commencing," assuming that Darwin's work, showing how the organic world would arise from natural causes, was not incompatible with the

---

[27] Darwin (1879). See the quotation of this letter in Kampourakis' contribution to this volume.

[28] In fact, Doedes misquoted Carlyle, who wrote in fact "wide as Immensity, deep as Eternity" (Carlyle, 1902, 2, p. 124). See the notes by the Darwin Correspondence Project to this letter (Doedes, 1873b).

possibility that the universe itself has "ever been." Finally, Doedes asked Darwin, if the question was "beyond the scope of man's intellect," how could he mention the Creator in the conclusion of the *Origin* or Chap. 2 of *The Descent of Man*? To our knowledge, Darwin did not reply to that. In any case, Doedes' argument is not necessary: showing how the diversity of life originated does not imply that the universe has ever been – if this is so, the possibility that it "arose" persists, as well as showing how anything did come into being.[29] If Leibniz were right to say that the *metaphysical* question "why is there something rather than nothing?" could only be answered by a necessary being,[30] Darwin would perhaps agree that this was well "beyond the scope" of his theory.[31]

In an 1878 letter to James Grant, Darwin wrote:

> The strongest argument for the existence of God, as it seems to me, is the instinct or intuition which we all (as I suppose) feel that there must have been an intelligent beginner of the Universe; but then comes the doubt and difficulty whether such intuitions are trustworthy (Darwin, 1878).

Finally, in a famous letter from 3 July 1881 to W. Graham, Darwin wrote that Graham had expressed his "inward conviction, though far more vividly and clearly" than he could have done, "that the Universe is not the result of chance." According to Darwin, one can see one of the passages where the complexity of the universe seemed challenging to explain without God. But also interesting is a note to this phrase, written by his son Francis Darwin:

> The Duke of Argyll ('Good Words,' Ap. 1885, p. 244) has recorded a few words on this subject, spoken by my father in the last year of his life. "... in the course of that conversation I said to Mr. Darwin, with reference to some of his own remarkable works on the 'Fertilisation of Orchids,' and upon 'The Earthworms,' and various other observations he made of the wonderful contrivances for certain purposes in nature—I said it was impossible

[29] Doedes himself acknowledges he is dealing with a *possibility*, since he writes: "Therefore it seems to me, that the impossibility of conceiving how this universe arose by chance, does not urge me to believe in God; because it may be possible that the materia has ever been."

[30] Leibniz (1714/1998), sections 7–8.

[31] Darwin's humility, if we may say, regarding the epistemological limits of his theory is impressive. But so it is his availability in talking about it; he replies to Doedes, a student, in a matter of days, and in the quoted letter to M. E. Boole, the widow of the logician Boole, Darwin added a P.S.: "I am grieved that my views should incidentally have caused trouble to your mind but I thank you for your Judgment & honour you for it, that theology & science should each run its own course & that in the present case I am not responsible if their meeting point should still be far off" (Darwin, 1866). One could still consider here a difference between something that *begins* in time and the metaphysical *origin* of something (in a different understanding of the word "origin" than Darwin's one): "To speak about the origin, is to speak about an act which is not limited to one specific moment in time. The origin must not be mistaken for the beginning. To speak about the beginning is to acknowledge something new and therefore to distinguish between what was before and what is now. It means establishing a difference in time, conceived as an ongoing, continuous experience. The origin does not confine us to one single instant in time. It is not one occurrence among others, but it is the constituting condition of all that is, at every moment, in the course of events which have happened in the time-space continuum. The origin cannot be adequately circumscribed to a scientific theory" (Maldamé, 2011).

to look at these without seeing that they were the effect and the expression of mind. I shall never forget Mr. Darwin's answer. He looked at me very hard and said, 'Well, that often comes over me with overwhelming force; but at other times,' and he shook his head vaguely, adding, 'it seems to go away'" (Darwin, 1887, p. 216, note).

7.6 Conclusions

Let us conclude that regarding religion, Darwin's "final" (or more precisely "fluctuating," as he wrote) position was agnosticism, implying that there are limits to scientific knowledge regarding reality. Blaise Pascal wrote that one should look for "the God of Abraham, Isaac, and Jacob, not the God of the philosophers and scientists" (Pascal, 1654/1991, p. 50). Unexpectedly, Darwin would perhaps agree with him: science cannot answer all possible questions, and evolution would prove neither God's existence nor the inexistence. As remembered by John Milton's *Paradise Lost*, a work beloved by Darwin, God has created men free – one could say, in particular, free to believe or not.

It is interesting to note that Darwin's position on this subject appears, as we saw, mainly in his correspondence – perhaps because he taught that a scientific text such as the *Origin* was not the proper place to debate God's existence. Nevertheless, the declarations he added to the book's later editions may also reflect a strategy to show that it is possible to understand his theory as compatible with the belief in God rather than an effort to prove God's existence.

In this sense, reading the *Origin* can be compatible with a religious view.[32] It is also consistent with an agnostic view of life – as the one Darwin himself professed at some moments. An atheist should find no problems with it (if bracketing metaphysical questions, as it is appropriate on a positive scientific method). One can thus see that some part of the polemic on the possibility of one being a religious person and accepting a Darwinian evolution is a false problem that should not have such a significant impact as the one seen not only in academia but also in the media and at schools.[33]

For natural theology, in particular, the question is far from simple. Although Darwin abandoned natural theology's arguments, such as Paley's watchmaker (a design argument), he did not discard all work associated with it in one way or

---

[32] As we saw, this *possibility* would be the case for Christianism – including Darwin's Anglican heritage, as well as the actual positions of the Catholic Church (cf. Haught, 2013) and some protestant churches. If accepting evolution would also be a *necessity* is a matter of debate in some cases.

[33] As we said, we are talking here about the compatibility between Darwinian evolution and certain views of Christianity. Other problems may arise regarding other religions, or Christians that rely on a literal reading of the Bible, proposing an *intelligent design* Creationism with no space for evolution – cf., for example, Numbers (2013) and Ruse (2009). Moreover, what to conclude metaphysically from evolution is not a consensus: does it imply a materialistic or a reductionist view, for instance? (see Haught, 2013).

another. The *Vestiges* prepared "the ground for the reception of analogous views." Gray's religious appreciations of adaptations were partially (even if not entirely) compatible with Darwin's. The history of science, and more generally the history of thought, does not proceed by "all or nothing" steps but involves multiple variations to use a Darwinian biological term.

In his autobiography and several letters, we see Darwin "bewildered" on how to explain this "wonderful universe" – a problem, as we saw, related to the cosmological argument. Darwin's 1842 first sketch of the *Origin* reads that there is "a simple grandeur" in the referred view of life – the word "simple" not appearing in the *Origin's* published versions. But that shouldn't throw us off the line of reasoning since Darwin also wrote there was "much grandeur" in this view.[34] Evolution's "simple" grandeur is indeed vast – but is it as great as immensity and as deep as eternity?

Acknowledgements I would like to thank Maria Elice Brzezinski Prestes for the opportunity to elaborate these reflections, both in the classroom and for this book, as well as for all her comments. I would also like to thank all my colleagues of the Philosophy and Biology study group at the Institute of Biosciences of the University of São Paulo for the discussion we had on this text. In particular, I would like to thank Roberto Rozenberg for this attentive reading of the text and his suggestions.

References

Brooke, J. H. (2009). "Laws impressed on matter by the Creator"?: The origin and the question of religion. In M. Ruse & R. J. Richards. (eds.), *The Cambridge companion to the origin of species* (pp. 256–274). Cambridge University Press.

Browne, J. (2006). *Darwin's origin of species: A biography*. Atlantic Books.

Campbell, G. D. (1862). Review of orchids and other works. *Edinburgh Review, 116*, 378–397.

Carlyle, T. (1902). *The French revolution* (Vol. 3 vols). George Bell and sons.

Chambers, R. (1844). *Vestiges of the natural history of creation* (1st ed.). Churchill.

Chignell, A., & Pereboom, D. (2020). Natural theology and natural religion. In E. N. Zalta (Ed.), *The Stanford encyclopedia of philosophy*. https://plato.stanford.edu/archives/fall2020/entries/natural-theology/

Darwin, C. R. (1845). Letter to J. D. Hooker, January 7 1845. Consulted on Darwin correspondence project, "letter no. 814". https://www.darwinproject.ac.uk/letter/?docId=letters/DCP-LETT-814.xml. Accessed on February 7, 2022.

Darwin, C. R. (1859). In J. Murray (Ed.), *On the origin of species by means of natural selection, or the preservation of favoured races in the struggle for life* (1st ed.). Consulted on Van Wyhe.

[34]Darwin wrote in the sketch: "There is much grandeur in looking at the existing animals either as the lineal descendants of the forms buried under thousand feet of matter, or as the coheirs of some still more ancient ancestor. It accords with what we know of the law impressed on matter by the Creator, that the creation and extinction of forms, like the birth and death of individuals should be the effect of secondary [laws] means. It is derogatory that the Creator of countless systems of worlds should have created each of the myriads of creeping parasites and [slimy] worms which have swarmed each day of life on land and water ⟨on⟩ [this] one globe" (Darwin, 1909, p. 51).

Darwin, C. R. (1860a, May 22). Letter to Asa Gray. Consulted on Darwin correspondence project. “letter no. 2814”. https://www.darwinproject.ac.uk/letter/?docId=letters/DCP-LETT-2814.xml. Accessed on February 7, 2022.
Darwin, C. R. (1860b). In J. Murray (Ed.), *On the origin of species by means of natural selection, or the preservation of favoured races in the struggle for life* (2nd ed.). Consulted on Van Wyhe (2002-).
Darwin, C. R. (1861a, June 5). Letter to Asa Gray. Consulted on Darwin correspondence project, “letter no. 3176”. https://www.darwinproject.ac.uk/letter/?docId=letters/DCP-LETT-3176.xml. Accessed on February 9, 2022.
Darwin, C. R. (1861b). *On the origin of species by means of natural selection, or the preservation of favoured races in the struggle for life* (3rd ed.). John Murray. Consulted on Van Wyhe (2002–).
Darwin, C. R. (1862). *On the various contrivances by which British and foreign orchids are fertilised by insects, and on the good Efect of intercrossing*. John Murray. Consulted on Van Wyhe (2002–).
Darwin, C. R. (1863, March 29). Letter to J. D. Hooker. Consulted on Darwin correspondence project, “letter no. 4065”. https://www.darwinproject.ac.uk/letter/?docId=letters/DCP-LETT-4065.xml. Accessed on March 4, 2022.
Darwin, C. R. (1866). Letter to M. E. Boole. Consulted on Darwin correspondence project, “letter no. 5307”. https://www.darwinproject.ac.uk/letter/?docId=letters/DCP-LETT-5307.xml. Accessed on February 3, 2022.
Darwin, C. R. (1868). *The variation of animals and plants under domestication* (Vol. 2, 1st ed.). John Murray. Consulted on Van Wyhe (2002–).
Darwin, C. R. (1871). *The descent of man, and selection in relation to sex* (Vol. 2, 1st ed.). John Murray. Consulted on Van Wyhe (2002–).
Darwin, C. R. (1873, April 2). Letter to N. D. Doedes. Consulted on Darwin correspondence project, “letter no. 8837”. https://www.darwinproject.ac.uk/letter/?docId=letters/DCP-LETT-8837.xml. Accessed on February 7, 2022.
Darwin, C. R. (1878, March 11). Letter to J. Grant. Consulted on Darwin correspondence project, “letter no. 11416”. https://www.darwinproject.ac.uk/letter/?docId=letters/DCP-LETT-11416.xml. Accessed on February 10, 2022.
Darwin, C. R. (1879). Letter to J. Fordyce. In: J. Fordyce, *Aspects of scepticism*, London: Elliot stock, 1883, p. 190. Consulted on Darwin correspondence project, “letter no. 12041”. https://www.darwinproject.ac.uk/letter/?docId=letters/DCP-LETT-12041.xml. Accessed on February 3, 2022.
Darwin, C. R. (1887). *The life and letters of Charles Darwin, including an autobiographical chapter, edited by his son, Francis Darwin* (Vol. 1). John Murray. Consulted on Van Wyhe (2002).
Darwin, F. (1909). *The foundations of the origin of species, a sketch written in 1842*. Cambridge University Press. Consulted on Van Wyhe (2002).
Darwin, C. R. (1958). *The autobiography of Charles Darwin, 1809–1882*. Edited and with appendix and notes by his grand-daughter Nora Barlow. Consulted on Van Wyhe (2002–).
Doedes, N. D. (1873a, March 27). Letter to Charles Darwin. Consulted on Darwin correspondence project, “letter no. 8828”. https://www.darwinproject.ac.uk/letter/?docId=letters/DCP-LETT-8828.xml. Accessed on February 9, 2022.
Doedes, N. D. (1873b, April 4). Letter to Charles Darwin. Consulted on Darwin correspondence project, “letter no. 8841”. https://www.darwinproject.ac.uk/letter/?docId=letters/DCP-LETT-8841.xml. Accessed on February 9, 2022.
Gray, Asa. (1863, March 22–30). Letter to Charles Darwin. Consulted on Darwin correspondence project, “letter no. 4056”. https://www.darwinproject.ac.uk/letter/?docId=letters/DCP-LETT-4056.xml. Accessed on February 9, 2022.
Haught, J. F. (2013). Darwin and Catholicism. In M. Ruse (Ed.), *The Cambridge encyclopedia of Darwin and evolutionary thought* (pp. 485–492). Cambridge University Press.
Huxley, T. H. (1896). *Darwiniana: Essays*. D. Appleton.

Huxley, T. H. (1899/1904). Agnosticism. In *Collected essays* (5th ed.). Science and Christian tradition.
Interdisciplinary Encyclopedia of Religion and Science. (2022). The Role of a Hypothetic Creator is Compatible with the Evolution of the Biological Species. https://inters.org/Darwin-Creator-Origin-Species. Accessed on February 10, 2022.
Kingsley, C. (1859, November 18). *Letter to Charles Darwin. Consulted on Darwin Correspondence Project, "Letter no. 2534"*. https://www.darwinproject.ac.uk/letter/?docId=letters/DCP-LETT-2534.xml. Accessed on 20 August 2023.
Leibniz, G. W. (1714/1998). Principles of nature and grace, based on reason. In R. S. Woolhouse & R. Francks (Eds.), *Philosophical texts* (pp. 258–266). Oxford University Press.
Lennox, J. G. (2013). Darwin and teleology. In M. Ruse (Ed.), *The Cambridge encyclopedia of Darwin and evolutionary thought* (pp. 152–157). Cambridge University Press.
Maldamé, J.-M. (2011). The various meanings of the word evolution in science, philosophy and theology. In G. Auletta, M. R. Leclerc, & A. Martínez (Eds.), *Biological evolution: Facts and theories. A critical appraisal 150 years after "the origin of species"* (pp. 557–571). G&B Press.
Moore, J. R. (1981). *The post-Darwinian controversies: A study of the Protestant struggle to come to terms with Darwin in Great Britain and America, 1870–1900*. Cambridge University Press.
Newman, J. H. (1868/1973). Letter to J. Walker of Scarborough. In C. S. Dessain & T. Gornall (Eds.), *The letters and diaries of John Henry Newman* (Vol. XXI). Clarendon Press.
Numbers, R. L. (2013). Creationism. In M. Ruse (Ed.), *The Cambridge encyclopedia of Darwin and evolutionary thought* (pp. 476–484). Cambridge University Press.
Ospovat, D. (1981). *The development of Darwin's theory: Natural history, natural theology, and natural selection, 1838–1859*. Cambridge University Press.
Paley, W. (1802/2006). *Natural theology or evidence of the existence and attributes of the deity, collected from the appearances of*. Oxford University Press.
Pascal, B. (1654/1991). Le Mémorial. In *Œuvres complètes* (Vol. 3). Desclée de Brouwer.
Pimentel, O. L. (2020). Charles Babbage: teologia, técnica e economia no Nono tratado de Bridgewater. In *Doctoral dissertation*. Universidade de São Paulo.
Reichenbach, B. (2021). Cosmological argument. In E. N. Zalta (Ed.), *The Stanford encyclopedia of philosophy*. https://plato.stanford.edu/archives/win2021/entries/cosmological-argument/
Ruse, M. (1979). *The Darwinian revolution: Science red in tooth and claw*. University of Chicago Press.
Ruse, M. (2009). Belief in god in a Darwinian age. In J. Hodges & G. Radick (eds.), *The Cambridge companion to Darwin* (pp. 368–389). Cambridge University Press.
Secord, J. A. (2000). *Victorian sensation: The extraordinary publication, reception, and secret authorship of vestiges on natural history of creation*. The University of Chicago Press.
Snyder, L. J. (2011). *The philosophical breakfast club: Four remarkable friends who transformed science and changed the world*. Random House.
Van Wyhe, J. (2002). The complete work of Charles Darwin online. http://darwin-online.org.uk/. Accessed on February 7, 2022.

João F. N. B. Cortese has a joint Ph.D. degree, in Epistemology and History of Science (Université de Paris 7 – Denis Diderot) and in Philosophy (University of São Paulo). He is currently a professor of Philosophy at the Institute of Biosciences of the University of São Paulo, as well as an associate researcher at the SPHERE Laboratory (CNRS / Université Paris Cité). His works deal with the History and Philosophy of Science, as well as Bioethics.

Chapter 8
Mr. Darwin's Beloved Barnacles: Using Cirripedes to Understand Evolution in "Origin of Species"

Marsha L. Richmond

Abstract In 1835, as a young naturalist on board the *Beagle* expedition exploring the delights of South American flora and fauna, Charles Darwin encountered a tiny new barnacle off the coast of Chile that he found most curious. Unlike all the usual shelled species found attached to rocks or even ship hulls, this one lived "naked," sheltered in the crevices of seashells. "Mr Arthrobalanus," as he dubbed the unusual little creature, continued to intrigue Darwin far beyond the initial discovery. Little did he know that some 10 years later he would embark on a 6-year-long taxonomic project that not only described and classified Mr. Arthrobalanus but also all other known cirripede species, both living and fossil. Nor could he imagine that in undertaking this endeavor, he would not only cement his reputation as an eminent naturalist but also test his developing ideas about the evolution of life on Earth. This chapter reveals how Darwin's study of barnacles sheds essential light on many foundational evolutionary tenets laid out in his next major work: *On the Origin of Species* (1859).

8.1 Introduction

In mid-January 1835, Charles Darwin was exploring the shoreline of the small island of Huafo (Guafo) near Chiloé in the Chonos Archipelago off the coast of Chile. As naturalist on board the HMS *Beagle*, charting the coasts off South America, Darwin was charged with collecting scientifically interesting plants, animals, and mineral specimens. On his evening expedition, he "extracted from the rock a good many fossil shells—" to take back to the ship to inspect later (Darwin, 1998, p. 279).[1] Once

[1] See also Darwin Online, Richard Keynes, "Charles Darwin's Ornithological and Animal Notes: An Introduction," http://darwin-online.org.uk/EditorialIntroductions/Keynes_Animal_notes_Intro.html

M. L. Richmond (✉)
Professor Emerita, Department of History, Wayne State University, Detroit, MI, USA
e-mail: marsha.richmond@wayne.edu

M. Elice Brzezinski Prestes (ed.), *Understanding Evolution in Darwin's "Origin"*, History, Philosophy and Theory of the Life Sciences 34,
https://doi.org/10.1007/978-3-031-40165-7_8

able to examine the shells more carefully, he discovered a creature living inside unlike any he had ever seen. Buried within the shell of a gastropod mollusk, he found a "most curious" tiny cirripede, as he related in his research notes. The shell was "completely drilled by the cavities formed by this animal"[2] (Fig. 8.1).

Darwin knew this organism was new to science, and he proceeded to describe it and tentatively identify it as a burrowing barnacle belonging to the Balanidae, sessile acorn cirripedes most frequently found attached to rocks and ship bottoms, and he also found eggs inside that resembled larva he previously observed in Crustacea.[3] This was curious indeed, given that at the time barnacles were classified as mollusks, not crustaceans.

Darwin filed his notes away, sending the shell back to London along with other specimens. Little did he know that "Mr Arthrobalanus," as he later dubbed this strange little barnacle, would eventually help guide him toward developing a theory of *evolution* or, as he described it, species transmutation. Nor could he know that some 10 years later he would invest 8 years in a taxonomic project to classify Mr. Arthrobalanus and all other known cirripedes, both living and fossil. This chapter shows how Darwin's study of barnacles sheds essential light on many foundational evolutionary tenets laid out in his next major work: *On the Origin of Species* (1859).

8.2 Mr. Arthrobalanus

Darwin returned to the little barnacle again in 1846, when, after completing works on the geology and natural history of the *Beagle* voyage, he reviewed specimens that remained undescribed. As he told his friend Joseph Dalton Hooker in October, he planned to:

> [...] begin some papers on the lower marine animals, which will last me some months, perhaps a year, & then I shall begin looking over my ten-year-long accumulation of notes on species & varieties which, with writing, I daresay will take me five years, & then when published, I daresay I shall stand infinitely low in the opinion of all sound naturalists—so this is my prospect for the future (Darwin Correspondence Project, letter no. 1003, n.d. [2 October 1846]).

Once completed, he aimed to begin his "big species book." This, however, was not how things worked out.

[2] Some of Charles Darwin's original drawings from the *Beagle* voyage are in DAR 29.3: 72. Darwin Archive, Reference Code: GBR/0012/MS DAR, Cambridge University Library, Cambridge; hereafter, CUL. See also Darwin Online, http://darwin-online.org.uk/content/frameset?itemID=F1840&viewtype=text&pageseq=1

[3] While a medical student at Edinburgh University in 1825, Darwin was introduced to marine organisms by Robert Grant and frequently naturalized along the coast of the Firth of Forth, examining marine invertebrates, especially paying attention to their reproductive systems. See Janet Browne, *Charles Darwin*, vol. 1: *Voyaging* (New York: Alfred A. Knopf, 1995).

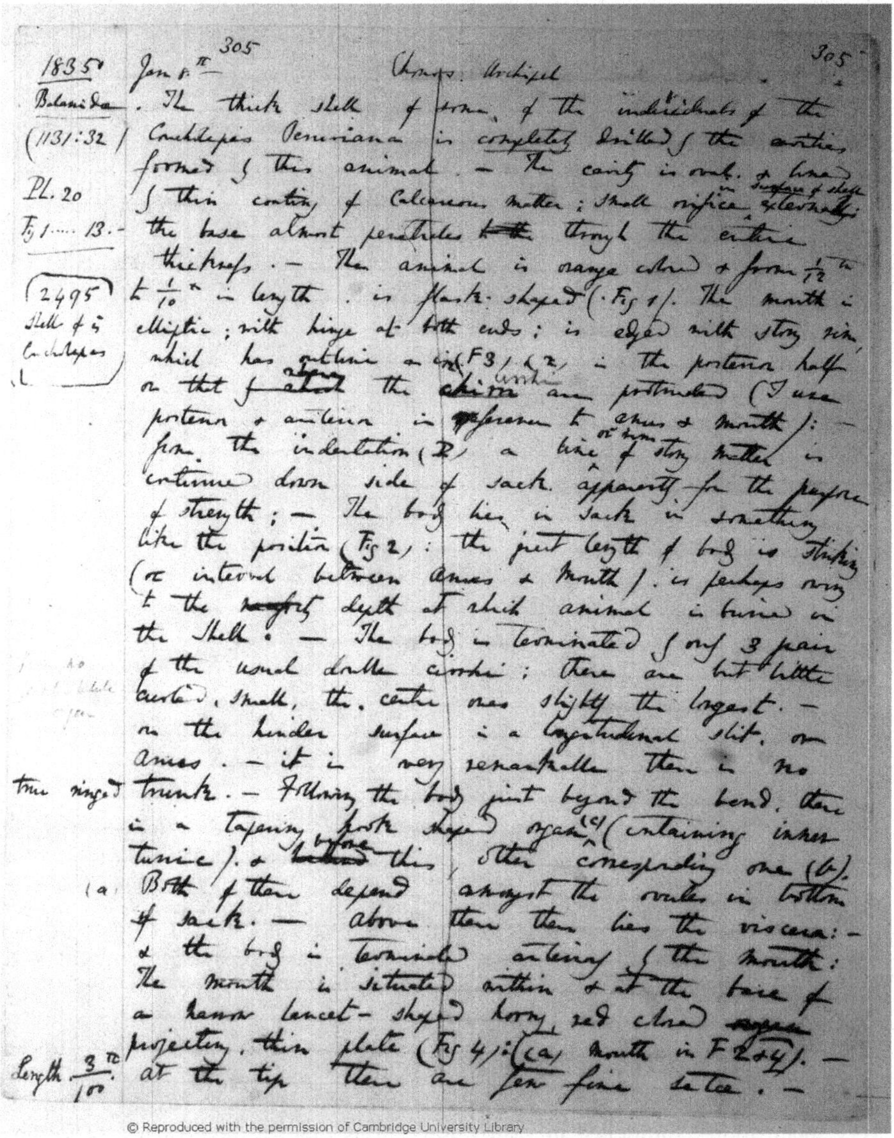

Fig. 8.1 Zoological diary: Chonos Archipelago, "Diary of observations on zoology of the places visited during the voyage". Darwin, C. R. 1835.01. Zoological diary: Chonos Archipelago. CUL-DAR31.305–314. (Edited by John van Wyhe (Darwin Online, http://darwin-online.org.uk/). Permission to reproduce requested from Cambridge University Library)

Darwin worked daily "in dissecting a little animal about the size of a pin's head from the Chonos Arch.," noting he "could spend another month on it, & daily see some more beautiful structure" (Darwin Correspondence Project, letter

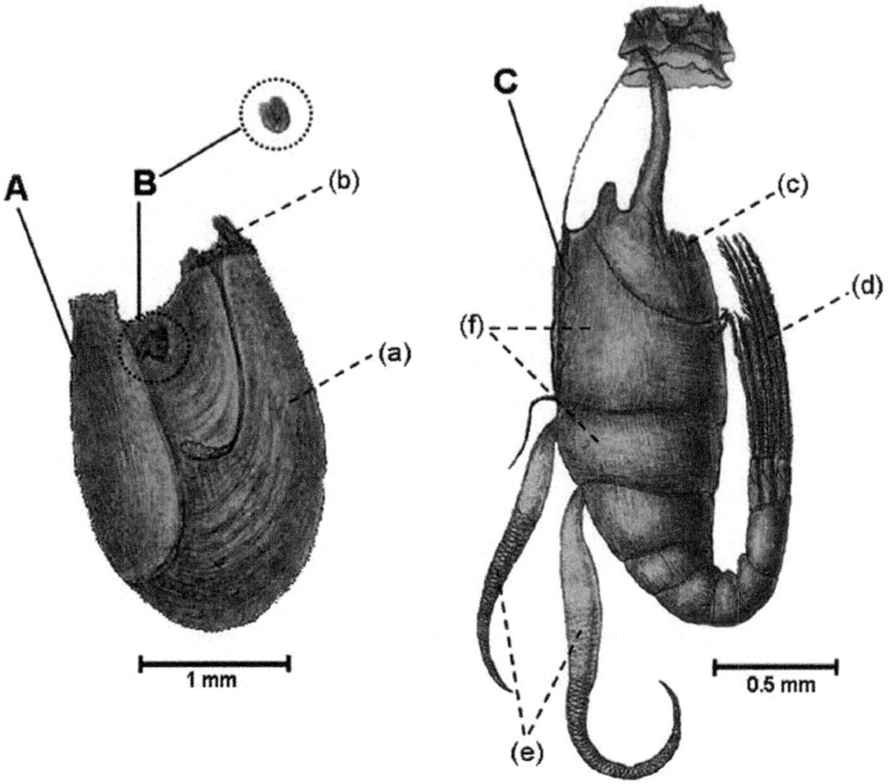

Fig. 8.2 Cryptophialus minutus Darwin, 1854. A. Lateral view of a female inside the mantle. B. Male (inside the circle) attached to free worn edge of the mantle. C. Lateral view of shell-less female removed from the mantle: (**a**) mantle or tunic encapsulating the female; (**b**) aperture ornaments and entrance to the internal mantle cavity; (**c**) maxillae; (**d**) three pairs of thoracic cirri; (**e**) dorsal body appendages; (**f**) body segments. (From Castilla, J. C. (2009). Darwin Taxonomist: Barnacles and Shell Burrowing Barnacles. Revista Chilena de Historia Natural 82: 477–483 (p. 480). Permission to reproduce granted by the author)

no. 1014, n.d. Darwin to Robert FitzRoy, 28 October [1846]). He was reading "heaps of papers on Cirripedia" and consulting with Hooker, saying "I have so very much more confidence in your observations than in mine own." "The more I read," he told his friend, "the more singular does our little fellow appear"; he soon gave "Mr Arthrobalanus" the more proper scientific name, *Cryptophialus minutus* (Fig. 8.2). Not an adept artist, Darwin was delighted when Hooker volunteered to prepare drawings of the larval specimen, which he found "quite beautiful" and begged for more, especially one "with legs retracted." More and more, Darwin understood how singular this little creature was, "for its relations are to various very distinct families" (Darwin Correspondence Project, letter no. 1012, n.d. [26 October 1846]). Classifying him would lead Darwin in a different direction.

The weeks spent on Mr. Arthrobalanus soon stretched into months and then almost a year, having acquired other specimens from naturalists. He gradually gained self-confidence in anatomical dissection and judging characteristics deemed important in tracing its relations to other barnacles.

Knowing that his next project was to write up his admittedly controversial views on species change, Darwin took note of Hooker's critical comment about someone whose taxonomic descriptions he did not trust. Paraphrasing Hooker's quip, Darwin repeated "that no one has hardly a right to examine the question of species who has not minutely described many" (Darwin Correspondence Project, letter no. 915, n.d. [10 September 1845]). He took this unintended criticism to heart and began to consider the value of not only describing Mr. Arthrobalanus but all living and fossil cirripedes – a daunting task but one that to date no naturalist had undertaken. Moreover, it was also scientifically important given the current confusion over whether barnacles should be ranked as mollusks or crustaceans. Finally, Darwin knew Hooker's view was right: such a study would provide an excellent way to test his ideas about species change over time and help him better understand varieties and variation as well as classification, which would provide him the professional gravitas by which to become a trustworthy naturalist before engaging in the "species question."

Darwin was not a total newcomer to principles of classification. Having adopted a view of species transmutation a short time after returning from the *Beagle* voyage, he grounded taxonomic relationships on genealogical principles. As he told the well-regarded taxonomist George Robert Waterhouse in July 1843:

> According to my opinion, [...] classification consists in grouping beings according to their actual relationship, ie their consanguinity, or descent from common stocks— In this view all relations of analogy &c &c &, consist of those resemblances between two forms, which they do not owe to having inherited it, from a common stock.— To me, of course, the difficulty of ascertaining true relationship ie a natural classification remain just the same, though I know what I am looking for.— This being the case viz. ignorance of a distinct object I think, we ought to look at classification as a simple logical process, i.e. a means of conveying much information through single words— (Darwin Correspondence Project, letter no. 684, n.d. [26 July 1843]).

Thus, while tracing inheritance from immediate ancestors formed a general tenet of Darwin's classificatory views, this was more difficult for a creature like Mr. Arthrobalanus, whose "relations are to various very distinct families." Darwin realized he needed additional rules to classify unusual species and to map out relations within a large genus, family, and class.

By November 1846, Darwin told Richard Owen he had "worked out pretty carefully my new articulated Balanus" and "become so much interested in the structure of the sessile Cirripedes" that he was "dissecting 5 or 6 of the other genera." His approach was to "work out mouths & cirri carefully, muscular structure & tunics of the sack, & a some of the structure of the viscera," although unsure whether this was necessary (Darwin Correspondence Project, letter no. 1033, n.d. 25 November

[1846]).[4] He also focused on generative organs, characters commonly useful to taxonomists. In this regard, he told Hooker about his surprising finding:

> I believe Arthrobalanus has no ovisac at all!, & that the appearance of one is entirely owing to the splitting, & tucking up to the posterior penis, of the inner membrane of sack.— I have just found a Cirripede with an indisputably abortive anterior penis; so that this chief anomalous feature (viz two penes) in Arthrobalanus is in some degree brought within bounds (Darwin Correspondence Project, letter no. 1022, n.d. [12 November 1846]).

Clearly Darwin's knowledge of cirripede anatomy was growing by leaps and bounds, and so was his proficiency at using hand lenses, microscopes, and contemporary microscopic technique – all necessary to dissect tiny creatures and immature specimens. Darwin enjoyed this work on what he called his "beloved Barnacles" (Darwin Correspondence Project, letter no. 1166, n.d.-a [26 March 1848]).

By 1848, Darwin decided to undertake a dedicated taxonomic study of the *Cirripedia*.[5] As he simply explained: "I had originally intended to have described only a single abnormal Cirripede, from the shores of South America, and was led, for the sake of comparison, to examine the internal parts of as many genera as I could precure" (Darwin, 1851).[6] In this he was encouraged by John Edward Gray, keeper of the zoological department of the British Museum, who loaned him the Museum's collection. He also relied on the kindness of collectors around the world, who sent him new and exotic specimens to describe.

8.3 The Cirripede Monographs, 1851 and 1854

Darwin did indeed benefit from the cirripede study. He discovered the vast amount of variation that existed between individuals, varieties, and species that supported his views about speciation. He learned firsthand from grappling with vexed questions

[4] Darwin asked Owen, one of the leading anatomists in England, to read and criticize his description, but in the event, he did not publish a separate paper but provided a description in his 1851 monograph on the *Cirripedia*. In the Preface to *Living Cirripedia*, Darwin explained his classificatory criteria: "I have given the specific or diagnostic characters, deduced from the external parts alone, in both Latin and English. As I found, during the progress of this work, that a similarly abbreviated character of the softer internal parts, was very useful in discriminating the species, I have inserted it after the ordinary specific character" (vol. 1, p. ix).

[5] In a letter to Louis Agassiz, thanking him for sending specimens, Darwin revealed that "when doubting whether to undertake a monograph of the class, or to confine myself to their anatomy, your sentence that 'a monograph on the Cirripedia was a pressing desideratum in Zoology' much helped to decide me" (Darwin Correspondence Project, letter no. 1205, 22 October 1848).

[6] In his *Autobiography*, Darwin gave a similar explanation for his decision: "To understand the structure of my new Cirripede I had to examine and dissect many of the common forms; and this gradually led me on to take up the whole group." Charles Darwin, *The Life and Letters of Charles Darwin, Including an Autobiographical Chapter*, ed. Francis Darwin, 3 vols. (London: John Murray, 1887), vol. 1, pp. 80–81; also Darwin Online, http://darwin-online.org.uk/content/frameset?itemID=F1452.1&viewtype=text&pageseq=1

about whether forms should be considered mere varieties or independent species and from suitably grouping forms into separate genera of barnacles. These were important lessons to learn for someone who intended soon to engage in the "species question."

8.3.1 The Evolution of Sex

An even more significant outcome of the cirripede study, however, was the opportunity it offered for considering the evolution of sex. Darwin had long reflected on the origin of sex, having sketched out ideas about the separation of the sexes from an ancestral hermaphrodite in Notebook D (1838, p. 162).[7] In the 1840s, he tested his views on how separate sexes could evolve by considering flowers with abortive stamens or pistils, which might indicate a transition toward independent sexes (Darwin Correspondence Project, letter no. 582, n.d. [21 November 1840]). Such views on how separate sexes might evolve helped Darwin become more receptive to a "curious point" relating to sex in *Cryptophialus minutus* (aka Mr. Arthrobalanus) that led to a major discovery (Darwin Correspondence Project, letter no. 1166, n.d.-b [26 March 1848]).

Most barnacles are hermaphrodites, possessing both male and female sexual organs. However, *Cryptophialus* was found to have separate sexes. This suggested that some forms might illustrate the emergence of two separate sexes or, in Darwin's terminology, become *bisexual.* On the lookout for such forms, Darwin discovered two other genera that indeed have separate males and females, *Ibla* and *Scalpellum.* In reporting on the "generative systems" of the genus *Ibla* (*Living Cirripedia* 1: 196–203), Darwin noted that not only were there separate sexes in *I. cumingii* but that, most surprising, the minute male was parasitic on the female. "In Ibla, the males and females are not organically united, but only permanently and immovably attached to each other. We have in this genus the additional singularity of occasionally two males parasitic on one female" (Darwin, 1851, p; 201).

This finding was particularly important for testing Darwin's views on how species change occurred, as he coyly implied in 1851:

> The individuals in every other genus (with the exception of Scalpellum), in the several families, in the three Orders of Cirripedia, are hermaphrodite or bisexual. Why, then, is Ibla

[7] "Theory of sexes (woman makes bud, man puts primordial vivifying principle) one individual secretes two substances, although organs for the double purpose are not distinguished, (yet may be presumed from hybridity of ferns) afterwards they can be seen distinct (in dioecious plants in their abortive sexual organs?): they then become so relates to each other as never to be able to impregnate themselves (this never happens in plants, only in subordinate manner in the plants which have male & female flowers on same stem. —) so that Molluscous hermaphroditism takes place. — thus one organ in each becomes obliterated, & sexes as in Vertebrates take place. — ∴ every man & woman is hermaphrodite: —" See Darwin Online, http://darwin-online.org.uk/content/frameset?itemID=F373&viewtype=text&pageseq=1

unisexual; yet, becoming, in the most paradoxical manner, from its earliest youth, essentially bisexual? Would food have been deficient, and was the seizure of infusoria by another and differently constructed individual, necessary for the support of the male and female organs? The orifice of the sack of the female is unusually narrow; would the presence of testes and vesiculæ seminales have rendered her thorax and prosoma inconveniently thick? Seeing the analogous facts in the six, differently-constructed species of the allied genus Scalpellum, I infer there must be some profounder and more mysterious final cause (Darwin, 1851, p. 203).

Even more perplexing was his discovery in *I. quadrivalvis*, a hermaphrodite, of independent or "complemental" males. His surprise is evident even in his published reference:

Considering the whole case, there seems no room to doubt the justness of the conclusion arrived at, under the former as well as under the present species, namely, that these little parasites are the males of the two species of Ibla to which they are attached;—wonderful though the fact be, that in one case, the male should pair with an hermaphrodite already provided with efficient male organs. It is to bring this fact prominently forward, that I have called such males, Complemental Males; as they seem to form the complement to the male organs in the hermaphrodite.... If the final cause of the existence of these Complemental Males be asked, no certain answer can be given; .... No explanation, as we have seen, can be given of the much simpler case of the mere separation of the sexes in Ibla Cumingii; nor can any explanation, I believe, be given of the much more varied arrangement of the parts of fructification in plants of the Linnean class, Polygamia (Darwin, 1851, p. 214).

Other species likewise contributed to his investigation of the "final cause" of such relations. In *Scalpellum*, he also found parasitic males. In *Alcippe*, which had separate sexes, he discovered females were inexplicably degenerate, assuming characters that resembled males. Did this mean that species could also undergo a transition from (using his terms) *bisexuality* to *unisexuality*? His delight at such discoveries is palpable in his correspondence. For instance, he credited his species theory with directing him toward such discoveries. As he told Hooker:

I have been getting on well with my beloved Cirripedia, & got more skilful in dissection: ... I have lately got a bisexual cirripede, the male being microscopically small & parasitic within the sack of the female; I tell you this to boast of my species theory, for the nearest & closely allied genus to it is, as usual, hermaphrodite, but I had observed some minute parasites adhering to it, & these parasites, I now can show, are supplemental males, the male organs in the hermaphrodite being unusually small, though perfect & containing zoosperms: so we have almost a polygamous animal, simple females alone being wanting. I never sh[d]. Have made this out, had not my species theory convinced me, that an hermaphrodite species must pass into a bisexual species by insensibly small stages, & here we have it, for the male organs in the hermaphrodite are beginning to fail, & independent males ready formed. But I can hardly explain what I mean, & you will perhaps wish my Barnacles & Species theory al Diabolo together. But I don't care what you say, my species theory is all gospel (Darwin Correspondence Project, letter no. 1174, n.d. 10 May 1848).

Darwin later pointed to the variation and transitions in the sexual systems of cirripedes to support his views about descent and modification in *Origin of Species*.

Darwin's views on evolution also benefited in several other ways from his study of cirripedes. For instance, barnacles reinforced his notion of the importance of crosses between individuals, thus supporting his view of cross-fertilization as a means (as he believed, in contrast to our modern understanding) of maintaining species.[8] In addition, most importantly, barnacles bolstered his conviction that genealogy (as established by the focus on embryonic development) provided the best evidence for establishing taxonomic relationships.

8.4 Understanding *Origin of Species* Through the Cirripedes

There are several important elements of Darwin's theory of evolution by natural selection that are well illustrated by examining passages on cirripedes in *Origin of Species* (Darwin, 1859/1964). These include Darwin's description of variation, intercrossing, importance of embryology, transition of organs, classification, and the imperfection of the geological record as an explanation for the lack of transitional forms.

8.4.1 The "Meaning" of Intercrossing

The first time Darwin referenced cirripedes in *Origin* was in Chap. 4: Natural Selection. In this chapter, he continued the discussion of the "struggle for existence," the topic of Chap. 3, by addressing variation. He opened by noting how ubiquitous variation is in nature, evidenced by the "endless number of strange peculiarities" in "domestic productions" (plants and animals domesticated for human uses) and, "to a lesser degree, those under nature" (Darwin, 1859/1964, p. 80). What was the importance of such variation?

Already Darwin had noted "that many more individuals are born than can possibly survive" and that "individuals having any advantage, however slight, over others, would have the best chance of surviving and of procreating their kind." The preservation of "favorable variations" and rejection of "injurious variations" is what Darwin called "natural selection" (Darwin, 1859/1964, p. 81). He continued:

> It may be said that natural selection is daily and hourly scrutinising, throughout the world, every variation, even the slightest; rejecting that which is bad, preserving and adding up all that is good; silently and insensibly working, whenever and wherever opportunity offers, at the improvement of each organic being in relation to its organic and inorganic conditions of life. We see nothing of these slow changes in progress, until the hand of time has marked the long lapse of ages, and then so imperfect is our view into long past geological ages, that we

[8] As he told the Danish biologist J.J.S. Steenstrup: "You will perhaps be interested by hearing that I once found a Balanus which had had its probosciform organ cut off & healed *absolutely imperforate*, & yet the ova had been impregnated & contained larvae; some of the neighbouring individuals in the cluster having certainly impregnated these ova." (Darwin Correspondence Project, letter no. 1330, 20 May [1850]).

> only see that the forms of life are now different from what they formerly were (Darwin, 1859/1964, p. 84).

After mentioning several examples, Darwin turns to a section "On the Intercrossing of Individuals."

As previously mentioned, Darwin adopted the "almost universal belief of breeders, that with animals and plants a cross between different varieties, or between individuals of the same variety but of another strain, gives vigour and fertility to the offspring; and on the other hand, that close interbreeding diminishes vigour and fertility." This led him to believe "that it is a general law of nature (utterly ignorant though we be of the meaning of the law) that no organic being self-fertilises itself for an eternity of generations; but that a cross with another individual is occasionally—perhaps at very long intervals—indispensable" (Darwin, 1859/1964, pp. 96, 97).

It was for this reason that cirripedes (and other hermaphrodites) seemingly posed a challenge to the supposed "general law of nature" of the necessity for intercrossing. That is to say, if an individual possesses both male and female sex organs, would they not always self-fertilize? He addressed this issue indirectly in volume 2 of *Living Cirripedia* (Darwin, 1854), where he mentioned "the singular case of some elongated specimens of Balanus balanoides, from Tenby, in South Wales" in which:

> In six out of these seven specimens, the probosciformed penis was quite short and abruptly truncated, as if from abortion. By cutting off the truncated apex, and cleaning the external tissue, I ascertained that it was imperforate, apparently in all the cases, and I am certain of this fact in several of the cases.... So that these three individuals certainly could not have impregnated their own eggs; nevertheless, within the shell of these very three, there were perfectly developed larvæ: I am led to conclude from this fact, that adjoining specimens in a perfect condition had, by means of their long probosciformed penis, effected the fecundation of their imperfect neighbours (Darwin, 1854, pp. 101–102).

While a promising observation, this was nonetheless a conjecture not based on direct evidence.

Darwin was thus delighted to receive news from C. S. Bate that a naturalist he knew had observed cirripedes in the act of copulation. Darwin immediately responded with a flurry of questions:

> The points on which I so much wish for more information, are.— Which was the species; if not known could I see a specimen of the kind. Was the probosciformed penis inserted into more than one individual? For about how long time was it inserted? Was it inserted deeply & at which end of valves? Especially did the recipient individual continue during the time exserting its cirri? Did it keep its opercular valves widely open for the reception of the organ? I am anxious to know whether the recipient was a willing agent or adulterer, or whether it was a case of rape by act.— If the recipient was in full vigour, I think it wd be impossible to insert anything without its consent. Were the specimens under water at time? (Darwin Correspondence Project, letter no. 2175F, n.d. 29 November 1857).

He told Bate he would like to contact the individual in order "that I might quote his authority."

In turn, Bate sent Darwin his correspondent's response. Richard Bishop replied that had he known his observations were new, he would have paid more attention. Still, he tried to answer Darwin's questions:

> 1st. The Species —Balanus communis?
> 2nd. Was the penis inserted into more than one individual? not positive, but I believe it was. It was extended to a length equal to about 3 times that of any single branch of the cirrus, & waved in every direction till it came in contact with the cirrus of a neighbouring balanus, when it was inserted, I believe into more than one. It should be observed that my group is not large, consisting of only 5 full sized animals & a few small ones.
> third The insertion did not occupy more than 2 or 3 seconds.
> 4th. Not inserted deeply as far as I could judge— at which end of valves was not noticed.
> 5 & 6. My own impression is decidedly, that the recipient individual was during the time exerting its cirri, with more than usual energy, & gave evidence of the intruder being a welcome guest, but on these points my memory is unfortunately not supported by that of a friend who joined me in noticing the act.
> 7. The specimens were under water at the time, I have not under other circumstances seen any movement of the cirri (Darwin Correspondence Project, letter no. 2179, n.d. 3 December 1857).

Darwin was delighted; even if this verification was not as robust as he would have liked, he trusted it enough to cite it tentatively in *Origin*:

> Of aquatic animals, there are many self-fertilising hermaphrodites; but here currents in the water offer an obvious means for an occasional cross. And, as in the case of flowers, I have as yet failed ... to discover a single case of an hermaphrodite animal with the organs of reproduction so perfectly enclosed within the body, that access from without and the occasional influence of a distinct individual can be shown to be physically impossible. Cirripedes long appeared to me to present a case of very great difficulty under this point of view; but I have been enabled, by a fortunate chance, elsewhere to prove that two individuals, though both are self-fertilising hermaphrodites, do sometimes cross. ... From these several considerations and from the many special facts which I have collected, ... I am strongly inclined to suspect that, both in the vegetable and animal kingdoms, an occasional intercross with a distinct individual is a law of nature (Darwin, 1859/1964, pp. 100–101).

On this basis, he concluded that "the difference between hermaphrodites and unisexual species, as far as function is concerned, becomes very small." Rather than being an exception to a "law of nature," cirripedes thus offered evidence of the evolution of reproductive systems that could offer organisms greater procreative success. The importance of this rule became more evident in future chapters.

8.4.2 Laws of Variation

Variation in organisms – including both external characters and internal organs and systems – was a major pillar of Darwin's theory of speciation by natural selection. His hypothesis envisioned change occurring through the natural tendency of individuals born with a "favorable variation" being more likely to survive and pass this variation onto their progeny. Over a long period of time, this would result in the gradual increase in those possessing this variation in the population, which could lead to speciation.

Thus, it is not surprising that Darwin discussed different aspects related to variation in Chaps. 1, 2, 3, 4, and 5 of *Origin*. Chap. 1 focused on "variation under domestication" and Chap. 2 on "variation under nature." Chap. 3 introduced the important premise of the "struggle for existence," which was the motive force behind selection, that is, all organisms experience a struggle to survive within their particular environment. To explain this, Darwin drew on the well-known principles governing human population increase and decline laid out 50 years earlier by the Reverend Thomas Robert Malthus (1766–1834) in *An Essay on the Principle of Population* (1798). Applying this principle to plants and animals, Darwin argued that organisms reproduce at rates governed by their inherited reproductive mechanisms. Moreover, all depend for their survival on securing a steady food supply. Thus, the population numbers of species are naturally determined by population size and the availability (or lack thereof) of food. This interdependent relationship sets up competition among individuals for access to adequate resources that enable them to survive and reproduce. The principle of a "struggle for existence" provided the motive force for change in species over time.

Just how this change occurs was developed in Chap. 4, in which Darwin explains "natural selection." Given a struggle for limited resources, those individuals who possess a variation that might provide them with any kind of advantage over those who do not have it would be more likely to survive and to pass it along to their children. Thus, variations and their inheritance were both key aspects of Darwin's theory. That's where the cirripedes came in.

Darwin, as we have seen, believed that "intercrosses between individuals of the same species" were important. As he explained:

> Intercrossing plays a very important part in nature in keeping the individuals of the same species, or of the same variety, true and uniform in character. It will obviously thus act far more efficiently with those animals which unite for each birth; but I have already attempted to show that we have reason to believe that occasional intercrosses take place with all animals and with all plants. Even if these take place only at long intervals, I am convinced that the young thus produced will gain so much in vigour and fertility over the offspring from long-continued self-fertilisation, that they will have a better chance of surviving and propagating their kind; and thus, in the long run, the influence of intercrosses, even at rare intervals, will be great (Darwin, 1859/1964, pp. 103–104).

Hence, Darwin's concern to show that hermaphroditism in cirripedes, who lived a sedentary existence attached to a host (rock, ship's hull, conch shell, etc.) did indeed have the ability to intercross with other individuals. The existence of minute males possessing long proboscis-like penises, in fact, well demonstrated both the ability to intercross and the action of natural selection favoring the gradual development of such a system of generation that conferred such an advantage.

In Chap. 5 – the final discussion of variation – Darwin highlighted what he called the "laws of variation." In this chapter, he focused on a series of factors that played a role in causing variation. As he explained:

> I have hitherto sometimes spoken as if the variations—so common and multiform in organic beings under domestication, and in a lesser degree in those in a state of nature—had been due to chance. This, of course, is a wholly incorrect expression, but it serves to acknowledge plainly our ignorance of the cause of each particular variation. Some authors believe it to be as much the function of the reproductive system to produce individual differences, or very slight deviations of structure, as to make the child like its parents. But the much greater

> variability, as well as the greater frequency of monstrosities, under domestication or cultivation, than under nature, leads me to believe that deviations of structure are in some way due to the nature of the conditions of life, to which the parents and their more remote ancestors have been exposed during several generations (Darwin, 1859/1964, p. 131).

He then laid out what kind of "conditions" could lead to variation, including "habit, use, and disuse." Each of these "has, in some cases, played a considerable part in the modification of the constitution, and of the structure of various organs." But, in addition, "innate differences" could arise in individuals during their development; given that "the whole organisation is so tied together during its growth and development, that when slight variations in any one part occur, and are accumulated through natural selection, other parts become modified" (Darwin, 1859/1964, p. 143).

Darwin called this law the "correlation of growth." At its core, the correlation of growth drew on a specialized area within natural history – embryology – not well understood by many naturalists let alone the general public. While one could gain a basic understanding of embryology by studying the works of specialists, deeper knowledge required familiarity with the use of microscopes and microscopic techniques as well as mastery of principles of development. Few naturalists could boast of proficiency in embryology, but Darwin was among them. Indeed, in 1853, his expertise was formally recognized by having been awarded the Royal Society of London's prestigious Royal Medal. In his notification, Hooker told him excitedly that while his geological work was mentioned, the award came largely on the basis of his cirripede monographs: "All along of the Barnacles!!!" (Darwin Correspondence Project, letter no. 1539, (n.d.)) The formal citation indicates Darwin's seminal discoveries, including "the larval condition of the Cirripeds" and recognition that the Lepadidae and Balanidae "pass through successive stages of metamorphosis." "The knowledge thus gained from the study of development," it noted, "is most sagaciously and happily applied by you to explain the homological relations between the Cirripeds and Crustaceans," an important new recognition (Parsons, 1853).

Thus, it is not surprising that in *Origin* Darwin drew on his expertise to explain how larval development of several cirripede species illustrated principles underlying variation. Certainly, not all of the premises Darwin cited to support species change were new; indeed, several were reappropriations of older "laws" in natural history to new purposes. For instance, he accepted the long-accepted notion of *balance* in nature and related *compensation* to maintain balance. Citing Goethe, he noted that "in order to spend on one side, nature is forced to economise on the other side" (Darwin, 1859/1964, p. 147). Thus, "natural selection is continually trying to economise in every part of the organisation. If under changed conditions of life a structure before useful becomes less useful, any diminution, however slight, in its development, will be seized on by natural selection, for it will profit the individual not to have its nutriment wasted in building up an useless structure" (Darwin, 1859/1964, pp. 147–148). He illustrated the principle of "economy in nature" by referencing the cirripedes:

> I can thus only understand a fact with which I was much struck when examining cirripedes, and of which many other instances could be given: namely, that when a cirripede is parasitic within another and is thus protected, it loses more or less completely its own shell or carapace. This is the case with the male Ibla, and in a truly extraordinary manner with the Proteolepas: for the carapace in all other cirripedes consists of the three highly-important anterior segments of the head enormously developed, and furnished with great nerves and muscles; but in the parasitic and protected Proteolepas, the whole anterior part of the head is reduced to the merest rudiment attached to the bases of the prehensile antennæ. Now the saving of a large and complex structure, when rendered superfluous by the parasitic habits of the Proteolepas, though effected by slow steps, would be a decided advantage to each successive individual of the species; for in the struggle for life to which every animal is exposed, each individual Proteolepas would have a better chance of supporting itself, by less nutriment being wasted in developing a structure now become useless.
>
> Thus, as I believe, natural selection will always succeed in the long run in reducing and saving every part of the organisation, as soon as it is rendered superfluous, without by any means causing some other part to be largely developed in a corresponding degree. And, conversely, that natural selection may perfectly well succeed in largely developing any organ, without requiring as a necessary compensation the reduction of some adjoining part (Darwin, 1859/1964, p. 148).

Thus, barnacles evidence both balance in nature and compensation.

Likewise, Darwin drew on barnacles to demonstrate another "law of variation," namely, "A part developed in any species in an extraordinary degree or manner, in comparison with the same part in allied species, tends to be highly variable" (Darwin, 1859/1964, p. 150). Stating that while he had a "long array of facts" to support his belief in this rule, it "almost invariably holds good in cirripedes." He thus provided only a single case, "as it illustrates the rule in its largest application":

> The opercular valves of sessile cirripedes (rock barnacles) are, in every sense of the word, very important structures, and they differ extremely little even in different genera; but in the several species of one genus, Pyrgoma, these valves present a marvellous amount of diversification: the homologous valves in the different species being sometimes wholly unlike in shape; and the amount of variation in the individuals of several of the species is so great, that it is no exaggeration to state that the varieties differ more from each other in the characters of these important valves than do other species of distinct genera (Darwin, 1859/1964, p. 151).

As he explained in *Living Cirripedia* (Darwin, 1854, p. 34), in *Pyrgoma*, "the opercular valves seem to have broken loose from all law, presenting a greater diversity in character than do all other species of Balaninæ and Chthamalinæ taken together."[9] Thus, the species well demonstrates this law's operation in nature.

In Chap. 6, having explained the theory of "descent with modification," Darwin turned to "difficulties on theory." This chapter addressed a number of points that seemingly posed problems for him. These included the "absence or rarity of transitional varieties," the transition in habits, and how organs "of extreme perfection"

[9]Darwin used *Pyrgoma* as an example of the rule that "when the species of a genus differed in some organ or part, which is usually constant in the species of the same genus, then that one or more of the species individually varied in some degree in this same organ or character" (Darwin Correspondence Project, letter no. 1749, 24 August [1855]).

(such as the eye) could arise through the selection of gradual improvements. Darwin had long agonized over how to explain the "evolution" of the eye – a prime example invoked by those advocating for special creation – as he readily admitted in *Origin*:

> To suppose that the eye, with all its inimitable contrivances for adjusting the focus to different distances, for admitting different amounts of light, and for the correction of spherical and chromatic aberration, could have been formed by natural selection, seems, I freely confess, absurd in the highest possible degree. Yet reason tells me, that if numerous gradations from a perfect and complex eye to one very imperfect and simple, each grade being useful to its possessor, can be shown to exist; if further, the eye does vary ever so slightly, and the variations be inherited, which is certainly the case; and if any variation or modification in the organ be ever useful to an animal under changing conditions of life, then the difficulty of believing that a perfect and complex eye could be formed by natural selection, though insuperable by our imagination, can hardly be considered real (Darwin, 1859/1964, pp. 186–187).

A correlated problem was the difficulty of explaining how, as he maintained, the function of an existing organ could transition to new purposes. Here, the cirripedes again provided an excellent, if somewhat specialized, illustration of balance and compensation:

> Pedunculated cirripedes have two minute folds of skin, called by me the ovigerous frena, which serve, through the means of a sticky secretion, to retain the eggs until they are hatched within the sack. These cirripedes have no branchiæ, the whole surface of the body and sack, including the small frena, serving for respiration. The Balanidæ or sessile cirripedes, on the other hand, have no ovigerous frena, the eggs lying loose at the bottom of the sack, in the well-enclosed shell; but they have large folded branchiæ. Now I think no one will dispute that the ovigerous frena in the one family are strictly homologous with the branchiæ of the other family; indeed, they graduate into each other. Therefore I do not doubt that little folds of skin, which originally served as ovigerous frena, but which, likewise, very slightly aided the act of respiration, have been gradually converted by natural selection into branchiæ, simply through an increase in their size and the obliteration of their adhesive glands. If all pedunculated cirripedes had become extinct, and they have already suffered far more extinction than have sessile cirripedes, who would ever have imagined that the branchiæ in this latter family had originally existed as organs for preventing the ova from being washed out of the sack? (Darwin, 1859/1964, pp. 191–192).

Interestingly, this anatomical interpretation was attacked by Richard Owen in his anonymous review of the *Origin*, which both enraged and hurt Darwin, who prided himself on this discovery. As he bemoaned to Huxley, "I never saw such an amount of misrepresentation," and to Alfred Russel Wallace he complained: "Talking of envy, you never read anything more envious & spiteful (with numerous misrepresentations) than Owen is in the Edinburgh Review. I must give one instance he throws doubts & sneers at my saying that the ovigerous frena of cirripedes have been converted into Branchiæ, because I have not proved them to be Branchiæ; whereas he himself admits, before I wrote, on cirripedes, without the least hesitation that these organs are Branchiæ" (Owen, 1860).[10]

[10]Darwin Correspondence Project, letter no. 2751, 9 April [1860]; and letter no. 2807 18 May 1860.

Darwin focused on the fossil record in geology in Chaps. 9, 10, 11, and 12 to address the difficulty of the absence of transitional forms. In Chap. 9, he explained why no "intermediate" forms were found in the geological record, noting that this was often simply because such forms had not yet been found. He himself, having failed to find sessile cirripeds in Secondary strata, had reluctantly concluded in *Fossil Cirripedia* that "this great group had been suddenly developed at the commencement of the tertiary series." But "my work had hardly been published, when a skilful palæontologist, M. Bosquet, sent me a drawing of a perfect specimen of an unmistakeable sessile cirripede, which he had himself extracted from the chalk of Belgium.... Hence we now positively know that sessile cirripedes existed during the secondary period; and these cirripedes [Chthamalus] might have been the progenitors of our many tertiary and existing species" (Darwin, 1859/1964, pp. 304–305).

In Chap. 13, Darwin focused on classification. There he repeated his long-held view, as discussed earlier, that grouping species into genera, families, class, etc. relied on more than "mere resemblance"; it reflected real historical relationship, as he said, the "propinquity of descent" (Darwin, 1859/1964, p. 413). Again, cirripedes offered a good illustration:

> With species in a state of nature, every naturalist has in fact brought descent into his classification; for he includes in his lowest grade, or that of a species, the two sexes; and how enormously these sometimes differ in the most important characters, is known to every naturalist: scarcely a single fact can be predicated in common of the males and hermaphrodites of certain cirripedes, when adult, and yet no one dreams of separating them. The naturalist includes as one species the several larval stages of the same individual, however much they may differ from each other and from the adult [...] (Darwin, 1859/1964, p. 424).

The chapter also includes a dedicated section on morphology and the important recognition "that the members of the same class, independently of their habits of life, resemble each other in the general plan of their organization, ... often expressed by the term 'unity of type.'

As Darwin emphasized, morphology "is the most interesting department of natural history, and may be said to be its very soul. What can be more curious than that the hand of a man, formed for grasping, that of a mole for digging, the leg of the horse, the paddle of the porpoise, and the wing of the bat, should all be constructed on the same pattern, and should include the same bones, in the same relative positions?" (Darwin, 1859/1964, p. 434). Moreover, embryology is an especially important subfield of morphology. Indeed, as Darwin's receipt of the Royal Medal well indicated, embryology often provided important clues about morphological relationships.

However, embryonic forms are often confounding. As Darwin explained, this is particularly true "when an animal during any part of its embryonic career is active, and has to provide for itself" (Darwin, 1859/1964, p. 440). The larvae of cirripedes served as particularly fine examples:

> So again the two main divisions of cirripedes, the pedunculated and sessile, which differ widely in external appearance, have larvæ in all their several stages barely distinguishable.

> The embryo in the course of development generally rises in organisation: . . . the larvæ in the first stage have three pairs of legs, a very simple single eye, and a probosciformed mouth, with which they feed largely, for they increase much in size. In the second stage, . . . they have six pairs of beautifully constructed natatory legs, a pair of magnificent compound eyes, and extremely complex antennæ; but they have a closed and imperfect mouth, and cannot feed: their function at this stage is, to search by their well-developed organs of sense, and to reach by their active powers of swimming, a proper place on which to become attached and to undergo their final metamorphosis. When this is completed they are fixed for life: their legs are now converted into prehensile organs; they again obtain a well-constructed mouth; but they have no antennæ, and their two eyes are now reconverted into a minute, single, and very simple eye-spot. In this last and complete state, cirripedes may be considered as either more highly or more lowly organised than they were in the larval condition. But in some genera the larvæ become developed either into hermaphrodites having the ordinary structure, or into what I have called complemental males: and in the latter, the development has assuredly been retrograde; for the male is a mere sack, which lives for a short time, and is destitute of mouth, stomach, or other organ of importance, excepting for reproduction (Darwin, 1859/1964, pp. 440–441).

Despite presenting such baffling variations, larvae nonetheless illustrated that "community in embryonic structure reveals community of descent" (Darwin, 1859/1964, p. 449).

The final chapter, Chap. 14, "Recapitulation and Conclusion," ties together the "one long argument" Darwin presents in the *Origin of Species* (Darwin, 1859/1964, p. 459). Certainly, his earliest thoughts about the possibility of species change over time were based on other species he examined on the *Beagle* voyage, most notably the Galapagos finches and even tortoises (Lack, 1983). Still, his chance discovery of a tiny curious creature – Mr. Arthrobalanus – discovered in a conch shell randomly picked up along a beach off the Chilean coast certainly must also be seen as critical for the shape of Darwin's mature theory of evolution. Once he reexamined it again more than a decade after returning from his South American voyage, it captured his attention and propelled him toward engaging in a deeper understanding of how life on Earth could evolve. Simply put, Darwin's 8-year study of cirripedes, as he recounted in his 1876 Autobiography, "was of considerable use to me, when I had to discuss in the *Origin of Species* the principles of a natural classification" (Darwin, 1887, pp. 117–118). Today, Darwin's beloved barnacles provide readers an engaging means by which to comprehend his theory of evolution.

References

Darwin Correspondence Project, "Letter no. 1003". (n.d.). https://www.darwinproject.ac.uk/letter/?docId=letters/DCP-LETT-1003.xml

Darwin Correspondence Project, "Letter no. 1014". (n.d.). https://www.darwinproject.ac.uk/letter/?docId=letters/DCP-LETT-1014.xml

Darwin Correspondence Project, "Letter no. 1012". (n.d.). https://www.darwinproject.ac.uk/letter/?docId=letters/DCP-LETT-1012.xml

Darwin Correspondence Project, "Letter no. 915". (n.d.). https://www.darwinproject.ac.uk/letter/?docId=letters/DCP-LETT-915.xml.

Darwin Correspondence Project, "Letter no. 684". (n.d.). https://www.darwinproject.ac.uk/letter/?docId=letters/DCP-LETT-684.xml
Darwin Correspondence Project, "Letter no. 1033". (n.d.). https://www.darwinproject.ac.uk/letter/?docId=letters/DCP-LETT-1033.xml
Darwin Correspondence Project, "Letter no. 1022". (n.d.). https://www.darwinproject.ac.uk/letter/?docId=letters/DCP-LETT-1022.xml
Darwin Correspondence Project, "Letter no. 1166". (n.d.-a). https://www.darwinproject.ac.uk/letter/?docId=letters/DCP-LETT-1166.xml
Darwin Correspondence Project, "Letter no. 1166". (n.d.-b). https://www.darwinproject.ac.uk/letter/?docId=letters/DCP-LETT-1166.xml
Darwin Correspondence Project, "Letter no. 582". (n.d.). https://www.darwinproject.ac.uk/letter/?docId=letters/DCP-LETT-582.xml
Darwin Correspondence Project, "Letter no. 1174". (n.d.). https://www.darwinproject.ac.uk/letter/?docId=letters/DCP-LETT-1174.xml
Darwin Correspondence Project, "Letter no. 2175F". (n.d.). https://www.darwinproject.ac.uk/letter/DCP-LETT-2175F.xml
Darwin Correspondence Project, "Letter no. 2179". (n.d.). https://www.darwinproject.ac.uk/letter/DCP-LETT-2179.xml
Darwin Correspondence Project, "Letter no. 1539". (n.d.). https://www.darwinproject.ac.uk/letter/?docId=letters/DCP-LETT-1539.xml
Darwin, C. R. (1851). *A monograph on the sub-class Cirripedia, with figures of all the species. Volume 1: The Lepadidae; or, pedunculated Cirripedes*. The Ray Society. http://darwin-online.org.uk/content/frameset?itemID=F339.1&viewtype=side&pageseq=1
Darwin, C. R. (1854). A monograph on the sub-class Cirripedia, with figures of all the species. In *The Balanidæ, (or sessile Cirripedes); the Verrucidæ, etc. etc. etc* (Vol. 2). The Ray Society. http://darwin-online.org.uk/content/frameset?itemID=F339.2&viewtype=text&pageseq=1
Darwin, C. R. (1887). In F. Darwin (Ed.), *The life and letters of Charles Darwin, including an autobiographical chapter*. John Murray.
Darwin, C. R. (1859/1964). On the origin of species by Charles Darwin: A facsimile of the first edition. In *With an introduction of Ernst Mayr*. Harvard University Press.
Darwin, C. R. (2001). In R. D. Keynes (Ed.), *Charles Darwin's beagle diary*. Cambridge University Press.
Lack, D. (1983). *Darwin's finches*. Cambridge University Press.
Owen, R. (1860). Darwin on the origin of species. *Edinburgh Review, 111*, 487–532.
Parsons, W. (Earl of Rosse). (1853). [Announcement of Darwin's Royal Medal]. *Proceedings of the Royal Society of London*, *6*: 355–6 [reprinted in Appendix II, "Darwin's Study of the Cirripedia", CCD 4: 388–409, on 406–407].

Marsha L. Richmond, Editor on the Darwin Correspondence, 1987–1993, has published on Darwin, genetics, cell theory, and women in biology. She is completing a book with Ida Stamhuis on Women in Genetics and working on a scientific biography of Theo Colborn and the discovery of endocrine disruption.

Chapter 9
Wallace, Darwin, and the Relationship Between Species and Varieties (1858)

Viviane Arruda do Carmo and Lilian Al-Chueyr Pereira Martins

Abstract In the Author's Introduction to the first edition of *Origin of Species*, Charles Robert Darwin (1809–1882) mentioned he was dealing with his ideas on the origin of species since he returned from the voyage of the *Beagle*. Besides that, he intended to publish them in 2 years. Nevertheless, it did not happen. Several years later, when he was still dealing with his manuscript, in 1858, Darwin received a memoir from Alfred Russel Wallace (1823–1913) with the same general conclusions as him. Due to this, he published *The Origin of Species* the following year. Before the *Origin* (1859), Darwin and Wallace published their ideas in the Linnean Society of London journal. This chapter discusses Wallace's ideas trying to detect to what extent they were similar to Darwin's in their papers published in 1858. The analysis concluded that despite the sequence differences, both authors' contributions are coherent. Although some terms or expressions are different, their connotation is the same. Both referred to the "struggle for existence" in nature, although Darwin did not use this expression. Wallace, contrary to Darwin, did not use the words "natural selection" but referred to a principle whose connotation is the same. Wallace and Darwin agreed that species exist first as varieties. Both of them admitted the principle of divergence. In short, their main ideas were similar, as they realized in 1858.

V. A. do Carmo (✉)
Municipal Department of Education, Group of History and Theory of Biology,
University of São Paulo, São Paulo, Brazil
e-mail: viviane.carmo@sme.prefeitura.sp.gov.br

L. A.-C. P. Martins
Department of Biology, FFCLRP-USP; Group of History and Theory of Biology,
University of São Paulo, São Paulo, Brazil
e-mail: lacpm@ffclrp.usp.br

M. Elice Brzezinski Prestes (ed.), *Understanding Evolution in Darwin's "Origin"*,
History, Philosophy and Theory of the Life Sciences 34,
https://doi.org/10.1007/978-3-031-40165-7_9

9.1 Introduction

Since 1837, Darwin had been dealing with a theory on the transmutation of species and registering his ideas in notebooks. In June 1842, he wrote a brief abstract of his theory in pencil on 35 pages, which he enlarged into 230 pages during the summer of 1844 (Barlow, 1958,[1] p. 120; Darwin, 1859, p. 1). Early in 1856, Charles Lyell (1797–1875) advised him to write his views as a whole. So he was doing. However, in 1858, Darwin received a letter with an essay from Alfred Russel Wallace (1823–1913), who was in the Malay Archipelago.

Concerning Wallace's essay, Darwin commented: "This essay contained the same theory as mine.[2] Mr. Wallace expressed the wish that if I thought well of his essay, I should send it to Lyell for perusal" (Barlow, 1958, p. 121).

On this occasion, Lyell and other close colleagues, such as the botanists Joseph Dalton Hooker (1817–1911) and Asa Gray (1810–1888), who had been following the development of Darwin's theory, proposed a joint presentation of the works by Darwin and Wallace to the Linnean Society of London (Carmo & Martins, 2008, p. 457). Darwin expressed his feelings concerning the situation: "I was at first very unwilling to consent, as I thought Mr Wallace might consider my doing so unjustifiable, for I did not then know how generous and noble his disposition was" (Barlow, 1958, p. 121).

In June 1858, Lyell and Hooker read both works at the Linnean Society, and in the same year, the *Journal of the Proceedings of the Linnean Society*[3] published them (Barlow, 1958, p. 121). Lyell and Hooker stated:

> These gentlemen having, independently and unknown to another, conceived the same very ingenious theory to account for the appearance and perpetuation of varieties and specific forms on our planet, may both claim the merit of being original thinkers in this important line of inquiry, but neither of them having published his views [. . .]. (Lyell & Hooker, 1858)

The following year, in the Introduction to the first edition of *Origin of Species*, Darwin gave Wallace due credit, mentioning that he had led him to bring forward the publication, which would still take 2 or 3 years to complete. Darwin emphasized that Wallace had reached "practically the same conclusions about the origin of species" as him (Darwin, 1859, pp. 1–2).

---

[1] Although Darwin started his autobiography in 1860, it appeared only in 1887, as part of Darwin's *Life and Letters* and later as a separate volume in 1958, edited by his granddaughter Norma Barlow. Opinions pro and against it have appeared (Colp Jr, 1985, p. 362). In the present chapter, we use Norma Barlow's edition (Barlow, 1958).

[2] It is worthwhile mentioning that Wallace also thought at the time that their theories were identical (Bulmer, 2005, p. 126). In an introductory note to a reprint of his Ternate paper (Wallace, 1891), Wallace wrote: "This [paper] sets for the main features of a theory identical with that discovered by Mr. Darwin many years before but not then published" (Wallace, 1891, p. 20, *apud*, Bulmer, 2005, p. 126).

[3] The communication of Darwin's and Wallace's works by Lyell and Hooker produced a low impact on the audience. According to Edward S. Rhayer, this happened because of the lack of understanding of its implications (Rayher, 1996, p. 160).

The original publication containing the works of Darwin and Wallace consists of an introductory letter by Lyell and Hooker followed by an extract from a chapter of Darwin's work on the *Origin of Species*, excerpts from a letter from Darwin to Asa Gray, and finally, Wallace's article.

Darwin and Wallace considered their ideas in 1858 identical. However, some authors, such as Kottler (1985) and Nicholson (1960), did not agree. They claimed marked differences between them (Bulmer, 2005, p. 126). In this chapter, we will revisit the works mentioned above, trying to detect their similarities and differences and to what extent.

9.2 Darwin's Views

In the extract from an unpublished work on species, sketched in 1839 and copied in 1844,[4] Darwin commented that organisms are constantly "at war" in nature, whose strength and intensity vary. He then referred to Thomas Malthus' (1766–1875) views[5] stating that while organisms reproduce in geometrical progression, resources, such as food supply, grow in arithmetical progression.[6]

Although organisms have an enormous power of multiplication, few reach the adult stage because there are limiting factors[7] that could act on the seeds, eggs, or chicks. Thus, the average percentage of each region's inhabitants remains constant (Darwin, 1858a, p. 47). In Darwin's words:

> Suppose in a certain spot there are eight pairs of birds, and that only four pairs of them annually (including double hatches) rear only four young, and that these go on rearing their young at the same rate, that at the end of seven years (a short life excluding violent deaths, for any bird) there will be 2048 birds, instead of the original sixteen. As this increase is quite impossible, we must conclude either that birds do not rear nearly half their young or the average life of a bird is, from accident, not nearly seven years. Both checks probably concur.

---

[4]Hooker read it in 1844 and communicated its contents afterward to Charles Lyell. These extracts are the ones published in the Journal of the Linnean Society of London (Darwin, 1858a).

[5]Darwin read Malthus' *Essay on natural populations* in the 1826 edition, where these ideas appear more detailed than in the first edition (Regner, 2004, p. 48). Darwin wrote: "In October 1838, that is, fifteen months after I had begun my inquiry, I happened to read for amusement Malthus on Population, and being well prepared to appreciate the struggle for existence which everywhere goes on from long-continued observation of the animals and plants, it at once struck me that under these circumstances favourable variations would tend to be preserved, and unfavourable ones to be destroyed" (Barlow, 1958, p. 120).

[6]Several historians of science agree that reading Thomas Malthus' *Essay on the principle of population*, published anonymously in 1798, brought elements that contributed to Darwin's formulation of the principle of natural selection (Bowler, 1989, p. 120). Some of them, such as Peter Vorzimer, Michael Ghiselin, and Ernst Mayr, attribute it to a catalyzing role in the development of Darwin's theory (Regner, 2004, p. 50). However, others consider that Darwin would propose it even without a reading, such as Anna Carolina K. P. Regner (2004).

[7]For Darwin, these limiting factors were food scarcity, long periods of drought, cold, and rain.

> The same kind of calculation applied to all plants and animals affords results more or less striking, but in very few instances more striking than in man. (Darwin, 1858a, p. 47)

Besides presenting hypothetical examples, Darwin also considered real ones:

> Many practical illustrations of this rapid tendency to increase are on record, among which, during peculiar seasons, are the extraordinary numbers of certain animals; for instance, the years 1826 to 1828, in La Plata, when from drought, some millions of cattle perished, the whole country actually *swarmed* with mice [. . .]. This astounding increase during three years must be attributed, to the greater number than usual surviving the first year, and then breeding, and so on till the third year, when their numbers were brought down to their usual limits on the return of wet weather. (Darwin, 1858a, p. 47)

If external conditions remain the same, recurrent struggles against other species or *external nature* maintain the average number of individuals in each region. On the other hand, if there is a change in the external conditions, a proportion of organisms will have slight changes. If these modifications continue occurring, they will act on their reproductive system, making the organization of these beings plastic (Darwin, 1858a, pp. 48–49). As more individuals are born than can survive, *selection* determines which will survive and which will perish (Darwin, 1858a, p. 49).

In addition to selection in nature, Darwin mentioned a second agent on most unisexual animals, tending to produce the same effect: "the fight of the males for the females." It could be for the battle, but it could also be in the case of birds, their songs, beauty, or the court like the rock-thrush of Guiana (Fig. 9.1). This "type of selection" is less rigorous since those who are not successful do not die but have fewer descendants (Darwin, 1858a, p. 50).

Darwin thought it was possible to compare the result of the struggle among the males to the one produced by the agriculturists who paid less attention to the careful selection of all their young animals and more to the occasional use of a choice mate (Darwin, 1858a, p. 50).

Fig. 9.1 Common rock-thrush (*Monticola saxatilis*). https://commons.wikimedia.org/wiki/File:Monsax.jpg

Some years later, in the abstract of the letter to Asa Gray, dated September 1857, Darwin referred to the selection by man and nature. He considered that the one made by man consciously or unconsciously in animals and plants since ancient times was responsible for the production of domestic breeds: "Man by his power of accumulating variations, adapts living beings to his wants – may be said to make the wool of one sheep good, for carpets, of another for cloth, &tc." (Darwin, 1858b, p. 51).

The second – "natural selection" – acts only through the accumulation of slight or more significant natural variations in all parts caused by external conditions, such as the differences between the offspring and their parents. According to Darwin, nature selects slight variations in practically unlimited time. *Natural Selection* selects for the good of each organic being. Only a few can survive. The useful variations will influence how they obtain food, fight with other living beings and disseminate their eggs and seeds. In his words: "The variety thus formed will either coexist with or, more commonly, will exterminate the parent form" (Darwin, 1858b, p. 52).

In addition to the principle of natural selection, Darwin considered the principle of divergence to play a relevant role in the origin of species. He commented:

> Each new variety or species, when formed, will generally take the place of and thus exterminate its less well-fitted parent. This I believe to be the origin of the classification and affinities of organic beings at all times; for organic beings always *seem* to branch and sub-branch like the limbs of a tree from a common trunk, the flourishing and diverging twigs destroying the less vigorous – the dead and lost branches rudely representing extinct genera and families. (Darwin, 1858b, p. 53)

Darwin wrote to Hooker 1 month before describing a few botanical calculations; he referred to the principle of divergence (Browne, 1980, p. 82).

> If it will all hold good, it is very important for me: for it explains, as I think, all classification, *i.e.* the quasi-branching and splitting up, etc., as you will perceive. But then comes in also, what I call a principle of divergence, which I think I can explain but which is too long, and perhaps you do not care to hear. (Letter from Darwin to Hooker, August 22, 1857, *Apud* Browne, 1980, p. 83)

9.3 Wallace's Views

In the essay written in the isle of Gilolo (Malay Archipelago) in 1858,[8] while recovering from a malaria crisis and sent to Darwin from Ternate, Wallace started by discussing the differences between varieties' production processes in the domestic and wild states. He considered that the idea that varieties produced under domestication tend to return to the ordinary form of the species that originated

[8] In a previous paper written in Sarawak, Borneo, published in 1855, departing from the observation of the geographical distribution of animals and plants, Wallace concluded that "Every species has come into existence coincident both in space and time with a pre-existing closed allied species" (Wallace, *Apud,* Bulmer, 2005, p. 125). Nevertheless, this paper aroused little interest and received no constructive criticism as his author expected (Beddall, 1972, p. 153).

them, which also applied to species in nature, was probably due to the absence or scarcity of facts occurring among wild animals. Such a view would have contributed to the belief in the stability of species.

At that time, scholars, such as Lyell, admitted that environmental changes could cause some morphological changes transmitted to the descendants. These changes happened within limits already stipulated by the Creator, but this was not enough to transmute one species into another. In most cases, species resist environmental changes and later go extinct (Bulmer, 2005, p. 126; Carmo, 2011, p. 51).

Wallace explained the purpose of the paper:

> But it is the object of the present paper to show that this assumption is altogether false, that there is a general principle in nature which will cause many *varieties* to survive the parent species and to give rise to successive variations departing further and further from the original type, and which also produces in domesticated animals, the tendency of varieties to return to the parent form. (Wallace, 1858, p. 54)

Wallace characterized the life of wild animals as a *struggle for existence*, searching for food, and protecting the offspring. In his words:

> The life of wild animals is a struggle for existence. The full exertion of their faculties and their energies is required to preserve their own existence and provide for that of their infant offspring. The possibility of procuring food during the least favourable seasons and of escaping the attacks of their most dangerous enemies, are the primary conditions which determine the existence both of individuals and of entire species. These conditions will also determine the population of a species [...]. (Wallace, 1858, p. 54)

Wallace sought explanations after observing that some closely related species are abundant and others are rare; large and carnivorous animals are less numerous than herbivores. He put aside higher fecundity as a possible factor for the difference in closely related species proportion since observations have shown that birds do not increase year-round in geometrical progression as they would if nothing were limiting this growth. In his words:

> Perhaps the most remarkable instance of an immense bird population is the passenger pigeon[9] (Fig. 9.2) of the United States, which lays only one, or at most two eggs and is said to rear generally but a young one. Why is this bird so abundant, while others producing two or three times as many young are much less plentiful? (Wallace, 1858, p. 55)

Wallace considered that one of the factors is food supply because birds cannot fly, migrate, or form large populations without feeding. Food availability explains the scarcity of woodpeckers in Britain and the abundance in the tropics. On the other hand, the house sparrow (Fig. 9.3) is more abundant than the redbreast (Fig. 9.4) because its food supply is constant and plentiful due to the preservation of the grass seeds during the winter and the provision of the food by the farms and post-harvest fields (Wallace, 1858, p. 56). The same explanation fits the fact that waterfowl are generally more numerous: "Their food never fails, the sea-shores and river-banks

[9] *Ectopistes migratorius*. This species, widespread and populous in the nineteenth century, became extinct at the beginning of the twentieth century due to hunting and habitat destruction.

Fig. 9.2 Passenger pigeon (*Ectopistes migratorius*). https://commons.wikimedia.org/wiki/File:Ectopistes_migratorius_(passenger_pigeon)_1_(15374997397).jpg

Fig. 9.3 House sparrow (*Passer domesticus*). https://commons.wikimedia.org/wiki/File:Passer_domesticus_male_(15).jpg

daily swarming with a fresh supply of small mollusca and crustacea" (Wallace, 1858, p. 56).

The Welsh-born English naturalist[10] believed that the same laws apply to mammals. In his words:

> Exactly the same laws will apply to mammals. Wild cats are prolific and have few enemies; why then are they never abundant as rabbits? The only intelligible answer is, that their supply of food is more precarious. It appears evident, therefore, that so long as a country

[10]Both of Wallace's parents were English, and he regarded himself as English. Although he was born in south Wales and lived there until age 6, when the family relocated back to England, he never learned the Welsh language. See Charles Smith: http://people.wku.edu/charles.smith/wallace/FAQ.htm#Welsh

Fig. 9.4 Redbreast (*Erithacus*). https://commons.wikimedia.org/wiki/File:Redbreast.jpg

> remains physically unchanged, the numbers of its animal population cannot materially increase. If one species does so, some others requiring the same kind of food must diminish in proportion. (Wallace, 1858, p. 56)

The animals that perish must be the weakest, the very young, the elderly, or the sick. The healthiest and most vigorous must feed themselves regularly and avoid enemies. There is a "struggle for existence" in which the weakest and least perfectly organized must always succumb. The number of animals that die is huge (Wallace, 1858, pp. 56–57).

According to Wallace, what happens with individuals of the same species must occur among the species of a group, among the several allied species that are best adapted to obtain food regularly and defend themselves from enemy attacks. At the same time, those unable to do so must decrease numerically and sometimes disappear. He concluded:

> 1st, that the animal population of a country is generally stationary, being kept down by a periodical deficiency of food and other checks; and, 2nd, that the comparative abundance or scarcity of the individuals of several species is entirely due to their organization and resulting habits, which, rendering it more difficult to procure a regular supply of food and to provide for their personal safety in some cases than in others, can only be balanced by a difference in the population which have to exist in a given area. (Wallace, 1858, p. 57)

Although Wallace did not mention Malthus' name, around 1844, he read Malthus' *Essay on the Principle of Population*, which he admired "for the summary of facts and logical induction to conclusions" (Raby, 2001, p. 21).

Even slight variations in a species will affect its habits and abilities, such as foraging for food, leading to population decline. Wallace explained:

> An antelope with shorter or weaker legs must necessarily suffer more from the attacks of the feline carnivora; the passenger pigeon with less powerful wings would sooner or later be affected in its powers of procuring a regular supply of food; and in both cases, the result must necessarily be a diminution of the population of the modified species. (Wallace, 1858, p. 58)

In this way, slight variations in a species, such as color change, greater or lesser development of hair, and more profound transformations, the increase in potency or the dimensions of the limbs or any external organ, could influence favorably or unfavorably the struggle for existence (Wallace, 1858, pp. 57–58).

If a species produces a variety slightly more capable of preserving its existence, that variety will acquire a superiority in number over time. Thus, for the Welsh-born English naturalist, varieties fell into two categories: those under the same conditions would never reach the population of the parental species and those that would get and keep numerical superiority in due course. However, suppose there were alterations in the physical conditions of a region; of all the individuals that make up the species, those who constitute the least numerous and weakly organized variety will suffer first and, in the case of more severe pressure, would soon become extinct. According to Wallace, if the exact causes continue acting, the parental species would suffer and then gradually decrease in number. In case of the recurrence of similar unfavorable conditions, it could be the subject of extinction. Thus, the superior variety would stand alone, and when favorable circumstances return, it will rapidly increase in number and replace extinct species and varieties (Wallace, 1858, p. 58).

The formed variety replaces the previous species and is a more developed and organized form. In all respects, it would be better adapted to guarantee its safety and prolong its existence and that of its race. Such a variety could not return to its original form since it would be inferior and unable to compete for its existence. Nevertheless, as time passes, the new, improved, and populous race can give rise to new varieties that exhibit several divergent form modifications.[11] In Wallace's words, "We have here *progression* and continued divergence deduced from general laws that regulate the existence of animals in the state of nature, and from the unquestionable fact that varieties do frequently occur" (Wallace, 1858, p. 58).

After criticizing Lamarck's theory in the last citation, Wallace stated: "The principles constantly at work in nature make the process of species modification more easily understandable" (Wallace, 1858, p. 61). In Wallace's words: "Neither did the giraffe acquire its long neck by desiring to reach the foliage [. . .]" (Wallace, 1858, p. 58). However, attributing changes in species to their volition and employing the word "desiring", Wallace misrepresented Lamarck's theory (Lamarck, 1809, v. 1, pp. 256–257; Martins, 1997, pp. 38–39; Martins, 2007, pp. 212–213).

[11] Wallace admitted that species tend to form varieties that outlive the parent species, giving rise to successive varieties that increasingly diverge from the original type (Wallace, 1858, p. 55). This idea was present in a previous Wallace publication (Wallace, 1858, p. 186; Carmo, 2011, p. 71).

Fig. 9.5 Domestic pig (*Sus domesticus*). https://commons.wikimedia.org/wiki/File:Domestic_pig_LCCN2017660728.jpg

Wallace thought that the production processes of varieties in nature and under domestication involve different states. In the first case, well-being and existence depend on the perfect functioning of all senses and physical conditions. Wild animals need more sight, hearing, and protection to survive and provide for their offspring. In the case of the domestic animal, it receives food and shelter from the attacks of its natural enemies and generally raises its offspring with human assistance. In this way, any variation does not represent an advantage and has the same chance of continuing. In Wallace's words (Figs. 9.5 and 9.6):

> Our quickly fattening pigs (Fig. 9.5), short-legged ship (Fig. 9.6), pouter pigeons, and poodle dogs could never have come into existence in a state of nature, because the very first step towards such inferior forms would have led to the rapid extinction of the race; still, less could they exist in competition with their wild allies. The great speed but slight endurance of the race horse, the unwieldy strength of the ploughman's team, would be useless in a state of nature. If turned wild on the pampas, such animals would probably soon become extinct, or under favourable circumstances, might each loose those extreme qualities which would never be called into action, and in few generations, would revert to a common type, which must be that in which the various powers and faculties are so proportioned to each other as to be best adapted to procure food and secure safety, – that in which be the full exercise of every part of its organization the animal can alone continue to live. (Wallace, 1858, p. 60)

Wallace then concluded that the production processes of varieties under domestication and in nature exhibit some differences. Varieties that constitute new species in nature did not return to their original form. They depart indefinitely from the original

Fig. 9.6 The horned sheep of Dorsetshire form a singularly well-marked race (1893). https://commons.wikimedia.org/wiki/File:Sheep,_breeds_and_management_(1893)_(14595217380).jpg

type, contrary to the ones produced under domestication that tend to revert to it if left to themselves.

9.4 Wallace's and Darwin's Agreements and Disagreements

In this section, we will consider the sequence, coherence, and terminology present in the works of Darwin and Wallace, published in 1858 in the *Journal of the Linnean Society of London.*

Wallace (1858) used the expression struggle for existence, whereas Darwin, (1858a) used *war in nature*, which idea is the same. Both considered that if there is a change in the external conditions, a proportion of organisms will have slight changes and the ones who had inherited more advantageous slight variations, that is, the more adapted, will survive. However, Wallace mentioned that the youngest, the aged, and the diseased would perish, while the healthier and more vigorous will succeed in the struggle for existence. Darwin only referred that the less adapted will perish. According to Darwin, if these *modifications* continue, they will act on the organic beings' reproductive system, making their organization plastic, whereas Wallace did not consider this aspect. Nor Darwin nor Wallace used the term "fit" in these papers; instead, they used the word *adapted.*

According to Bulmer (2005), Wallace indicated that he would embrace the Darwinian view. Still, he did not because he considered the weakest, young or old, accidental rather than a heritable weakness. Thus, Wallace proposed Malthus' doctrine to explain the population decrease in these cases. We understand that although Wallace mentioned these conditions as possible causes of the failure in the struggle for existence, he also admitted, like Darwin, that hereditary variations were fundamental, as they would determine the success or failure of organisms' survival. When mentioning the weakest and less perfect in the organization, we consider that Wallace was not excluding hereditary differences.

When addressing the struggle for existence, Wallace referred only to animals; besides this group of living beings, Darwin also mentioned plants.

According to Malthus' ideas, Darwin and Wallace referred to the tendency for increasing populations unless there were some limiting factors. One of these limits is the food supply. However, Wallace did not mention Malthus' name. Therefore, we agree with Bulmer (2005) that Wallace presented the Malthusian argument (although he did not call it by that name) when he admitted that density control factors limit the population of each species (Bulmer, 2005, p. 126) in the same way as Darwin.

Besides *selection in nature,* Darwin (1858a) also considered the *fight between males* to possess the female. In the case of birds, their songs and beauty determine the female's choice. He also compared this selection to the one made by the agriculturists. He did not use the expression *sexual selection*, though. Wallace did not refer to this selection in his paper (Wallace, 1858).

In the letter to Gray (1858b), Darwin mentioned two kinds of selection: natural selection that occurs in nature and the selection made by man. In this particular letter, he used the expression "natural selection," which did not appear in his previous work (Darwin, 1858a). As we have seen, Wallace discussed the difference between the two sorts of selection in detail, arguing that what happens with the varieties is different: domestic varieties can return to the original form, but this never happens with wild species that depart indefinitely from the original type. Darwin (1858a) considered that besides the principle of natural selection, there is another one in nature: the principle of divergence, which has an essential role in the origin of species. Each new variety or species formed in general will exterminate its less adapted parental form, soon substituting it. Wallace also considered such a principle. According to Wallace (1858), the struggle for existence occurs between the same and different species.

According to Martin Fichman (2004), Darwin relied heavily on the analogy between human selection and natural selection when addressing evolution, while Wallace found this analogy suspect and misleading. Still, according to the historian, Wallace considered it a weakness of Darwin to extensively utilize the evidence between variation and selection between domestic animals and plants. He devoted his career to showing that the evolution theory could be supported only by evidence of natural variation (Fichman, 2004, p. 104). We disagree with Fichman's statement, as we believe Darwin used the analogy of artificial and natural selection in another sense. Darwin wanted to show that just as man selects and accumulates the characteristics that he deems desirable in forming new varieties, natural selection carries out the same process in nature by selecting variations that are helpful for survival. However, Darwin made it clear that, unlike man selection, natural selection favors the essential characteristics for the struggle for survival. Wallace has referred to domestic animals in this work, intending to show that their variations are not as crucial for maintaining life as wild animals because their living conditions are entirely different. In his later works, Wallace referred to artificial selection in the same way as Darwin.

As Darwin himself had admitted, neither the extract of his manuscript nor the letter to Gray intended to be published because their writing was poor (Barlow, 1958, pp. 121–122). On the other hand, the situation in Wallace's essay was different. For these reasons, Wallace's article is more precise regarding sequence and coherence. He made explicit the aim of the paper. He offered plausible arguments to show the difference between what happens concerning the production process of varieties produced under domestication and in the natural state, among other things. We agree with Nelson Papavero and Christian Santos that Wallace's essay is considerably better written and developed than Darwin's (Papavero & Santos, 2014, p. 175).

Wallace and Darwin used examples found in nature and hypothetical examples to illustrate their ideas. They also agreed that the species exist first as varieties that depart from the original type.

9.5 Final Remarks

The analysis developed based on the works of Darwin and Wallace, published in 1858, showed that their ideas were very similar, as they stated at that time. However, we also pointed out some differences in the previous section.

More marked differences between both theories became apparent between the late 1860s and early 1870s, especially after Darwin's death (1882). These mainly concerned the scope of natural selection and sexual selection. According to Wallace, unlike Darwin, natural selection did not explain the origin of sensation and consciousness and man's emergence from the lower animals. On the other hand, Wallace restricted sexual selection to characteristics used in the fight between males for possessing the female, such as defense weapons, horns, and spurs. Nevertheless, these and other aspects discussed in other works (Carmo, 2006; Carmo & Martins, 2006, 2009; Hidalgo & Carmo, 2007) are beyond the scope of this chapter.

We agree with Barbara Beddall that Wallace's and Darwin's contributions stimulated each other making them co-discoverers (Beddall, 1988, p. 64).

This case study shows aspects of the nature of science teachers can explore in the classroom. One is that natural philosophers or scientists can independently arrive at the same ideas simultaneously as Darwin and Wallace. Besides Wallace's and Darwin's, we can mention other examples in the history of science, such as differential and integral calculus by Isaac Newton (1643–1727) and Gottfried Wilhelm Leibniz (1646–1716).

References

Barlow, N. (Ed.). (1958). *The autobiography of Charles Darwin, 1809–1882, with original omissions restored*. Collins.

Beddall, B. G. (1972). Wallace, Darwin, and Edward Blyth: Further notes on the development of evolution theory. *Journal of the History of Biology, 5*(1), 153–158.

Beddall, B. G. (1988). Darwin and divergence: The Wallace connection. *Journal of the History of Biology, 21*(1), 1–68.

Bowler, P. (1989). *Evolution: The history of an idea*. University of California Press.

Browne, J. (1980). Darwin's botanical arithmetic and the "principle of divergence". *Journal of the History of Biology, 13*(1), 53–59.

Bulmer, M. (2005). The theory of natural selection of Alfred Russel Wallace. *Notes & Records of the Royal Society, 59*, 136–152.

Carmo, V. A. (2006). *Concepções evolutivas de Charles Darwin no Origin of Species e de Alfred Russel Wallace em Darwinism: um estudo comparativo* (M.Sc. thesis). Pontifícia Universidade Católica de São Paulo.

Carmo, V. A. (2011). *Episódios da história da biologia e o ensino da ciência: as contribuições de Alfred Russel Wallace* (Ph.D. thesis). Universidade de São Paulo.

Carmo, V. A., & Martins, L. A.-C. P. (2006). Charles Darwin, Alfred Russel Wallace e a seleção natural: Um estudo comparativo. *Filosofia e História da Biologia, 3*, 377–392.

Carmo, V. A., & Martins, L. A.-C. P. (2008). Algumas concepções evolutivas de Darwin *no Origin of Species* e de Wallace em *Darwinism*. In R. de, A. Martins, C. C. Silva, J. M. H. Ferreira, & L. A.-C. P. Martins (Eds.), *Filosofia e história da ciência no Cone Sul: Seleção de trabalhos do 5° Encontro* (pp. 455–461). Associação de Filosofia e História da Ciência do Cone Sul.

Carmo, V. A., & Martins, L. A.-C. P. (2009). Alfred Russel Wallace e o princípio de seleção natural. *Filosofia e História da Biologia, 4*, 209–233.

Colp, R., Jr. (1985). Notes on Charles Darwin's "autobiography". *Journal of the History of Biology, 18*(3), 357–401.

Darwin, C. (1858a). Extract from an unpublished work on species by C. Darwin, Esq., consisting of a portion of a chapter entitled, "on the variation of organic beings in a state of nature; on the natural means of selection; on the comparison of domestic races and true species". *Journal of the Linnean Society of London, 3*, 46–50.

Darwin, C. (1858b). Abstract of a letter from C. Darwin, Esq., to prof. Asa gray, Boston, U.S., dated down, September 5, 1857. *Journal of the Linnean Society of London, 3*, 50–53.

Darwin, C. (1859). *On the origin of species by means of natural selection or the preservation of Favoured races in the struggle for life*. John Murray.

Fichman, M. (2004). *An elusive Victorian: The evolution of Alfred Russel Wallace*. University of Chicago Press.

Hidalgo, J. M., & do Carmo, V. A. (2007). Wallace e a origem do homem: Suas concepções e as interpretações historiográficas. *Filosofia e História da Biologia, 2*, 227–244.

Kottler, M. J. (1985). Charles Darwin and Alfred Russel Wallace: Two decades of debate over natural selection. In D. Kohn (Ed.), *The Darwinian heritage* (pp. 367–432). Princeton University Press.

Lamarck, J.-B. M. (1809). *Philosophie zoologique* (Vol. 1). Dentu/Chez L' Auteur.

Lyell, C., & Hooker, J. D. (1858). On the tendency of species to form varieties; and on the perpetuation of varieties and species by natural means of selection. By Charles Darwin and Alfred Wallace. Letter from Charles Lyell and Joseph D. Hooker to the Linnean Society of London. *Journal of the Linnean Society of London, 3*, 45–46.

Martins, L. A.-C. P. (1997). Lamarck e as quatro leis da variação das espécies. *Episteme: Filosofia e História da Ciência em Revista, 2*(3), 33–54.

Martins, L. A.-C. P. (2007). *A teoria da progressão dos animais de Lamarck*. Booklink/FAPESP.

Nicholson, A. J. (1960). The role of population dynamics in natural selection. In S. Tax (Ed.), *Evolution after Darwin* (Vol. 1, pp. 477–522). Chicago University Press.

Papavero, N., & Santos, C. F. M. S. (2014). Evolucionismo darwinista: Contribuições de Alfred Russel Wallace à teoria da evolução. *Revista Brasileira de História, 34*(67), 158–180.
Raby, P. (2001). *Alfred Russel Wallace: A life*. Princeton University Press.
Rayher, E. S. (1996). *Confusion and Cohesion in Emerging Sciences: Darwin, Wallace, and Social Darwinism* (Ph. D. thesis). McGill University.
Regner, A. C. K. P. (2004). A teoria darwiniana da seleção natural. In R. de, A. Martins, L. A.-C. P. Martins, C. C. Silva, & J. M. H. Ferreira (Eds.), *Filosofia e História da Ciência no Cone Sul: 3° Encontro* (pp. 47–63). Associação de Filosofia e História da Ciência do Cone Sul.
Wallace, A. R. (1858). On the tendency of varieties to depart indefinitely from the original type. *Journal of the Linnean Society of London, 3*, 53–62.
Wallace, A. R. (1891). *Natural selection and tropical nature*. Macmillan.

Viviane Arruda do Carmo graduated in Biology, Pharmacy, and Pedagogy. She has done a master's in History of Science and Ph.D. in Education. She is also a researcher of the *History and Theory of Biology Group*/USP. Being a specialist in Alfred Russel Wallace, she develops research in the History of Science and Teaching and has published several works in this area. In addition, she is a founding member of the *Brazilian Association of Philosophy and History of Biology* (ABFHiB). She has held several positions on the Board of Directors as Secretary and Treasurer.

Lilian Al-Chueyr Pereira Martins researches history, philosophy of life sciences, and its epistemic interfaces, including science education. She has devoted herself to studying spontaneous generation since Antiquity, evolutionary theories in the nineteenth and twentieth centuries, and classical genetics. Being the first president of the Brazilian Association for History and Philosophy of Biology (ABFHiB) and author of the book *Lamarck's Theory on Animal Progression* (A teoria da progressão dos animais de Lamarck, 2007), she is currently the editor of the Brazilian journal *Philosophy and History of Biology* (Filosofia e História da Biologia).

Chapter 10
There Have Been Few Such Naturalists Before, but Still…: Darwin's Public Account of Predecessors

Maria Elice Brzezinski Prestes

Abstract This chapter explores Darwin's inclusion of the "Historical Sketch" in the third edition of *On the Origin of Species*, addressing criticisms from his contemporaneous readers regarding his book's absence of historical context. The Sketch, however, poses challenges for today's readers due to its lengthy list of mostly unfamiliar authors. Additionally, the chronological presentation fails to effectively highlight the similarities and differences between these authors' ideas and Darwin's own. A research-based teaching exercise is included in this chapter, with insights from Sketch and excerpts from his *Notebooks*, to clarify Darwin's line guide and motivate the reading. The activity revealed his perception that, while some naturalists believed in species modification, none *simultaneously* proposed the two fundamental principles of his (and Wallace's) theory: common descent and natural selection. Additionally, it rectified historical errors and reassessed the crucial difference Darwin considered when comparing his to Lamarck's theories, that is, the commitment with necessary progression and the lack of a unified origin for all living beings in the French naturalist theory. Viewing Sketch as a meta-scientific discourse encourages readers to explore the Nature of Science aspects in Darwin's work.

10.1 Introduction

On the Origin of Species can be surprising to readers who are unfamiliar with primary texts in the history of science. Although distant from today's citation conventions, the text draws attention to the numerous and wide range quotations. However, in the first edition of the book, Darwin remained silent about those who already believed in species modification before him and during his lifetime, except

M. Elice Brzezinski Prestes (✉)
Laboratory of the History of Biology and Teaching, Department of Genetics and Evolutionary Biology, Institute of Biosciences, University of São Paulo, São Paulo, Brazil
e-mail: eprestes@ib.usp.br

M. Elice Brzezinski Prestes (ed.), *Understanding Evolution in Darwin's "Origin"*,
History, Philosophy and Theory of the Life Sciences 34,
https://doi.org/10.1007/978-3-031-40165-7_10

for acknowledging that Alfred Russel Wallace "has arrived at almost exactly the same general conclusions" (Darwin, 1859, p. 10). Darwin did not mention other predecessors who explored the idea of species change generating new species. The omission prompted criticism, as there were numerous proponents of advanced ideas during that time, including theologians who questioned traditional interpretations of religious scriptures and speculated on species changes (Browne, 2006, p. 62; 78).

> One by one, Victorian thinkers claimed the right to investigate the world around them without recourse to either God's miraculous powers, or the Bible's word, or the church's doctrinal authority. (Browne, 2006, pp. 62–63)

Responding to criticisms, 2 years later, the *Origin* third edition (1861) included a new text, the "Historical Sketch of the Progress of the Opinion, Prior to This Work, on The Origin of Species." This Sketch underwent revisions in subsequent editions, with augmentations and the exclusion of the word "recent" from its title in the sixth edition (Darwin, 1872).[1] Except where noted, in this chapter, references to the text will be made to the "final" 6th London edition, of 1876.[2]

However, for today's readers, the Sketch proves challenging as it primarily consists of extensive names of authors, with only a few more informative paragraphs. The list of authors is composed chiefly of unknown names, diminishing the interest in reading the text thoroughly. Furthermore, Darwin's chronological presentation fails to effectively highlight the overall points of similarity and divergence he was identifying between the ideas of these authors and his own.

The initial paragraphs are the most engaging, as readers are naturally curious about Darwin's thoughts on famous predecessors, such as Lamarck, Erasmus Darwin, his grandfather known for his transformational ideas, or Richard Owen, who was considered a leading naturalist in Britain (Browne, 2006, pp. 93–94). The mention of Aristotle also may call attention to twenty-first-century readers, particularly those unfamiliar with Aristotle's biological work.[3] However, the reading quickly becomes tiresome. Consequently, the 34 names[4] – according to Darwin's account – and their specific ideas remain obscure and with little significance.

Interestingly, the Historical Sketch didn't hold much appeal for historians and philosophers of science either (Johnson, 2020, p. ix). However, considering historiographical procedures that enhance the reliability of an autobiographical account, as Curtis Johnson, we argue that the Historical Sketch deserves more attention.

[1] A fascinating comparison of editions in the Online Variorum of Darwin is available for free at: http://darwin-online.org.uk/Variorum/1861/1861-xiii-c-1872.html

[2] Freely available at: http://darwin-online.org.uk/content/frameset?viewtype=text&itemID=F401&pageseq=1

[3] Darwin describes an example in which Aristotle refers to adaptation in the formation of teeth. However, it is worth calling attention that, the search for transformist ideas in Aristotle was pointless, since such an idea was incompatible with the Greek notion of a perfect, eternal cosmos, that is, with no end, and no origin at all.

[4] Darwin's sum of named authors appears in a footnote of the Historical Sketch, declaring 30 in the 3rd edition (Darwin, 1861, p. xviii) and 34 from the 4th edition (Darwin, 1866, p. xx) until the last 6th edition (Darwin, 1876).

Before proceeding further, it is important to clarify two points. The first regards the object of Darwin's Sketch. He was addressing his predecessors, those who had expressed similar ideas prior to the publication of the *Origin*.[5] It is crucial to differentiate this recognition from that expressed by the notion of "precursors", usual in common sense and the "old" historiography of science, portraying them as geniuses ahead of their own time. If there were precursors, said Georges Canguilhem, the history of science would lose all meaning (1970, p. 21).[6] Since the 1960s, "the precursor virus" was considered a form of non-history, linked with the amateurish history that lacks the epistemological principles of the field.[7] Discussing the very object of the history of science, Canguilhem was shifting historians' questions from the traditional triad of *who*, *why*, and *how* to the understanding of "What is the history of science the history of?" – "*De quoi* l'histoire des sciences est-elle l'histoire?" (Canguilhem, 1970, p. 9). He emphasized that more than just *the science of the past*, the historian of science turned to what makes up the *whole past of current science*. In line with the prevailing sentiment of the era to identify the object of the history of science, these criticisms propelled ongoing transformations in the field. Eventually, in the 1980s, the historiography of science came to be characterized as "contextualist" (Smocovitis, 1996) or, simply, a "new" historiography.

Second, the purpose of this chapter is not to contrast what Darwin wrote in the Sketch and the historical records he mentioned or failed to mention.[8] Instead, the

---

[5]The terms "predecessor" and "antecedent" are mentioned only five times in the entire book (6th ed., 1876), specifically in reference to earlier geological periods or species and not in relation to authors or individuals. In the Historical Sketch, it seems that Darwin clearly avoided using similar expressions. For example, he states that Mr. Matthew recognized "the full force of the principle of natural selection" (Darwin, 1876, p. xvi). The closer notion of anticipation is mentioned next in a passage where Darwin claims to have sent a letter to Mr. Matthew, "fully acknowledging" that "he had anticipated" his ideas (Darwin, 1861, p. xv), but it was removed from subsequent editions of *Origin*.

[6]Georges Canguilhem discussed the concept of the "precursor virus" in a conference held at the *Société Canadienne d'histoire et de philosophie des Sciences*, in Montreal on October 28, 1966. The text was later prepared for publication in the book *Études d' histoire et de philosophie des sciences* in 1968. Canguilhem had explored the theme in seminars at the Institute of the History of Science and Technology at the University of Paris in the academic years 1964–1965 and 1965–1966, as mentioned in a footnote (Canguilhem, 1970, p. 9). Toward the end of the text (p. 22), he references Alexandre Koyré's *The Astronomical Revolution* (1961), which had already highlighted the potential harm caused by the notion of the precursor to the historian of science. Canguilhem's attention to specifying the "here and there" in which he exposed his arguments underscores the impact of this criticism on the historiography of science during the 1960s.

[7]According to Canguilhem (1970, p. 20), the term "the precursor virus" was coined by J. T. Clark, who argued for its elimination from historical analysis in the chapter "The Philosophy of Science and History of Science" he authored for the canonical book *Critical problems in History of Science*, edited by Marshall Clagett in 1962. The historian Quentin Skinner also criticized the notion of the precursor as a historical absurdity, a tendency to seek ideal-type approximations that "produces a form of non-history that is almost entirely devoted to pointing out earlier 'anticipations' of later doctrines and crediting each writer in terms of that foresight" (Skinner, 1969, p. 11).

[8]Several historians of science have made significant contributions to the understanding of Darwin's predecessors. A few notable examples shedding light on this topic are, on Lamarck, Martins (1993,

focus shifts in another direction, aiming to acknowledge Darwin's own representation of whom he considered his predecessors. The interest lies in understanding which ones Darwin intended to present to the world through his public testimony in *On the Origin of Species*.

The Historical Sketch combines a list-like format with a discursive writing style, featuring chronological paragraphs dedicated to specific authors. While most paragraphs are concise and focus on presenting the essential elements of authorship, work title, and date, there are also paragraphs that offer deeper analysis and additional information regarding the mentioned authors' ideas.

What captures the reader's attention is Darwin's meticulous inclusion of dates, although sometimes only indirectly determined. In instances where the title of a work is absent, Darwin explicitly mentions the idea and the date when the author endorsed it. What does it signify? Engaging in class discussions on this topic can lead to fascinating reflections. By posing such questions, students' interest can be sustained until the very last line of the Historical Sketch.

The following section outlines a pedagogical strategy modeled for conducting an investigative reading designed to stimulate curiosity and develop a focused approach to Darwin's predecessors, pointed out by himself in the Historical Sketch.[9]

10.2 A Model of Active Reading

The proposed pedagogical activity serves as a model of active reading. Aligned with constructivist pedagogy, it employs the content analysis technique as formulated by Laurence Bardin (1977/1995). The main science content learning objective of the activity is to promote students' skills to identify and compare Darwin's predecessors' positions regarding species change as he presented them in the Historical Sketch. Additionally, the activity aims to familiarize students with the qualitative research commonly used in Science Education, which contrasts the quantitative research used in natural sciences, preparing students for the diachronic reading of a primary source of the history of science.

To facilitate collaborative learning, the activity encourages students to work in groups. They can form small groups of up to four students or engage in discussions as a whole class group, with guidance from the instructor. Working together in groups allows students to share ideas, perspectives, and insights, fostering a richer

1997, 2007, 2015a, b), in European Natural History, Corsi (2005, and this volume), and, in general, Hodge (2013) and Sloan (2019).

[9]The pedagogical exercise described in this chapter was implemented in three instances of the course "The Origin of Species, by Charles Darwin," offered to undergraduate biology students and graduate science education students at the University of São Paulo in 2018, 2019, and 2021. It was developed in the "teaching with research" model, using a qualitative approach. Data were collected on multiple occasions, including pre- and post-tests, providing empirical evidence that served to the analysis presented in this study.

learning experience. The instructor plays a crucial role in guiding the groups and facilitating discussions. He or she can provide guidance, pose thought-provoking questions, and encourage active participation from all students. By guiding the groups, the instructor ensures that the activity remains focused, productive, and aligned with the learning objectives.

The activity will be presented in two parts. The first, analytical in nature, combines the processes of codification and categorization; the second, synthetical, addresses the flowchart construction.

10.2.1 Part 1: Codification and Categorization

In the context of the content analysis research methodology, codification involves decoupage of the text into its meaninfull constitutive elements, called units or indexes, that allow for a precise description of the relevant characteristics of the text content. This process involves selecting and extracting data from the text and organizing it in a structured manner. Categorization, on the other hand, involves classifying the constitutive elements of a set based on determined criteria. It is a process of differentiation and regrouping of data according to specific criteria.

Indexes and categories can be established prior to data collection based on the theoretical framework or previous research, or they can be developed during the analysis phase based on emerging patterns and themes.

When indexes and categories are established prior to data collection, it means that the researcher has a predetermined set of codes and categories based on existing knowledge and theories. This approach provides a structured framework for analyzing the data and ensures consistency in the analysis process. It allows researchers to focus on specific aspects of the data that align with their research objectives.

On the other hand, indexes and categories can also be developed during the analysis phase. This approach is often used in exploratory research or when there is limited existing knowledge about the topic. Nevertheless, developing indexes and categories during the analysis phase can have significant heuristic value in classroom activities, particularly when engaging with complex texts. By immersing and actively analyzing text content, readers can independently identify patterns, themes, and relevant aspects that contribute to its deeper understanding. Furthermore, this flexible approach encourages students to explore the text from different perspectives, enabling the discovery of new insights and connections.

Both approaches have their merits and can be employed depending on the research context and objectives, the nature of the study, the available knowledge on the topic, and the pedagogical objectives of the activity.

In the exercise described in this chapter, a mixed approach of coding and categorization processes was employed. Authors names for analysis were pre-selected by the instructor to create a representative sample that encompasses the diversity of positions regarding species changes in Darwin's text. The author's ideas on the origin of species and for the means of species change were identified during the analysis of the text itself. Additionally, the set of means by which species change identified in the text were contrasted and complemented with the list of means provided by Darwin in the preface of the second edition of his other book, *The Descent of Man* (1882). The comparison helps in gaining a broader understanding of Darwin's views and provides additional context for the analysis of the Historical Sketch.

Reading and Coding

The analysis process begins with an initial comprehensive reading of the entire text, which, in the case of the 1876 edition of the *Origin*, comprises 18 pages. The purpose of this initial reading is to establish familiarity with the content and obtain a general understanding of its scope. This fluctuating reading allows the reader to immerse in the text and form initial impressions and orientations (Bardin, 1977/1995, p. 96).

To facilitate the exercise, students are encouraged to enumerate the 31 paragraphs of the text, including footnotes. Afterward, in small groups, students discuss their observations and insights gained from the fluctuating reading. This discussion helps consolidate their general understanding of the text and prepares them for the subsequent analytical reading and codification phase.

During the analytical reading directs to identify the types of information Darwin provides about each predecessor. Students are encouraged to focus on the nine selected authors below, listing all the relevant information they encounter, and engaging in a close re-reading of each paragraph to uncover details as they delve deeper into the text.

These identified items form the basis for the coding process. Following the principles outlined by Bardin (1977/1995, pp. 97–98), the list of codes should strive to meet four key characteristics: exhaustiveness, representativeness, homogeneity, and relevance. Students should aim to list as many codes as possible, and the level of detail in the coding will determine the potential for diverse and in-depth analyses.

As the codes are compiled, students are encouraged to groupe them according to the criteria discussed and selected by the students themselves. This grouping and organization of codes will form the categories of analysis.

In the exercise described, these were the authors indicated for analysis:

1. Alfred Russel Wallace
2. Author of Vestiges of Creation (Robert Chambers)
3. Charles Darwin (referring to himself)
4. Geoffroy Saint-Hilaire
5. Isidore Geoffroy Saint-Hilaire
6. Jean-Baptist Lamarck
7. Omalius d'Halley
8. Patrick Matthew
9. Richard Owen

These are the categories and respective codes (types of information) that emerged in the exercise:

I. AUTHORS

1. Author's name
2. Profession
3. Work title or idea
4. Date
5. Quotation length (number of words)

II. DARWIN'S ASSESSMENT OF PREDECESSORS

6. Merits of work, compliments
7. Work difficulties, limitations, criticism

III. AUTHORS' GENERAL POSITION ON SPECIES CHANGE

8. Immutable species: single creation
9. Immutable species: multiple creations
10. Unlimited variability
11. Limited variability
12. Separate origins
13. Common origin

IV. AUTHOR'S PROPOSED MEANS OF SPECIES CHANGE

14. Inherited effects of use and disuse
15. Progressive development (law, impulse)
16. Direct and prolonged action of changed conditions of life
17. Natural selection
18. Crossing of existing forms
19. Sexual selection
20. Correlated growth
21. Impulse to adapt
22. Degeneration of the same type
23. Reversion of structures

V. EXTRA OBSERVATIONS

24. Provenience of author's ideas
25. Agency
26. Gradual or leap changes
27. Spontaneous generation
28. Darwin's metascientific comments

To facilitate data organization and analysis, it is recommended to construct a table where each row represents a predecessor author, and the columns are labeled with the categories and their corresponding codes. This table serves as a powerful heuristic tool that can be used for various subsequent investigations and analyses of past evolutionary theories. As an illustration, Figs. 10.1 and 10.2 show the five categories and corresponding codes plotted in an Excel table. Figure 10.1 shows that the table also includes three extra columns for

	A	B	C	D	E	F	G	H	I	J	K	L	M	N	O	P
1	INDEXING DATA		AUTHORS						DARWIN'S ASSESSMENT		AUTHOR'S GENERAL POSITION ON SPECIES CHANGE					
2	Number	Sketch Page	Transcription	Author's name	Profession	Work title or idea	Date	Number of words	Merits	Difficulties	Single creation	Multiple creations	Unlimited variability	Limited variability	Common origin	Separate origin
3																

Fig. 10.1 Quotation, Darwin's assessment, and the author's position categories and respective codes

Q	R	S	T	U	V	W	X	Y	Z	AA	AB	AC	AD	AE	AF
AUTHOR'S PROPOSED MEANS OF SPECIES CHANGE											EXTRA OBSERVATIONS				
Not addressed	Inherited effects of use and disuse	Progressive development (law, impulse)	Direct and prolonged action of changed conditions of life	Natural selection	Crossing of existing forms	Sexual Selection	'Correlated' growth	Impulse to adapt	Degeneration of the same type	Reversion of structures	Provenience of author's ideas	Agency	Gradual or Leap Changes	Spontaneous Generation	Darwin's metascientific comments

Fig. 10.2 Means or causes of species change and extra observation categories and respective codes

numbering the analyzed authors, the page of the Historical Sketch where they are mentioned, and the transcript of the paragraphs dedicated to them by Darwin.

Inserting Data into the Table

When filling in the table, it is crucial to transcribe *the exact words used by Darwin*[10] to avoid any anachronistic interpretations and the use of present concepts. This ensures that students are capturing Darwin's original perceptions and ideas rather than imposing their own interpretations onto the text. Additionally, it's important to include codes for ideas that might not be explicitly stated but can be inferred from the text, taking care of differentiating them from the exact words used by Darwin.

Expanding the search for information beyond the Historical Sketch to chapters of the *Origin* and other works by Darwin, as well as reliable secondary sources on the history and philosophy of science, can provide a broader context and deeper understanding. However, it's important to exercise caution when using online sources, particularly websites or blogs by non-professionals in the field.

By following these guidelines and being diligent in the transcription and interpretation of Darwin's words, students can accomplish the science content learning objective of the activity mentioned above, that is, to be able to identify and compare Darwin's predecessors' positions regarding species change as he presented them in the Historical Sketch and establish a solid foundation for further analysis and inferences in the subsequent steps of the research process.

[10]Which is easily accomplished by the free online availability of the text modality of the work, at http://darwin-online.org.uk/content/frameset?viewtype=text&itemID=F401&pageseq=1

10.2.2 Part 2: Flowchart

The second part of the activity concern the construction of a visual representation of the comparison of authors named by Darwin in the Sketch. The proposal is to create a flowchart, manually or digitally, with the objective of comparing Darwin and Wallace's core concepts with the ideas of selected predecessors based on chosen indexes. The visual representation of the flowchart will respond to the question: When comparing his ideas with those of precedent authors, what does Darwin point out that is unique to his and Wallace's theory?

Here's how the flowchart can be constructed:

1. Start by creating boxes for each of the following authors: Alfred Russel Wallance, Charles Darwin, Geoffroy Saint-Hilaire, Isidore Geoffroy Saint-Hilaire, Jean-Baptist Lamarck, Omalius d'Halley, Patrick Matthew, Richard Owen, the Vestiges of creation author (Robert Chambers); and code terms: common origin, correlated growth, crossing of existing forms, degeneration of the same type, direct and prolonged action of changed conditions of life, divine creation, does not address the means, few naturalists, great majority of naturalists, immutable species, impulse to adapt, inherited effects of use and disuse, limited variability, multiple creations, natural laws, natural selection, progressive development (law, impulse), reversion of structures, separete origins, sexual selection, single creation, species, species change, unlimited variability.
2. Connect the boxes with arrows and lines to represent the inferences and relationships between them. The connections should reflect the analysis of the data and highlight the unique aspects of Darwin and Wallace's theory. For example:
    - Connect the concepts of species change among each other and to the authors' boxes to show their respective positions and proposals.
    - Connect the concepts related to immutable species among each other and to the authors.

As the groups construct their flowchart, the instructor should engage them in making their inferences explicit and discuss the connections between the boxes. It is worth asking them to explain the reasoning behind each inference and clarify any unclear links. If the Historical Sketch does not provide enough information, suggest relevant works by Darwin or other sources that can shed light on the connections being sought. In reporting some of the cases, Sect. 10.4 includes sources that have been consulted.

Figure 10.3 represents one example that resulted from interaction with students.

At this point, students have made a detailed examination of the alleged predecessors listed by Darwin, along with the recognition of how Darwin considered their contributions, acquiring valuable insights into the development of Darwin's own ideas and theory. By studying the mechanisms proposed by these predecessor authors to explain species change, students could compare them to the mechanisms proposed by Darwin (and Wallace), identifying similarities, differences, and unique aspects of their theory.

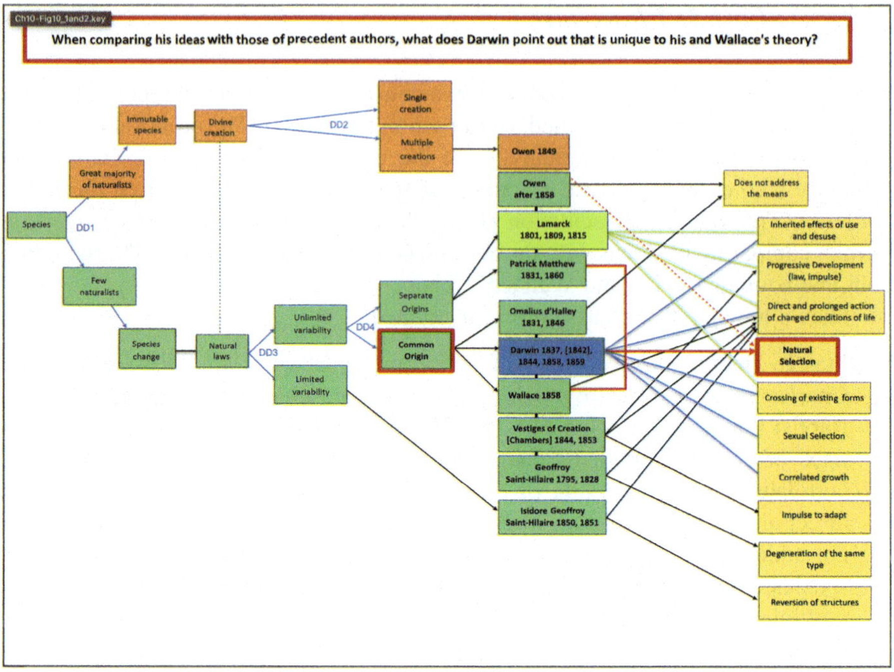

Fig. 10.3 Example of a flowchart with concepts and authors from the Historical Sketch (6th edition of The Origin of Species, 1876)

The following sections of this chapter expand on some of the topics highlighted by the flowchart that have the potential to increase students' understanding of the fact of evolution and the *means* by which it occurs, drawing on what Darwin interpreted from his predecessors.

10.3 Discussion

10.3.1 Dichotomous Reasoning

Dichotomous reasoning, in its philosophical sense, follows the approach that involves dividing a concept or category into two distinct differences. This method of reasoning, as described by Aristotle and referenced in Lennox (2001, 2006), requires that each term, referred to as the genus, be divided into two and only two sub-terms, known as the species. There is no allowance for a third option or alternative.

By employing dichotomous reasoning, individuals are forced to make a clear choice between two mutually exclusive possibilities. This approach can be useful in analyzing and categorizing information, as it allows for a clear differentiation between distinct alternatives or positions, such as common origin by natural selection and special creation, a strategy constantly used by Darwin.

In his book, Darwin aimed to challenge the prevailing doctrine of creation by providing alternative explanations for natural phenomena. According to David Depew, throughout the book, Darwin consistently argues against the "ordinary doctrine of creation" and likely had specific individuals in mind when formulating his arguments (Depew, 2009, p. 241). Janet Browne noted that Darwin presented instances that were "quite inexplicable" if one were to consider independent acts of creation as the explanation (Browne, 2006, p. 69).

To illustrate this approach, consider Darwin's investigation of a plant species found in two geographically distant territories in Chap. 11 of the *Origin* first ed. As employed in a Teaching Learning Sequence developed by Tatiana Tavares Silva (2013), Darwin proposed two possible explanations for this occurrence:

1. The plant species could have been independently created in each of the different places where it was found.
2. The species in question has a single origin, and its dispersal mechanisms, such as seed dispersal, allowed it to reach different locations on the planet.

Darwin then sought evidence to support the plausibility of the second explanation, close to his theory. He conducted three experiments to explore this possibility:

(a) Buoyancy of plant structures: Darwin examined whether the plant structures could float or be transported through various means, such as wind or water currents.
(b) Viability of seed transportation by sea currents: He investigated whether seeds could remain viable and survive transportation by ocean currents.
(c) Viability of seed germination after prolonged immersion in saltwater: Darwin tested whether seeds could still germinate after being immersed in saltwater for an extended period.

By conducting these experiments and presenting the evidence, Darwin aimed to provide strength to the plausibility of his arguments based on natural mechanisms, such as seed dispersal, as explanations for the distribution of species.

The use of dichotomous reasoning in the flowchart allows for a clear and structured analysis of the concepts and ideas under consideration. In this exercise, four dichotomous divisions, labeled DD1, DD2, DD3, and DD4, were identified and examined, as amplified in Fig. 10.4.

Dichotomous division 1 – species change or not change. DD1 represents the first dichotomous division, which revolves around the idea of species origin status. The options are either species don't change or species change, meaning they do originate new species. In the Historical Sketch, Darwin (1876, p. xiii) highlights that until recently, the prevailing belief among naturalists was that species were immutable, while a few naturalists endorsed that species change.

Dichotomous division 2 – one act or multiple acts of creation. Taking the immutable species side, there are two and only two historical theological options for the

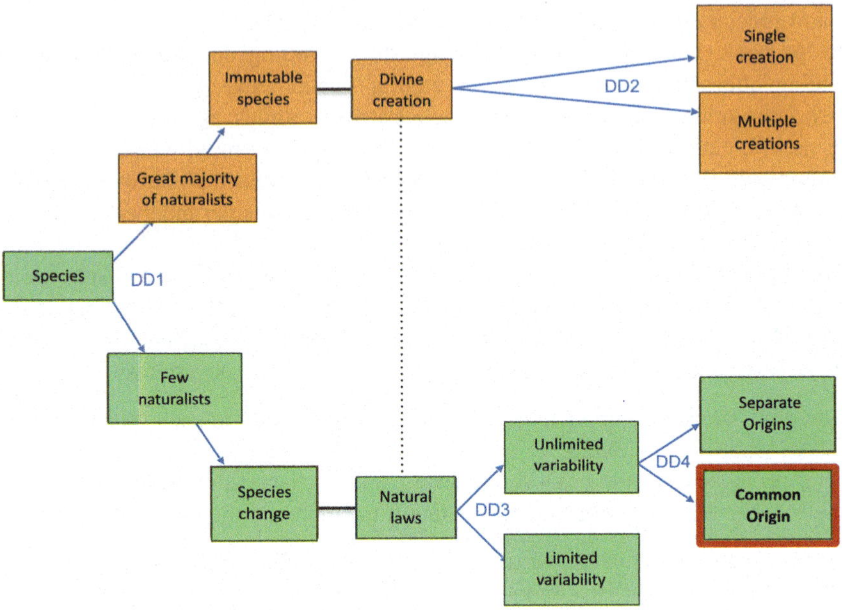

Fig. 10.4 The amplified piece of the flowchart with dichotomous divisions DD1–DD4

divine creation to occur: species were created all at once, synchronically in time and space, or created multiple times and in distinct places. Having been created at once or various times, species have existed ever since precisely like they were made. Darwin (1876, p. xii) mentions Richard Owen as having expressed his belief in the second conception, separate creations, in 1849.

Dichotomous division 3 – limited or unlimited changes. Moving back to the proponents of species change, DD3 examines the scope of change. There are two historically held possibilities that have divided naturalists into: those who believe in universal change, where all living beings are subject to change at any time and place, and those who believe in a limited capacity of species change. While most of the naturalists mentioned in the Sketch conceived of universal change, Darwin (1876, p. xix) indicates Isidore Geoffroy Saint-Hilaire as an example of someone who advocated for limited variability, emphasizing the dependence of species change on surrounding circumstances.

Dichotomous division 4 – one or multiple origins. Moving back to those who advocate for unlimited variability, DD4 addresses the question of the natural origin of species between two possibilities: those who believe in a common origin for all species and those who propose separate origins in different times and places. Among the selected naturalists for the flowchart, Darwin, Omalius d'Halley, and Wallace support the idea of a common origin, while Lamarck,

Owen (after 1858), and Patrick Matthew argue for separate origins. This division highlights a key distinction Darwin saw between his and Lamarck's theory.

10.3.2 Historical Backgrounds and Sources

In the dichotomous division DD1, the focus is on the concept of immutability and its relationship to divine creation. According to Richards and Ruse (2016), proponents of immutability argue that species, as products of divine creation, are already perfect and well-adapted to their original environments. This perspective implies that species do not undergo changes or transformations over time.

On the other hand, the idea of species change and the emergence of new species necessitates a different understanding of their origin. This perspective suggests that species evolve through natural processes governed by natural laws, as discussed by Brooke (2009). This viewpoint does not necessarily exclude the possibility of a divine role, as it allows for the notion that nature and its laws may have been created by a higher power.

Therefore, in the context of DD1, the dichotomy lies between the belief that immutability is a result of divine creation and the notion that species change is a consequence of natural laws and processes.

For discussions on DD2, it is worth referring to the impact of the astonishingly growing number of species revealed by exploring the new worlds, both in the West and the East, drastically reshaping our understanding of living beings. Naturalists of the seventeenth century acknowledged that even well-adapted species could thrive "in countries other than their own" (Browne, 1983, p. 11). Migration, dispersion, and various modes of transportation could explain the widespread distribution of species across the globe. Simultaneously, the sheer number of species challenged the traditional notion of Noah's Ark. An alternative perspective emerged, suggesting a nonliteral interpretation of the Sacred Scriptures, speculating that the deluge may have been a localized event in Europe and the Middle East. It was proposed that humans, as well as animals and plants from the rest of the world, remained "unaffected by the catastrophe" (Browne, 1983, p. 13). Instead of migration or post-deluge dispersion of species originating from Mount Ararat, it was proposed that "there had been not one but many creations," suggesting that animals and humans in the New World had existed from the beginning (Browne, 1983, p. 14).

In the eighteenth century, the debate continued with the known rivalry between Linné and Buffon, leading them to opposing viewpoints. Expanding on the concept of "centers of creation" (Herbert, 2005, p. 302), Linné proposed that "all living beings, including mankind, [...] originated on a high mountain as the primaeval waters began to recede" (Browne, 1983, p. 18). The observed climatic stratification of plants on a mountain led to it being considered a microcosm representing "an entire world in miniature" (p. 19). On the other hand, Buffon, taking into account the difficulties species faced in traveling long distances and through inhospitable lands, revived the hypothesis of species being created in various locations (Browne, 1983, p. 23).

The conundrum persisted into the nineteenth century, with Charles Lyell incorporating Linné's concept of "centres of creation" in his *Principles of Geology*. He used this framework to explain the distribution of plants on a group of islands, postulating the existence of a "primitive focus, or center," with a specific type of vegetation from which a smaller number of species dispersed to the surrounding islands (Herbert, 2005, p. 302).

Overall, the exploration of new worlds and the increasing number of discovered species challenged traditional beliefs about the origin and distribution of life forms. The dichotomy between a single divine creation and multiple creations in different places became a central aspect of the ongoing debate, with various naturalists proposing different explanations to account for the observed diversity and distribution of species.

Isidore Geoffroy Saint-Hilaire's position advocating limited variability of DD3 can be traced back to Darwin's writings in Sketch. In his Lectures delivered in 1850, as Darwin transcribed in French, Geoffroy Saint-Hilaire briefly provides his reasons for believing that specific characteristics are fixed, for each species, as long as it persists in the same circumstances, and they are modified only if the surrounding conditions change. "In summary, the observation of wild animals already demonstrates the *limited variability* of species" (Darwin, 1876, pp. xviii–xix, emphasis added).

Regarding the common or multiple of natural origin of species expressed by DD4, the Sketch provides the position advocated by Patrick Matthew on the side of separate origins, "at successive periods."[11] Matthew's perspective suggests that the world underwent near depopulation during successive periods, followed by re-stocking with new forms. Darwin states that Matthew had even proposed an alternative explanation that new forms could generate "without the presence of any mould or germ of former aggregates." This inclusion of Matthew's position in the Sketch highlights the presence of alternative ideas about the origins of species and the mechanisms of change. Matthew's emphasis on separate origins at successive periods stands in contrast to the concept of a common origin advocated by Darwin and other naturalists.

Let's now turn to the case of Lamarck in two separate lines of argument. One is the examination of whether the theory of Lamarck proposed one common or multiple origins of all living beings. Another line is to determine how Darwin conceived Lamarck's theory regarding this issue. In both lines, though, there is not a clear-cut interpretation. But following Darwin's strategy, we seek to give some

[11] Investigating the reasons by which Darwin states the differences between Matthew's view and his own "are not of much importance" reveals significant social aspects of the nature of science. The priority dispute between Patrick Matthew and Charles Darwin is a curious aspect of their relationship and is specifically addressed by Johnson (2020, pp. 110–125). Additionally, Dagg (2018) offers a comprehensive comparison of the thoughts of Matthew, Darwin, and Wallace.

evidence of the plausibility that among their differences, there was the question about the common or multiple origins.

Indeed, Darwin was explicit about the differences between his own theory and that of Jean-Baptiste Lamarck. One of the most quoted passages is in his notebooks, where Darwin simply states "my theory very distinct from Lamarck's" (Darwin, 1837–1838, p. 213), without explicitly explaining what the specific differences are. However, through various other writings, Darwin provided further insights into the contrast between his theory and Lamarck's. According to Janet Browne (2002, p. 61), Darwin highlighted the fundamental difference in their views regarding goal-directedness or any internal driving force that directed adaptative changes toward specific directions. Darwin's criticism of Lamarck's ideas can be seen in his notebooks as well. In *Notebook B*, he expressed "Lamarck's 'willing' doctrine absurd" (Darwin, 1837–1838, p. 216). In 1993 Lilian Martins clearly argued that Lamarck never even suggested that there was an influence of desire [*désir*] in the formation of new organs, but of the physiological "need" [*besoin*] to be satisfied (Martins, 2007, p. 303, 344, 421) . By using the term "will" in his summary of Lamarck's ideas, Lyell made the mistake that was repeated in Darwin. Additionally, when discussing a German edition of his *Origin*, where natural selection was translated to something like "choice of life style" (which for him implied a Lamarckian interpretation), Darwin expressed his rejection of the Lamarckian doctrine, as he considered habits of life to be far from the sole determination of evolution (Darwin in Browne, 2002, p. 142).

Regarding Lamarck's own works, there is some evidence he conceived of living things as having separate origins – albeit all natural, by the physical laws of universal attraction and repulsive action. For Lamarck, the Great Author created "the order of things" (nature), governed by its own laws, nature itself gave rise to all living beings successively, without divine intervention, through spontaneous generation (Martins, 2007, pp. 58–59). Assumptions pointing to separate origins can be found in Lamarck's monumental work, *Histoire naturelle des animaux sans vertèbres* (1815–1822). Throughout volume two, passages like the following can be found:

> [...] I will prove that there is *no real chain which generally binds together* the productions of nature, and that it can only be found *in certain branches* of the series which they form [...]. (Lamarck, 1835, p. 51, emphasis added)

The difficulty with the passage, though, is that it may not be clear whether Lamarck referred to morphological proximity (that is, taxonomic) or origin proximity (that is, by descent) when he mentions unique connections between certain branches.[12]

[12] In "Lamarck Revisited," Ernst Mayr showed that the understanding of the actual or suspected continuity between higher taxa, which Lamarck grappled with and revised throughout his work, could only begin to be clarified much later in the twentieth century. It was during this time that the distinction between "well-defined and sharply delimited" zoological groups like birds, bats, or beetles, and the artificial and somewhat subjective taxonomic categories (e.g., family, order, class) became clearer in terminology (Mayr, 1972, pp. 83–84).

On the other hand, passages found in Lamarck's work indicate a common origin. Based on them, Lilian Martins argues that this is the position of the French naturalist, as she expresses in this summary:

> In a remote past there were no living beings; then nature created the first (simplest) ones. These first living beings created by nature were produced from spontaneous generation. From these first ones, with time and favorable circumstances, others emerged and increased their complexity, giving rise to an animal and plant scale with large groups in different degrees of perfection. This scale also presents some ramifications. (Martins, 2007, p. 63)

Sometimes, Lamarck's text affirms the separation of animals and plants but brings their base as an exception, like here:

> [. . .] while I have established a *positive distinction between vegetables and animals*, and have shown that, even when vegetables seem to be linked to animals by some point in their series, instead of forming together a chain or a ladder graduated, *they would always present two separate branches, very distinct*, and *only brought together at their base*, in relation to the simplicity of organization of the beings which are found there. (Lamarck, 1835, p. 110, emphasis added)

What is tricky, though, is how to cohere those lines with other passages that strongly affirm the radical separation of animals and plants. For example:

> Now, as there exist two very distinct sorts of living bodies, namely, vegetables and animals, let us examine the essential characteristics of these first, and showing the line of separation which nature has established between these two sorts of beings, let us prove that *plants cannot unite with animals at any point in their series to form a real chain*. (Lamarck, 1835, p. 73, emphasis added)

Besides, one can speculate about these words by contrast. Could they have been uttered by someone whose picture of nature was entirely founded on common descent, like Darwin? Hardly so.

However, the focus of this exercise was to explore evidence rather than delve into the debate itself. Experts are better suited to argue whether Lamarck oscillated between the notion of a single origin and separate origins or if he embraced only one of these models. In the context of inferring what Darwin attributed to Lamarck, the dichotomous arrangement seemed to support the defense that, when Darwin stated "my theory so different from Lamarck's," in 1838, he was referring to the contrast with his cornerstone concept, the absolute common descent of all living beings.

It is also important to acknowledge that Darwin recognized other differences between his theory and Lamarck's. Perhaps the most significant discrepancy was the law of progressive development of living beings, clearly expressed in the Sketch. In this passage, after citing three aspects accepted by both, Darwin introduces the fourth preceded by the unquestionable "but," denouncing the disagreement:

> [. . .] he attributed something to the *direct action of the physical conditions of life*, something to the *crossing of already existing forms*, and much to *use and disuse, that is, to the effects of habit*. To this latter agency he seems to attribute all the beautiful adaptations in nature – such as the long neck of the giraffe for browsing on the branches of trees. *But* he likewise believed in *a law of progressive development*; [. . .] (Darwin, 1876, p. xiv, emphasis added)

Indeed, Lamarck conceived a law of progression in the animal scale, from the simple to the complex, and of degradation when the scale is inverted from the complex to the simplest (Martins, 2007, p. 59). For Darwin, however, the criticism of progress in species change goes back a long way:

> Degradation and complication see Lamarck: *no tendency to perfection*: if room, [even] high organism would have greater power in beating lower one, thought ⟨?⟩ to be selected for a degraded end. (Darwin, 1909, p. 47, emphasis added)

At this point, the uniqueness of the theory of Darwin and Wallace, in contrast to the representative sample of predecessors explored in this exercise, becomes evident. While several naturalists advocated for a single origin of living beings, and Patrick Matthew also contemplated the concept of natural selection, it was only Darwin and Wallace who integrated both the principles, common descent and natural selection, into a unified theory.

10.4 Final Comments

The teaching exercise described has achieved several objectives. It successfully conveyed Darwin's perception of his predecessors and contemporaries who held transformist ideas and provided a diachronic understanding of the revolutionary core of his theory (common descent and natural selection).[13] It clarified differences between Lamarck's and Darwin's theories: Lamarck accepted progressive development, Darwin not; possibly, Lamarck conceived multiple origins, Darwin not; finally, only Darwin conceived natural selection, the main, though not exclusive means by which species evolve. At the same time, the exercise reiterated various means of evolution accepted by Lamarck and Darwin, including the inherited effects of use and disuse. The exercise also identified the type of reasoning Darwin employed in the Historical Sketch, a metascientific text, as a deductive reasoning through dichotomous division.

However, it is important to acknowledge the limitations of the exercise. The content analysis tool and the resulting flowchart focused on understanding what Darwin intended to communicate rather than critically evaluating the reliability of his testimony. Evaluating the accuracy of Darwin's opinions or confronting and challenging the information he presented about his predecessors were beyond the scope of this exercise and required specialized research in the history of science.

The exercise also generated several research topics that emerged during the lessons. These include investigating the reasons behind Darwin's decision to include a Historical Sketch only in the 3rd edition of the *Origin* and his selection of specific

[13] The difficulties in understanding natural selection and evolution are indeed well-documented in pedagogical research. Theoretical and empirical work on this topic provides valuable insights into the conceptual obstacles that learners face and help inform effective teaching strategies: Gregory, 2009; Kampourakis, 2013; Deniz & Borgerding, 2018; Harms & Reiss, 2019.

authors to mention.[14] The exercise also raises questions about the extent to which Darwin should have addressed Wallace's contribution in more detail in the Sketch, considering the joint publication and their shared theory. Exploring the implications of citation numbers and their impact on the strength of Darwin's argument is another topic for further discussion.

Nonetheless, the exercise has provided clarity regarding the place of Darwin and Wallace's ideas among intellectuals and naturalists of the mid-nineteenth century. It has laid the foundation for future investigations in the history or philosophy of biology. The curiosity generated among students about these topics serves as a compelling reason to continue using Darwin's *Origin of Species* in the teaching of evolution and the history of science, promoting a grounded understanding of the nature of science.

References

Bardin, L. (1977/1995). *Análise de Conteúdo* (p. 70). Edições.

Brooke, J. H. (2009). "Laws impressed on matter by the creator"? The origin and the question of religion. In M. Ruse & R. J. Richards (Eds.), *The Cambridge companion to the "origin of species"* (pp. 256–274). Cambridge University Press.

Browne, J. (1983). *The secular ark: Studies in the history of biogeography*. Yale University Press.

Browne, J. (2002). *Charles Darwin: The power of place*. Princeton University Press.

Browne, J. (2006). *Darwin's origin of species: A biography*. Grove.

Canguilhem, G. (1970). Introduction. L'Objet de l'Histoire des Sciences. In *Études d'Histoire et de Philosophie des Sciences* (2nd ed., pp. 9–23). Vrin.

Corsi, P. (2005). Before Darwin: Transformist concepts in European natural history. *Journal of the History of Biology, 38*(1), 67–83.

Dagg, J. L. (2018). Comparing the respective transmutation mechanisms of Patrick Matthew, Charles Darwin and Alfred Wallace. *Biological Journal of the Linnean Society, 123*, 864–878.

Darwin, C. R. (1837–1838). Notebook B: [Transmutation of species]. CUL-DAR121. Edited by John van Wyhe (Darwin Online, http://darwin-online.org.uk/).

Darwin, C. R. (1859). *On the origin of species by means of natural selection, or the preservation of favoured races in the struggle for life* (1st ed.). John Murray.

Darwin, C. R. (1861). *On the origin of species by means of natural selection, or the preservation of favoured races in the struggle for life* (3rd ed.). John Murray.

Darwin, C. R. (1866). *On the origin of species by means of natural selection, or the preservation of favoured races in the struggle for life* (4th ed.). John Murray.

Darwin, C. R. (1872). *The origin of species by means of natural selection, or the preservation of favoured races in the struggle for life* (6th ed.). John Murray.

Darwin, C. R. (1876). *The origin of species by means of natural selection, or the preservation of favoured races in the struggle for life*. 6th edition, with additions and corrections. [First issue of final definitive text]. John Murray.

Darwin, C. R. (1882). *The descent of man, and selection in relation to sex* (2nd ed.). John Murray. http://darwin-online.org.uk/content/frameset?itemID=F955&viewtype=text&pageseq=1.

[14] Pietro Corsi's work (e.g., Corsi, 2005), including the chapter in this volume, provides a sample of the numerous, and still very little explored authors who, before Darwin, were supporters of the idea that species change and produce new species.

Darwin, F. (Ed.) (1909). *The foundations of* The origin of species, *a sketch written in 1842*. University Press. http://darwin-online.org.uk/content/frameset?itemID=F1555&viewtype=text&pageseq=1.
Deniz, H., & Borgerding, L. A. (2018). *Evolution education around the globe*. Springer.
Depew, D. J. (2009). The rhetoric of the *origin of species*. In M. Ruse & R. J. Richards (Eds.), *The Cambridge companion to the "origin of species"* (pp. 237–254). Cambridge University Press.
Gregory, T. R. (2009). Understanding natural selection: Essential concepts and common misconceptions. *Evolution: Education and Outreach, 2*, 156–175.
Harms, U., & Reiss, M. (Eds.). (2019). *Evolution education re-considered: Understanding what works*. Springer.
Herbert, S. (2005). *Charles Darwin: Geologist*. Cornell University Press.
Hodge, J. (2013). Darwin's book: On the origin of species. *Science & Education, 22*, 2267–2294.
Johnson, C. N. (2020). *Darwin's "historical sketch": An examination of the "preface" to the origin of species*. Oxford University Press.
Kampourakis, K. (Ed.). (2013). *The philosophy of biology: A companion for educators*. Springer.
Lamarck, J. B. P. A. (1835). *Histoire naturelle des animaux sans vertebres* (2nd ed.). Baillière.
Lennox, J. (2001). *Aristotle's philosophy of biology: Studies in the origins of life sciences*. Cambridge University Press.
Lennox, J. (2021). Aristotle's biology. The Stanford Encyclopedia of Philosophy. Edward N. Zalta (Ed.). https://plato.stanford.edu/archives/fall2021/entries/aristotle-biology/.
Martins, L. A.-C. P. (1993). *A teoria da progressão dos animais de Lamarck.* Campinas (Mestrado em Genética) – Universidade Estadual de Campinas.
Martins, L. A.-C. P. (1997). Lamarck e as quatro leis da variação das espécies. *Epistéme. Filosofia e História da Ciência em Revista, 2*(3), 33–54.
Martins, L. A.-C. P. (2007). *A teoria da progressão dos animais, de Lamarck*. Booklink.
Martins, L. A.-C. P. (2015a). A herança de caracteres adquiridos nas teorias "evolutivas" do século XIX, duas possibilidades: Lamarck e Darwin. *Filosofia e História da Biologia, 10*(1), 67–84.
Martins, R. A. (2015b). *Aristóteles e o estudo dos seres vivos*. Lulu Press.
Mayr, E. (1972). Lamarck revisited. *Journal of the History of Biology, 5*(1), 55–94.
Richards, R. J., & Ruse, M. (2016). *Debating Darwin*. The University of Chicago Press.
Silva, T. T. (2013). *Darwin na sala de aula: Replicação de experimentos históricos para auxiliar a compreensão da teoria evolutiva*. Dissertação (Mestrado em Ensino de Ciências – Universidade de São Paulo.
Skinner, Q. (1969). Meaning and understanding in the history of ideas. *History and Theory, 8*(1), 3–53.
Sloan, P. (2019). Evolutionary Thought Before Darwin. *The Stanford Encyclopedia of Philosophy*. Edward N. Zalta (Ed.). https://plato.stanford.edu/archives/win2019/entries/evolution-before-darwin/.
Smocovitis, V. B. (1996). *Unifying biology: The evolutionary synthesis and evolutionary biology*. Princeton University Press.

Maria Elice Brzezinski Prestes researches the history and philosophy of biology, including its approach to Science Education. She focused on the generation of living beings, especially on the modes of observation and experiments in the seventeenth and eighteenth centuries. Currently, she is developing studies on Darwin's and Aristotle's works. She is a Sênior Professor at the Department of Genetics and Evolutionary Biology of the Institute of Biosciences of the University of São Paulo, where she coordinates the Laboratory of History of Biology and Teaching (LaHBE). She is a former president of the Brazilian Association of Philosophy and History of Biology (ABFHiB) and the current editor of the Brazilian journal *Filosofia e História da Biologia*.

Chapter 11
You Too Can Find "Grandeur in This View of Life": A Linguistic Remedy for Resisting the Desire to Abandon Darwin's *Origin of Species*

Bárbara Jiménez-Pazos

Abstract Given the scientific, socio-cultural, philosophical, and anthropological implications of Charles Darwin's theory of evolution, it should be within reach of every motivated reader to discover its theoretical intricacies by reading the original source, *On the Origin of Species*. That said, many students, researchers, or interested readers in general admit that the dense prose and complex explanations have led them to abandon their reading of it. Therefore, in this chapter, I propose a linguistic remedy, through semantic analysis of some of the most emphatically expressive parts of the work, to avoid renouncing its reading in a fit of desperation. To do this, I analyze exclamatory passages from the *Origin* that denote scientific-aesthetic emotions, especially Darwin's admiration, passion, and respect for nature and the study of it. My aim is to awaken in the reader a similar interest by paying detailed attention to Darwin's sensibility and expressiveness throughout these passages. The desired result would be for the reader to empathize with the author and his enthusiasm for transmitting the grandeur that characterizes the view of life explained throughout the *Origin* and, therefore, to wish to go on reading it, despite the cognitive fatigue Darwin's arguments might sometimes cause.

11.1 The *Origin*: Something More than a Treatise on Evolutionary Biology

It is undeniable that Charles Darwin's *On the Origin of Species* (1859) is one of the most influential works in the history of science since it explains the workings of the mechanism that allows living beings to evolve in space and time, in accordance with ever-changing resources and environment: natural selection. This mechanism, the

B. Jiménez-Pazos (✉)
University of the Basque Country, San Sebastián, Spain
e-mail: barbara.jimenez@ehu.eus

M. Elice Brzezinski Prestes (ed.), *Understanding Evolution in Darwin's "Origin"*, History, Philosophy and Theory of the Life Sciences 34,
https://doi.org/10.1007/978-3-031-40165-7_11

basis of the Darwinian theory of evolution, throws light onto the biological identity of living and extinct species, explaining the relationship between lineages that emerge from one or various common ancestors. Even though Darwin was careful not to include explicit conclusions about human origins in the *Origin*, there are enough theoretical indications in the text to infer the animal descent of humankind (Radick, 2013):

> [I]n the distant future I see open fields for far more important researches. Psychology will be based on a new foundation, that of the necessary acquirement of each mental power and capacity by gradation. Light will be thrown on the origin of man and his history. (Darwin, 1859, p. 488)

Allusions such as in the above quotation indicate that the animal origin of human beings and subsequent parentage with the rest of the animal kingdom was a hypothesis that was very present in Darwin's mind.

Acceptance of the theory of evolution involved assimilating implications of great scientific, socio-cultural, philosophical, and anthropological relevance which were partly unforeseen prior to the publication of the *Origin* (November 24, 1859). In fact, their permeation into social ideology after its publication proved difficult. Darwin's theory removed the need for a divine creator of *ad hoc* fixed animal species, placing humans in the species chain as just one more link, without divine privileges that separated them from the rest of the animal kingdom. The human's place – now a human animal – in nature went from being on a vantage point with privileged views over the animal kingdom to being down below, on the battlefield for survival. Therefore, the conceptual power of this theory was colossal, as was its capacity for social and cultural repercussions.

The eager social anticipation, moved by the perspective of the imminent revelation of a revolutionary theory, was huge. It is, therefore, understandable that the first edition of the *Origin* sold out on the first day of its publication. Furthermore, given that the work exposed the plausibility of the natural origin of human beings, as opposed to the divine – thus breaking with what we can call human bio-anthropological identity as conceived up to that point –, it is understandable that its reading was at the time, as it is today, a challenge and an essential task for students, researchers or readers in general. Ultimately, the *Origin* is an icon in the history of science, and thus, recommended reading in many educational centers and faculties.

Despite such editorial success, Darwin was very critical of his own writing style.[1] In a letter to his editor, John Murray, referring to the proofreading of the *Origin* some months before its publication, admits that he found "[. . .] the style incredibly bad, &

[1] I recommend reading "Logical and Persuasive Structures in Charles Darwin's Prose Style" by Charles Kay Smith (1970) since, in addition to bringing to light and studying the logical and persuasive structures in Darwin's prose, he quotes a large part of the founding works about Darwin's language style and writing which, due to restrictions of space, are not included in this paper. Similarly, I recommend more recent works, especially those by Gillian Beer, from whom I would highlight specifically *Darwin's Plots* (2009). See also Depew (2008).

most difficult to make clear & smooth." After a few lines, he wonders about the reason, responding tentatively and apologizing:

> How I could have written so badly is quite inconceivable, but I suppose it was owing to my whole attention being fixed on general line of argument, & not on details. All I can say is that I am very sorry. (Darwin Correspondence Project, Letter no. 2469)

Darwin's guilty conscience regarding his prose style could equally be extended to other works, such as the proofreading of the 1858 Darwin-Wallace "On the tendency of species to form varieties; and on the perpetuation of varieties and species by natural means of selection," published a year before the *Origin*. In a letter to Joseph D. Hooker, Darwin complains and explains the possible reason for his "bad" writing in the paper:

> I am disgusted with my bad writing. I could not improve it, without rewriting all, which would not be fair or worthwhile, as I have begun on better abstract for Linn. Soc. My excuse is that it never was intended for publication. (Darwin Correspondence Project, Letter no. 2311)

Whether they be the causes of a supposedly bad prose style or not – on the one hand, the fact that he focused on the general argument of the *Origin* rather than the details, and, on the other hand, that he did not conceive the Darwin-Wallace paper as a potential publication – Darwin's thoughts at that time conveyed concern.

Is it possible that Darwin's complaints about his bad writing style could be justified?[2] I have met many readers of the *Origin*, of any age, who have abandoned their reading in an act of frustration, finding it too arduous. Moreover, there is a generalized opinion – mostly, perhaps, among those with no prior knowledge in biology – concerning the excessive length of the work,[3] Darwin's intricate explanations, and his cumbersome and tedious style that make the conclusions confusing. Nevertheless, whether or not Darwin was as bad a writer as he claimed – and some resentful receivers of his work would probably confirm – it is not an issue to be judged here.

That said, assuming there are, indeed, difficulties in maintaining continuity in the reading of the *Origin*, in what follows, I propose and develop a linguistic exercise consisting in the semantic analysis of the most emphatically expressive paragraphs in the book, as a palliative remedy for the cognitive fatigue that Darwin's prose might cause. In addition to strictly scientific explanations, the *Origin* includes discreet wake-up calls to the reader using occasional exclamation marks. As we will see later, the passages that contain one or more exclamation marks are

---

[2] George Levine would probably answer "no" to this question. Levine argues (2011) in favor of the argumentative and stylistic dexterity of Darwin's prose in the *Origin*, which he considers to be a literary work in addition to a scientific one.

[3] Whether the length of the *Origin* is excessive or not is a judgment that depends on each reader. That said, it is true that it is a demanding read, given that Darwin's theory covers multiple areas of knowledge, requiring a considerable number of pages for the detailed development of its arguments. I suggest reading *Darwin's Origin of Species: A Biography* (2006) by Janet Browne for detailed data about the editorial history of the *Origin*.

invitations to participate in the scientific awe that inspired Darwin at the time and which he thus enshrined in the descriptions and arguments of his *opus magnum*.

I, therefore, believe that the *Origin*, in addition to causing a scientific revolution regarding the origin of the species with wide-ranging and profound scientific, philosophical, anthropological, and socio-cultural implications, also contains stylistic features that testify to the emotional-conceptual relationship Darwin had with nature, seen in the light of his own scientific activity. A semantic journey through a selection of the exclamatory passages of the *Origin* will allow the reader to discover additional aspects of Darwin's personality beyond his evident capacity for practicing science, such as his scientific-aesthetic sensibility. This kind of sensibility shows up in the work with exclamations that reveal his awe and admiration for nature, stimulated by a capacity for perceiving beauty in the perfection and complexity of its mechanisms; or his awareness of the insignificance of human existence in the face of the grandeur of nature. The aim of this linguistic strategy of approaching the passion Darwin feels for nature is that the readers become "infected" with Darwinian enthusiasm and renounce the desire to abandon their reading when overcome by exhaustion or cognitive weariness.

To achieve this aim, below I develop an analysis of Darwin's scientific and aesthetic sensibility manifested in his text through the use of exclamatory language.

11.2 A Map of Darwin's Awe for Nature

Exclamations are one of the linguistic features Darwin uses to express his emotions when faced with extraordinary or striking natural facts, objects, processes, etc., as well as with notable elements from his own research and that of other experts. The exclamatory language throughout the pages of the *Origin* transmits both intellectual euphoria and scientific humility, expressed in sentiments like awe, admiration, respect, and passion that Darwin seems to feel when studying nature.

Therefore, to develop an analysis of Darwin's scientific and aesthetic sensibility, I have focused on all the sentences ending with an exclamation mark ('!') throughout the 14 chapters that make up the first edition of the *Origin*. The search for the exclamation mark in the digitalized text of the *Origin* shows 46 exclamatory sentences.[4] I have arranged them all, according to their semantic similarities, into 6 groups (G), transcribed them in Tables 11.1, 11.2, 11.3, 11.4, 11.5, and 11.6, and listed them individually in the order in which they appear in the book. Where there were two or three consecutive exclamations in the text, they have been kept on a single line in the table. The tables also refer to the chapter number and the title of the sub-section (some abbreviated in accordance with space limitations) where

[4] All quotations from the *Origin* have been taken from *The Complete Work of Charles Darwin Online* (Van Wyhe, 2002).

Table 11.1 Number (N), chapter (Ch), and exclamatory sentences of Group 1 (G1): Miscellany about frequency, probability, plausibility, and predictability of natural facts

N	Ch	[Title of the chapter's sub-section] Exclamatory sentence
1–2	I	[CAUSES OF VARIABILITY] How many animals there are which will not breed, though living long under not very close confinement in their native country! [...] but how many cultivated plants display the utmost vigour, and yet rarely or never seed! (p. 8)
8	III	[COMPLEX RELATIONS] Hence it is quite credible that the presence of a feline animal in large numbers in a district might determine, through the intervention first of mice and then of bees, the frequency of certain flowers in that district! (p. 74)
12	III	[STRUGGLE BETWEEN SAME SPECIES] How frequently we hear of one species of rat taking the place of another species under the most different climates! (p. 76)
15	IV	[INTERCROSSING OF INDIVIDUALS] every hybridiser knows how unfavourable exposure to wet is to the fertilisation of a flower, yet what a multitude of flowers have their anthers and stigmas fully exposed to the weather! (p. 97)
19	V	[USE AND DISUSE] Mr. Wollaston has discovered the remarkable fact that 200 beetles, out of the 550 species inhabiting Madeira, are so far deficient in wings that they cannot fly; and that of the 29 endemic genera, no less than 23 genera have all their species in this condition! (p. 135)
20	V	[ACCLIMATIZATION] The case of the Jerusalem artichoke, which is never propagated by seed, and of which consequently new varieties have not been produced, has even been advanced – For it is now as tender as ever it was – As proving that acclimatization cannot be effected! (p. 142)
21	V	[REVERSIONS TO LONG-LOST CHARACTERS] Call the breeds of pigeons, some of which have bred true for centuries, species; and how exactly parallel is the case with that of the species of the horse-genus! (p. 166)
23	VII	[INSTINCTS COMPARABLE WITH HABITS] How unconsciously many habitual actions are performed, indeed not rarely in direct opposition to our conscious will! (p. 208)
24	VII	[DOMESTIC INSTINCTS] How rarely, on the other hand, do our civilised dogs, even when quite young, require to be taught not to attack poultry, sheep, and pigs! (p. 215)
25	VIII	[DEGREES OF STERILITY] So that certain individual plants and all the individuals of certain species can actually be hybridised much more readily than they can be self-fertilised! (p. 250)
34	XI	[SINGLE CENTRES OF CREATION] What a strange anomaly it would be, if, when coming one step lower in the series, to the individuals of the same species, a directly opposite rule prevailed; and species were not local, but had been produced in two or more distinct areas! (p. 353)
35	XI	[MEANS OF DISPERSAL] Even in this case, how small would the chance be of a seed falling on favourable soil, and coming to maturity! (p. 364)
37	XII	[FRESH-WATER PRODUCTIONS] [T]he plants were of many kinds, and were altogether 537 in number; and yet the viscid mud was all contained in a breakfast cup! (pp. 386–387)
39	XIII	[RULES AND DIFFICULTIES IN CLASSIFICATION] So with plants, how remarkable it is that the organs of vegetation, on which their whole life depends, are of little signification, excepting in the first main divisions; whereas the organs of reproduction, with their product the seed, are of paramount importance! (p. 414)
41	XIII	[RUDIMENTARY ORGANS] Nothing can be plainer than that wings are formed for flight, yet in how many insects do we see wings so reduced in size as to be utterly incapable of flight, and not rarely lying under wing-cases, firmly soldered together! (p. 451)

Table 11.2 Number (N), chapter (Ch), and exclamatory sentences of Group 2 (G2): About scientific inaccuracy or ignorance

N	Ch	[Title of the chapter's sub-section] Exclamatory sentence
3	I	[ORIGIN OF DOMESTIC VARIETIES] One author believes that there formerly existed in Great Britain 11 wild species of sheep peculiar to it! (p. 19)
5–6	II	[DOUBTFUL SPECIES] Under genera, including the most polymorphic forms, Mr. Babington gives 251 species, whereas Mr. Bentham gives only 112 – a difference of 139 doubtful forms! [. . .] how many of those birds and insects in North America and Europe, which differ very slightly from each other, have been ranked by one eminent naturalist as undoubted species, and by another as varieties, or, as they are often called, as geographical races! (p. 48)
7	III	[COMPLEX RELATIONS] Nevertheless so profound is our ignorance, and so high our presumption, that we marvel when we hear of the extinction of an organic being; and as we do not see the cause, we invoke cataclysms to desolate the world, or invent laws on the duration of the forms of life! (p. 73)
9	III	[COMPLEX RELATIONS] When we look at the plants and bushes clothing an entangled bank, we are tempted to attribute their proportional numbers and kinds to what we call chance. But how false a view is this! (p.74)
32	X	[EXTINCTION] But how utterly groundless was my astonishment! (pp. 318–319)

Table 11.3 Number (N), chapter (Ch), and exclamatory sentences of Group 3 (G3): First-person singular descriptions

N	Ch	[Title of the chapter's sub-section] Exclamatory sentence
4	I	[DOMESTIC PIGEONS] [F]or instance, I crossed some uniformly white fantails with some uniformly black barbs, and they produced mottled brown and black birds; these I again crossed together, and one grandchild of the pure white fantail and pure black barb was of as beautiful a blue colour, with the white rump, double black wing-bar, and barred and white-edged tail-feathers, as any wild rock-pigeon! (p. 25)
22	VI	[TRANSITIONAL HABITS] [A]nd on the plains of La Plata, where not a tree grows, there is a woodpecker, which in every essential part of its organisation, even in its colouring, in the harsh tone of its voice, and undulatory flight, told me plainly of its close blood-relationship to our common species; yet it is a woodpecker which never climbs a tree! (p. 184)

exclamatory sentences or extracts of sentences appeared. The result is a map of Darwin's awe for nature. Let's go through it:

The semantic analysis of the 46 exclamatory sentences which make up the map of Darwin's awe for nature began by identifying whether they referred to trivial, generic objects or states of objects, in which case they could be mere, almost inertial, rhetorical resources; or whether they referred to something significant to Darwin's theory and could, therefore, be regarded as genuine exclamations. As we will see later in more detail, I discovered that they do represent remarkable, peculiar, or extraordinary aspects of nature and its study that are very pertinent to Darwin's theory. I then went on to identify the semantic similarities of the 46 sentences, categorizing six groups (G) which received consecutive numbers according to the order of their first appearance throughout the chapters of the *Origin*:

Table 11.4 Number (N), chapter (Ch), and exclamatory sentences of Group 4 (G4): About the passing of time

N	Ch	[Title of the chapter’s sub-section] Exclamatory sentence
10–11	III	[COMPLEX RELATIONS] What a struggle between the several kinds of trees must here have gone on during long centuries, each annually scattering its seeds by the thousand; what war between insect and insect [...] all striving to increase, and all feeding on each other or on the trees or their seeds and seedlings, or on the other plants which first clothed the ground and thus checked the growth of the trees! Throw up a handful of feathers, and all must fall to the ground according to definite laws; but how simple is this problem compared to the action and reaction of the innumerable plants and animals which have determined, in the course of centuries, the proportional numbers and kinds of trees now growing on the old Indian ruins! (pp. 74–75)
13–14	IV	[NATURAL SELECTION VS. MAN’S SELECTION] How fleeting are the wishes and efforts of man! How short his time! And consequently how poor will his products be, compared with those accumulated by nature during whole geological periods. (p. 84)
26	IX	[VAST LAPSE OF TIME] But how often do we see along the bases of retreating cliffs rounded boulders, all thickly clothed by marine productions, showing how little they are abraded and how seldom they are rolled about! (p. 283)
27	IX	[VAST LAPSE OF TIME] And what an amount of degradation is implied by the sedimentary deposits of many countries! (p. 284)
28	IX	[VAST LAPSE OF TIME] So that the lofty pile of sedimentary rocks in Britain, gives but an inadequate idea of the time which has elapsed during their accumulation; yet what time this must have consumed! (p. 284)
29–30	IX	[VAST LAPSE OF TIME] What an infinite number of generations, which the mind cannot grasp, must have succeeded each other in the long roll of years! Now turn to our richest geological museums, and what a paltry display we behold! (p. 287)

G1: *Miscellany about the frequency, probability, plausibility, and predictability of natural facts*: This group contains the 16 exclamations that most manifestly describe how frequent, probable, plausible, and predictable – and therefore how striking or unintuitive – a miscellany of natural facts or situations can be, as well as the results of the research into them carried out by Darwin and other renowned naturalists.

G2: *About scientific inaccuracy or ignorance*: This group includes six exclamations about: on the one hand, inaccurate or erroneous research of other naturalists; and, on the other hand, human – and Darwin’s own – ignorance, in the face of striking or unknown facts, phenomena or situations that could lead to the attribution of false causes to the observed scientific object in question.

G3: *First-person singular descriptions*: In the two exclamatory paragraphs in this group, Darwin dispenses with the inclusive “we” and imposes himself onto the text to describe and enumerate mainly visually perceptible aspects of the object of study in the first person singular.

G4: *About the passing of time*: This group includes nine exclamations that describe the effects of the passing of the centuries in nature, compare humans’ brief time

Table 11.5 Number (N), chapter (Ch), and exclamatory sentences of Group 5 (G5): About the clarity acquired by otherwise inexplicable facts

N	Ch	[Title of the chapter's sub-section] Exclamatory sentence
16–17–18	IV	[INTERCROSSING OF INDIVIDUALS] How strange are these facts! How strange that the pollen and stigmatic surface of the same flower, though placed so close together, as if for the very purpose of self-fertilisation, should in so many cases be mutually useless to each other! How simply are these facts explained on the view of an occasional cross with a distinct individual being advantageous or indispensable! (p. 99)
40	XIII	[MORPHOLOGY] How inexplicable are these facts on the ordinary view of creation! (p. 437)
42	XIV	[RECAPITULATION OF CIRCUMSTANCES IN FAVOUR OF NATURAL SELECTION] How strange it is that a bird, under the form of woodpecker, should have been created to prey on insects on the ground; that upland geese, which never or rarely swim, should have been created with webbed feet; that a thrush should have been created to dive and feed on sub-aquatic insects; and that a petrel should have been created with habits and structure fitting it for the life of an auk or grebe! But on the view of each species constantly trying to increase in number, [. . .] these facts cease to be strange. . . (pp. 471–472)
43–44	XIV	[RECAPITULATION] How inexplicable on the theory of creation is the occasional appearance of stripes on the shoulder and legs of the several species of the horse-genus and in their hybrids! How simply is this fact explained if we believe that these species have descended from a striped progenitor, in the same manner as the several domestic breeds of pigeon have descended from the blue and barred rock-pigeon! (p. 473)
45	XIV	[RECAPITULATION] On the view of each organic being and each separate organ having been specially created, how utterly inexplicable it is that parts, like the teeth in the embryonic calf or like the shrivelled wings under the soldered wing-covers of some beetles, should thus so frequently bear the plain stamp of inutility! (p. 480)
46	XIV	[CONCLUSION-EFFECTS OF NATURAL SELECTION ON NATURAL HISTORY] When we no longer look at an organic being as a savage looks at a ship, as at something wholly beyond his comprehension; when we regard every production of nature as one which has had a history; when we contemplate every complex structure and instinct as the summing up of many contrivances, each useful to the possessor, nearly in the same way as when we look at any great mechanical invention as the summing up of the labour, the experience, the reason, and even the blunders of numerous workmen; when we thus view each organic being, how far more interesting, I speak from experience, will the study of natural history become! (pp. 485–486)

on earth with quasi-eternal nature, or transmit multiple reflections regarding the temporal immensity of the geological periods.

G5: *About the clarity acquired by otherwise inexplicable facts*: The nine exclamations in this group relate to the clarity and simplicity with which knowledge explains facts that were previously inexplicable, together with the satisfaction generated by such explanations.

G6: *About the representativeness of living and non-living productions*: This group includes four exclamatory paragraphs in which Darwin reflects on the possibility

Table 11.6 Number (N), chapter (Ch), and exclamatory sentences of Group 6 (G6): About the representativeness of living and non-living productions

N	Ch	[Title of the chapter's sub-section] Exclamatory sentence
31	IX	[ABSENCE OF INTERMEDIATE VARIETIES] The Malay archipelago is one of the richest regions of the whole world in organic beings; yet if all the species were to be collected which have ever lived there, how imperfectly would they represent the natural history of the world!(pp. 299–300)
33	XI	[PRESENT DISTRIBUTION] Notwithstanding this parallelism in the conditions of the old and new worlds, how widely different are their living productions! (p. 347)
36	XI	[GLACIAL PERIOD] A list of the genera collected on the loftier peaks of Java raises a picture of a collection made on a hill in Europe!(p. 375)
38	XII	[RELATIONS OF INHABITANTS] [T]here is a considerable degree of resemblance in the volcanic nature of the soil, in climate, height, and size of the islands, between the Galapagos and cape de Verde archipelagos: But what an entire and absolute difference in their inhabitants! (p. 398)

that collections of fossils or living species could represent the flora and fauna of a particular place.

The semantic study of the 46 exclamations that has enabled their inclusion into the six preceding categories suggests that: First, the text of the *Origin*, and therefore Darwin too, are not aesthetically or emotionally indifferent. With that in mind, it is necessary to clarify that, in the context of this chapter, the term "aesthetic" does not only refer to the visible aesthetic characteristics of nature but, above all, to the intrinsic beauty of essential qualities such as the excellence of its mechanisms – functional, structural, and adaptive – reflected in the selective perfection of natural selection or the unstoppable erosive force of imposing geological masses. And second, that the application of exclamations is restricted almost exclusively to the scientific field, emphasizing fundamental explanations of Darwin's research; exclamations are, therefore, essential for the formulation of his theories.

This dual semantic feature of the exclamations – aesthetic-emotional and scientific-explanatory – is displayed symmetrically in the text of the *Origin* through a language that is also dual. On the one hand, there is a marked scientific-technical language, with an application that is predominantly explanatory, about the myriad of extraordinary aspects of nature and the research into them. Although this type of language is the leading vehicle for transmitting ideas in all the exclamatory passages of the map of Darwin's awe, it assumes a notable predominance in groups G1, G2, and G6, possibly as a result of the argumentation thread in the paragraphs containing the exclamations.

On the other hand, there is also an aesthetic-emotional language, whose presence in the text is, in most cases, secondary; subtly interwoven with the scientific-technical language, and whose application is predominantly descriptive (non-explanatory). Its use would be motivated by aspects of nature that are also extraordinary but touched deeper corners of Darwin's scientific sensibility. As a consequence of this, Darwin's vocabulary becomes more subjective, literary, and even occasionally poetic. Groups such as G3, G4, and G5 contain exclamatory passages in which the aesthetic-emotional language is more evident.

There is no chapter of the *Origin* free from exclamation marks. This fact may indicate that despite the scientific-technical language being understandably predominant in the work, the use of this linguistic resource – which, let us not forget, is a medium for transmitting emotions – is equally a constituent part of Darwin's explanations. In other words, Darwin doesn't renounce showing his feelings regarding the facts, observations or conclusions described in the work that most caught his attention as a naturalist.[5] As will be seen below, this linguistic peculiarity is a distinctive stamp of Darwin's use of language in the *Origin* that adds personality and proximity to the text.

11.3 In Search of Nature's Grandeur Using the Map of Darwin's Awe

Every aim involves seeking out the most appropriate path to achieve it. That said, difficulties may appear along the way, as might new, favorable impulses of motivation and aspiration. Darwin's main aim in the *Origin* – and one of the main ones in his journey through the history of science – was to explain all possible aspects of the mechanism that makes species change, causing some to perish and others to emerge: natural selection. The path taken to achieve this aim was that of research into, and more importantly, *in* nature. From the results in Tables 11.1, 11.2, 11.3, 11.4, 11.5, and 11.6, I would go as far as to say that Darwin discovered an added motivation on his path – that being his delight in the awe he felt at new discoveries; in other words, his enthusiasm and passion for nature and the study of it.

As was put forward in Sect. 11.1, this paper aims to encourage any weary reader of the *Origin* not to abandon their reading, by following the path of the semantic analysis of Darwin's use of exclamatory language. The added motivation for achieving this aim aspires to be that same awe and passion for nature that inspired Darwin in his day, and that drove him to end the *Origin* with these beautiful words: "There is grandeur in this view of life, with its several powers, having been originally breathed into a few forms or into one; and that, whilst this planet has gone cycling on according to the fixed law of gravity, from so simple a beginning endless forms most beautiful and most wonderful have been, and are being, evolved" (Darwin, 1859, p. 490). I propose to search for the grandeur characteristic of Darwin's view of life explained in the pages of the *Origin* before he delivers those final inspirational

[5] Levine (2006) and White (2009) study Darwin's emotional attachment to (Levine), or detachment from (White), his objects of study.

lines. To do this, in the following three sub-sections I will analyze some of the most impressive "landforms"[6] from the map of Darwin's awe set out in Sect. 11.2.

11.3.1 On the Origin of Darwin's Two Languages: Scientific-Technical and Aesthetic-Emotional

The living conditions that nature provides are unique, and human efforts to imitate and reproduce the same conditions are ineffective in most cases. What better way to transmit this message about nature's inimitable vital energy than by analyzing the laws that determine the reproduction of animals and plants in confinement, such as that reflected in exclamations 1–2 (G1)? The reproductive inefficiency of animals in a relatively relaxed domesticated state, and that healthily grown plants rarely produce seeds, are curious and singular facts that led Darwin to exclaim in awe a few pages after beginning the *Origin*. Exclamations 1–2 are good examples that the *Origin* conceals an emotional subsoil discreetly embedded with priority explanatory aims. Therefore, it is the scientific-technical language that predominates in this case.

Nevertheless, let us now look at the difference in type of emotional expressiveness between exclamations 1–2 and 13–14 (G4). In 13–14, Darwin exclaims about the fleetingness of humanity and, consequently, about how poor its achievements are in terms of selection in comparison with those of nature, whose longevity allows it to engage in a task of constant selection throughout entire geological periods. Darwin's aesthetic sentiments, excited by the idea of the perfection that the products of long-lived nature can acquire, are not so hidden beneath the cloak of the scientific-technical language of the descriptions, but are shown here much more explicitly than in 1–2. Exclamations 13–14 are good examples of how, on occasion, the aesthetic-emotional language emerges from the text of the *Origin* and imposes itself on the scientific-technical language.

The message being transmitted in both groups of exclamations – the hegemony of nature against human inferiority – is the same, but it is evident that exclamations 13–14 have different descriptive power from numbers 1–2. These later exclamations are characterized by not having the rhetorical flourishes of the former, characteristics which, rather than describing the awe arising from the knowledge of a curious scientific fact (such as that explained in 1–2), describe, above all, Darwin's aesthetic sentiments at the functional beauty of the perfect selective mechanism of nature.

What motivates this dual use of language in the *Origin*? Is this doubly descriptive fact the manifestation of a dual attitude and experience – scientific and aesthetic – in nature? Although it is not my intention to investigate Darwin's underlying aesthetic theory in the *Origin* in any depth, the notions of "aesthetic attitude" and the

[6]For reasons of spatial economy, I will analyse a very limited selection of the exclamations of Darwin's "landscape" of awe that best respond to the explanatory requirements of the argumentation thread in the article.

distinction between "internal" and "external" aesthetic experiences, posed by theories of taste and aesthetic values, could shine light onto the crux of these matters. Therefore, it may be helpful to observe the following: First, the genuinely aesthetic attitude in the strict sense of aesthetic theories, i.e., the willingness to appreciate an object disinterestedly (Kant, 1790; Schopenhauer, 1819) and with no purpose beyond that of enjoying it in itself, is not dominant in the text of the *Origin*. Second, if the distinction between the concepts of "internalism" (Dewey, 1934; Beardsley, 1958) and "externalism" (Beardsley 1982; Dickie, 1988) of theories of aesthetic experience were projected onto the exclamations in the *Origin* (according to which an internal aesthetic experience would appeal to internal characteristics of the experience, i.e., general phenomenological characteristics; while an external aesthetic experience would appeal to the external characteristics of the experience, usually characteristics of the object experienced), one would have to opt for the external experience, since what predominates in the *Origin* is always the desire to explain the natural characteristics of the object.

The conclusions produced from the semantic analysis of the 46 exclamations carried out in Sect. 11.2 reveal that the application of the exclamations is almost exclusively restricted to the field of science, although the text of the *Origin* – and therefore Darwin – are not aesthetically-emotionally indifferent. My hypothesis, then, is that Darwin's dual scientific-technical and aesthetic-emotional language is the lexical expression of his also dual capacity for, on the one hand, the perception and scientific explanation of nature, and on the other hand, the subsequent experience of aesthetic feelings of awe and admiration, in constant interdependence. In light of the previously outlined aesthetic notions, I intend to show below that while there are no indications in the *Origin* that Darwin adopts in advance a genuinely aesthetic attitude or predisposition before nature, the results that Darwin obtains in the – external – intellectual experience of explaining the mechanisms of nature, cause in him aesthetic emotions of awe, enthusiasm, and admiration before it. This shows that his aesthetic experience is inseparable from an intellectual experience that is genuinely explanatory and scientific.

That said, not all the exclamations reflect the same degree of the scientifically, aesthetically, and emotionally satisfactory view of life that Darwin transmits throughout the *Origin*. As will be seen below, certain matters being described, such as the sub-theme of the passing of time in the case of exclamations 13–14,[7] stimulate Darwin's scientific-aesthetic sensibility to a greater degree. The exclamations dedicated to different aspects of the influence on nature of the passing of time are, for this reason, those which would best clarify Darwin's sentiments about the

[7]The passing of time is not the central theme of the extract including exclamations 13–14, but a secondary theme; the central theme is made explicit in the title of the sub-section in the *Origin* in which they appear: the power of natural selection compared with human selection. I have classified them in G4 because the subject of time is a fundamental part of Darwin's explanations about the difference between natural selection and artificial selection, as can be seen in exclamations 13–14 and in the lines that follow them in the text of the *Origin* (p. 84).

grandiose characteristics of nature and those which, consequently, will be given priority in testing the hypothesis mentioned above.

11.3.2 The Passing of Time, or the Sublime in Darwin: The Hidden Beauty of Rocks

The temporal immensity represented by the extent of geological ages was, in fact, a recurring subject considered worthy of exclamation in works by outstanding naturalists of the time. In addition to revising the biblical conception of time on earth in favor of a revolutionary and scientifically more helpful geological conception, this awoke – in all those who internalized it – sentiments of sublimity on a par with those described in the verses of renowned romantic poets.

The passing of the centuries in nature is a fact that also awoke in Darwin the quasi-romantic expressivity characteristic of one of his closest predecessors – and prominent scientific figure of reference – Alexander von Humboldt, who in *Views of Nature* (1850) exclaimed about the unceasingly regenerative power of nature in comparison with perishable human generations:

> Thus pass away the generations of men! – thus perish the records of the glory of nations! Yet when every emanation of the human mind has faded – when in the storms of time the monuments of man's creative art are scattered to the dust an ever new life springs from the bosom of the earth. Unceasingly prolific nature unfolds her germs, – regardless though sinful man, ever at war with himself, tramples beneath his foot the ripening fruit! (von Humboldt, 1850, p. 173)

The expressive and semantic similarity between this extract from Humboldt and exclamations 13–14 from Darwin is overwhelming: in both cases, nature is exalted as an unbeatable regenerative force, with a language usage that does this force justice.

This expressive eloquence inspired by the inexhaustible power of nature is also perceptible in other passages from Chap. 9 of the *Origin*, such as those containing the exclamations 26, 27, 28, and 29–30 (G4) dedicated to the number of centuries that have passed in those sedimentary rock masses in constant, but very slow erosion. Exclamation 29 is especially noteworthy, in which Darwin transmits the overwhelming sensation of imaginative impotence when faced with the incommensurability of time and of the infinite generations that have passed, and been succeeded by others.

Over and above this explicit message of emotional awe, exclamation 29 contains an additional emotional undertone: Darwin suggests an awareness of the humble position of human beings in the brief time that we have formed part of the history of the earth and thus, a respect regarding a quasi-infinite and sublime nature. However, this characteristically romantic description of the sentiment of the sublime is immediately superimposed by exclamation number 30, which is a call to recover the central subject of the chapter – namely the imperfection of the fossil record, a poor sample of what Darwin's study of geology leads him to deduce as time's magnificent endeavor in nature.

The exclamations above indicate that there is beauty in old rocks. Still, it is hidden among their sedimentary strata, in its slow formation and erosion, in the fragments of natural history held within fossils; and ultimately, among the pages of the incomplete book of the earth's geological history. Darwin's capacity to perceive this beauty lies in his ability to *read* and *interpret* this geological book. This fact leads us to conclude that observing sedimentary rocks provokes Darwin to exclaim in awe because he *knows* the key to the workings of this rocky landscape: the inconceivable time needed for its formation, development, and deterioration.

Nowadays, we are fortunate enough to know the scientific and aesthetic satisfaction Darwin experienced in his day on contemplating such geological giants, thanks to the exclamatory legacy of the *Origin* and its characteristic prose; a prose made up of explanations of natural facts and descriptions of emotional states in an uninterrupted conversation. This hybrid language could be the proof, just as the hypothesis posed in Sect. 11.3.1 proposes, that there is a circular relationship between the study and subsequent acquisition of new knowledge about nature and Darwin's emotional reaction: the study of the geological landscape, and all natural facts in general, produce intense emotions in Darwin, and these encourage him to go on studying and discovering new facts, these evoking new emotions.

11.3.3 Knowledge, or the Key to Finding Nature's Grandeur: Darwin's Entangled Banks

The description of natural facts and emotional states in constant feedback extends throughout the *Origin*, as can be seen in exclamations 10–11 (G4),[8] which are some of the most epic in tone of all those in Darwin's map of awe. Here Darwin allows himself to be swept along by the satisfaction of *knowing* the mysteries of the struggle for life among innumerable plants and animals over the centuries. There is no action without a reaction in the natural world, and every event influences a future event in nature, this being described as a complex network of interdependence between the

[8] Once again (see footnote 7), the passing of time in nature is not the central theme of exclamations 10–11, but the complex relationships between species in the fight for existence. That said, the mentions "during long centuries" and "in the course of centuries" are proof that the passing of the centuries is not only a necessary principle for Darwin's hypothesis surrounding the interrelationship between species, but, probably, a source of inspiration that adds additional poetic expressivity to his descriptions.

It is no coincidence that G4 is one of the groups with some of the most analyzed exclamations. As was pointed out at the beginning of sub-section 11.3.2, the passing of time is one of the subjects that arouses most scientific-aesthetic interest in Darwin. Consequently, the explanations of natural processes that require the mentioning of the time factor to a greater extent are those that better reflect the grandiose view of life that Darwin sustains throughout the *Origin*.

plant and animal kingdoms that will determine the diversity and proportion of the flora in a specific area.

These exclamations appear in a paragraph with descriptive characteristics very similar to the aforementioned last lines of the *Origin*, where Darwin masterfully summarizes the laws which have made the appearance of multiple natural, beautiful, and marvelous forms possible from one simple beginning. This is Darwin's grandiose view of life, described as a relationship of co-dependence between animals and plants, guided by natural laws in the entangled bank: "It is interesting to contemplate an entangled bank, clothed with many plants of many kinds, with birds singing on the bushes, with various insects flitting about, and with worms crawling through the damp earth, and to reflect that these elaborately constructed forms, so different from each other, and dependent on each other in so complex a manner, have all been produced by laws acting around us" (ibid, p. 489). If we analyze exclamation number 9 (G2) – which, despite being shown in Table 11.2 as separate from numbers 10–11 (G4), belongs to the same paragraph as these – we will see that, in fact, Darwin also mentions the example of an entangled bank to explain the organic interconnection that determines the "beautiful diversity and proportion of kinds" (ibid., p. 74) of vegetation mentioned some lines previously. Here Darwin exclaims against the tendency to attribute the diversity and proportion of trees and bushes to chance when, in fact, it depends on the complex and uninterrupted interaction between plants and animals.

Both references to entangled banks show an aesthetic-emotional and scientifically satisfactory view of life: in the last paragraph of the *Origin*, natural forms are described as beautiful and marvelous; in the paragraph that contains exclamations 9 and 10–11, the diversity and proportion of trees are beautiful in the expert eyes of Darwin. The key to the perception of said beauty in both cases is the knowledge Darwin possesses about the workings of natural mechanisms that have made possible the development of countless entangled banks, or ecosystems, which have accommodated among their branches the complex yet perfect relationship between animal and plant species over centuries. This conclusion leads us to suppose that Darwin's allusion to the entangled bank in the *Origin* could be a metaphor[9] that represents the totality of nature working in perfect harmony. It could even be surmised that, by extending the semantic possibilities of the metaphor, it perhaps represents the passionate and respectful link Darwin has with nature and the study of it. This relationship would be supported, as we have just suggested, by the firm basis of knowledge Darwin has about natural mechanisms that, since time immemorial, have populated the earth with beautiful, marvelous and elaborately constructed forms, just as he confirms in the last lines of the *Origin*.

The satisfaction provoked by knowing which natural mechanisms have made possible the appearance of a perfect natural order in all corners of the earth is reaffirmed in exclamations 16–17–18, 40, 42, 43–44, 45, and 46 (G5). Here Darwin repeatedly exclaims about how strange and inexplicable some natural facts are in

[9] For a detailed analysis of metaphors in the *Origin*, see Beer (2009).

light of the special creation theory – according to which every living being has been independently created – and how simply they can be explained within the framework of the accumulation of adaptive improvements mechanism that natural selection provides: "How strange are these facts! How strange that... [...] How simply are these facts explained on the view of..." (16–17–18; here there is no explicit allusion to special creation); "How inexplicable are these facts..." (40); "How strange it is... [...] these facts cease to be strange" (42); "How inexplicable... [...] How simply is this fact explained..." (43–44); "[...] how utterly inexplicable it is that..." (45). This pattern of exclamations and responses is repeated in every case (the responses to exclamations 40 and 45 can be found in their surrounding lexical context) except for exclamation number 46, which deserves special analysis.

From number 42 onward, all the exclamations as a whole serve to prepare the arguments that guide the reader toward the final lines of the *Origin*. In these passages of recapitulation and conclusion, Darwin shares his satisfaction at knowing the secret that shines a light on many obscure facts: natural selection. This satisfaction acquires special force in exclamation 46, which appears in one of the two paragraphs that best reflect Darwin's scientific-aesthetic sensibility, and also in the last paragraph of the *Origin*. Four pages before the end of the work, as if aiming to speak directly to the reader, Darwin's language becomes personal, intimate, and persuasive. Darwin tries to convince the reader from his own personal experience ("I speak from experience"[10]), that the conception of living beings, with their complex structures and instincts, as products of nature with an evolutionary history of accumulation of adaptive improvements – as opposed to products of special creation – will make the study of natural history much more interesting. In fact, Darwin once again stresses this idea at the beginning ("It is interesting to contemplate an entangled bank...") and at the end of the last paragraph of the *Origin* ("There is grandeur in this view of life...").

This call to knowledge for the sake of the readers' interest in nature is forged throughout the 46 exclamations in the *Origin* with a clear strategy: to convince them that the view of life orchestrated by the explanatory possibilities of the mechanism of natural selection is much more grandiose than that offered by the special creation theory. If nature, we might ask, so beautiful and complex, had been created by an omniscient and almighty mind, what then would be the sense in being in awe of it? What's more, under the creationist hypothesis, one would have to explain why nature – and the world – is not much more beautiful, perfect and fair.

Darwin seems to want to say that there is a lot of beauty to be discovered in nature, and the key to its perception and enjoyment can ultimately be found within the pages of the *Origin*.

[10] With the exception of the exclamations in G3, this is the only exclamatory extract from Tables 11.1, 11.2, 11.3, 11.4, 11.5, and 11.6 in which Darwin expresses himself in the first person singular.

11.4 There's More Awe Still to Find in the *Origin*: Further Perspectives of Research

It may be that being in possession of this aesthetically optimistic message beforehand encourages potential readers of the *Origin* to begin reading, to carry on, and to become "infected" by the passion Darwin feels for nature and the pleasant awe he experiences in studying it. Of course, this is a highly desirable collateral aim. However, the main aim of this paper is to alleviate the weariness produced by Darwin's supposedly dry prose in readers that have felt the temptation to abandon their reading or who, at some point, did give up in a fit of desperation. The memory of said weariness may be challenging to ignore, but I hope it can be overcome with this analysis of Darwin's scientific and aesthetic sensibility.

To summarize, I have approached Darwin's scientific-aesthetic sensibility by semantically analyzing some of the 46 exclamations in the *Origin*. To this end, in Sect. 11.2, I have presented the 46 exclamatory passages in the map of Darwin's awe (Tables 11.1, 11.2, 11.3, 11.4, 11.5, and 11.6), and I have explained how I classified all of them into six thematic groups depending on their semantic similarities. The semantic analysis of the exclamations shows that, although the text of the *Origin*, and therefore Darwin, are not aesthetically-emotionally indifferent, the application of the exclamations is almost exclusively restricted to the field of science, being a constituent part of the formulation and explanation of Darwin's theories. This dual semantic feature of the exclamations – aesthetic-emotional and scientific-explanatory – is shown symmetrically in the *Origin* with a language use that is also dual; a type of language that is predominantly scientific-technical, together with a different kind of language that is aesthetic-emotional, distinguishable by the additional use of vocabulary with a literary tenor. Both are discreetly interwoven in the exclamatory passages of the *Origin*. What motivates the use of this dual prose in the *Origin*?

In Sect. 11.3, I have proposed a hypothesis according to which Darwin's dual scientific-technical and aesthetic-emotional prose would be the lexical manifestation of the inseparable link between his aesthetic experience and his scientific experience. This hypothesis is based, on the one hand, on the conclusions obtained from the semantic analysis of all the exclamations in Sect. 11.2; along with, on the other hand, basic notions about aesthetic attitude and experiences posed by the theories of taste and aesthetic values.

The source of both types of language is nature and what occurs in it. Its striking and astonishing facts are marvelous in Darwin's eyes. In addition, however, there are aspects of nature, such as the passing of time, that especially stimulate Darwin's scientific-aesthetic sensibility and thus satisfactorily show the interdependence between the scientific explanation of nature and subsequent aesthetic feelings of awe and admiration. Consequently, I have tested my hypothesis first by analyzing the exclamations that Darwin dedicates to the influence of the passing of time on the geological landscape. The passing of time in nature expressed by ancient sedimentary rocks provoked in Darwin feelings of sublimity when faced with an awareness of human insignificance in the natural history of the earth, in comparison with grandiose nature, operational since time immemorial. But what is the key for perceiving sublimity in nature? Second, the analysis of exclamations that Darwin dedicates to the satisfaction in explaining natural facts in the light of newly acquired knowledge shows that Darwin exclaims in awe and astonishment in the *Origin* before sedimentary rock masses, entangled banks, and, ultimately, the multiplicity of the natural processes he studies because he *knows* how to explain the hidden details of their complex, perfect, and therefore beautiful mechanisms. Knowledge is, for this reason, the portal through which to access the perception of nature's beauty and grandiosity.

Nevertheless, Darwin's awe is not exhausted in the 46 exclamations in the *Origin*. A further, essential task for defining in more detail the characteristics of Darwin's scientific-aesthetic emotional response would be to analyze the evolution of the 46 exclamations throughout the six editions of the *Origin* that Darwin corrected, extended, and reduced over the years. The evolution of the use of exclamations over time would be a good indicator that, similarly, Darwin's scientific-aesthetic sensibility evolved over time. As an example of said evolution, the two previously analyzed exclamations, 13–14 (G4), become three ("and consequently how poor will his products be, compared with those accumulated by Nature during whole geological periods!") in the third (Darwin, 1861, p. 88), fourth (Darwin, 1866, p. 95), fifth (Darwin, 1869, p. 96), and sixth (Darwin, 1872, p. 65) editions of the *Origin*. This simple example indicates that Darwin's capacity for awe may have changed and even grown, like in this case, as the different editions of the work progressed.

Likewise, an exciting aspect to analyze would be to see whether Darwin's awe grew alongside his increasing knowledge of nature. If Darwin refined and extended his knowledge of nature over time, as can be observed in the different editions of the *Origin*, and knowledge is the pathway to perceiving intense emotions – just as we concluded in Sect. 11.3 –, then it is to be expected that Darwin's awe would grow in line with the newer editions of the work. It would be precipitous to confirm such growth by referring solely to the number of times an exclamation mark is used in the six editions of the *Origin* – there is a growing tendency in usage from the second edition but a declining one in the fifth and sixth editions: 46 times in the second, 49 in the third, 52 in the fourth, 44 in the fifth, and 45 in the sixth. What new aspects of nature provoke a growth in the use of exclamations in the third and fourth editions? Why does the use of exclamations in the fifth and sixth editions decrease to numbers approximating those of the first edition? Of course, usage numbers in themselves do not answer these questions, nor will they provide any conclusions until a semantic analysis of the extracts containing the exclamations is carried out.

If we wanted to expand this perspective of analysis even further, there are more research possibilities in the *Origin* that would provide new clues about Darwin's scientific-aesthetic sensibility. For example, a preliminary search for the question mark ('?') in the first edition of the work shows that Darwin used it 107 times, mainly to add emphasis, rhythm, rhetoric and even sarcasm to his explanations. Additional proof of Darwin's vehement passion for nature can be found in questions such as: "As man can produce and certainly has produced a great result by his methodical and unconscious means of selection, what may not nature effect?" (Darwin, 1859, p. 83). Does this question not also transmit a clear message about the superiority of nature over human beings in terms of the power of selection? In fact, this question appears in the same paragraph as exclamations 13–14 do a few lines earlier, and, as we have seen, these carry this very same idea about the supremacy of nature.

In one way or another, whether it be with exclamations or questions, Darwin insists on expressing the idea around which his emotional response is structured, and without which the argumentative tapestry of the *Origin* could not have been constructed in such a masterful way: there is grandeur in nature, and you too can perceive it; I would dare to add, by way of the linguistic remedy presented in this paper.

References

Beardsley, M. C. (1958). *Aesthetics*. Hackett.

Beardsley, M. C. (1982). *The aesthetic point of view*. Cornell University Press.

Beer, G. (2009). *Darwin's plots. Evolutionary narrative in Darwin: George Eliot, and nineteenth-century fiction*. Cambridge University Press.

Browne, J. (2006). *Darwin's origin of species: A biography*. Atlantic Books.

Darwin, C. R. (1859; 1860; 1861; 1866; 1869). *On the origin of species by means of natural selection, or the preservation of favoured races in the struggle for life*. John Murray.

Darwin, C. R. (1872). *The origin of species by means of natural selection, or the preservation of favoured races in the struggle for life*. John Murray.

Darwin Correspondence Project. (n.d.-a). Letter no. 2469. https://www.darwinproject.ac.uk/letter/?docId=letters/DCP-LETT-2469.xml

Darwin Correspondence Project. (n.d.-b). Letter no. 2311. https://www.darwinproject.ac.uk/letter/?docId=letters/DCP-LETT-2311.xml

Darwin, C. R., & Wallace, A. R. (1858). On the tendency of species to form varieties; and on the perpetuation of varieties and species by natural means of selection. *Journal of the proceedings of the Linnean Society of London, Zoology, 3*(9), 45–62.

Depew, D. J. (2008). The rhetoric of the origin of species. In M. Ruse & R. J. Richards (Eds.), *The Cambridge companion to the "Origin of Species"* (pp. 237–255). Cambridge University Press.

Dewey, J. (1934). *Art and experience*. Putnam.

Dickie, G. (1988). *Evaluating art*. Temple University Press.

Kant, I., (1790) [2000]. *Critique of the power of judgment* (P. Guyer & E. Matthews, Trans.). Cambridge University Press.

Levine, G. (2006). *Darwin loves you: Natural selection and the re-enchantment of the world*. Princeton University Press.

Levine, G. (2011). *Darwin the writer*. Oxford University Press.

Radick, G. (2013). Darwin and humans. In M. Ruse (Ed.), *The Cambridge encyclopedia of Darwin and evolutionary thought* (pp. 173–181). Cambridge University Press.

Schopenhauer, A. (1819 [1969]. *The world as will and representation* (vol. 1, E. Payne, Trans.). Dover.

Smith, C. K. (1970). Logical and persuasive structures in Charles Darwin's prose style. *Language and Style 3, 4*, 243–273.

Van Wyhe, J. (Ed.) (2002). *The complete work of Charles Darwin online* http://darwin-online.org.uk.
Von Humboldt, A. (1850). *Views of nature or contemplations on the sublime phenomena of creation; with scientific illustrations* (E. C. Otté & Henry G. Bohn, Trans.). Henry G. Bohn.
White, P. (2009). Darwin's emotions: The scientific self and the sentiment of objectivity. *Isis. A Journal of the History of Science Society, 100*(4), 811–826.

Bárbara Jiménez-Pazos is a Philosophy lecturer in the Department of Philosophy at the University of the Basque Country, San Sebastián (Spain). She carried out her undergraduate (2007–2011), postgraduate (2011–2012), and Ph.D. (2012–2016) studies in Philosophy at the University of the Basque Country, before going on to the University of Leeds (United Kingdom) to complete a two-year post-doc research stay (2018–2020) in the School of Philosophy, Religion and History of Science, Centre for History and Philosophy of Science.

Part III
Spreading the New Theory to the World

Chapter 12
Origin's Chapter I: How Breeders Work Their Magic

Gregory Radick

Abstract Darwin begins his "one long argument" not in the natural world of the deep past but – surprisingly and, for some readers, disappointingly – on the present-day world of the farm, providing a detailed look at domesticated plants and animals as well as the humans who breed them. Darwin's opening chapter divides roughly into two halves. In the first half, Darwin surveys the amazing variability of plants and animals under domestication and some of the main causes of that variability. In the second half, he turns from variation to selection: the picking for breeding of the individual dogs or pigeons or cabbages or whatever which are the fastest or the strongest or the most colorful or the tastiest or whatever, in the hopes that those attractive variations will be inherited by offspring. It is mainly through selection of inheritable variation, Darwin argues, that humans have created so many and such diverse domesticated varieties of plants and animals, suited so splendidly to human needs and desires. A large part of being able to follow Darwin's reasoning through the densely fact-packed paragraphs in this chapter lies with seeing where, in his view, it's all going: namely, the laying down of the foundations for an analogical argument linking the selective breeding-into-being by humans of new varieties, or "artificial selection," with a comparable process in nature capable of producing not merely new varieties but new species – what he will call, to flag the analogy, "natural selection." It also helps to be alert to a number of ideas that Darwin held but which modern biologists reject, such as the notion that domestication itself induces animals and plants to become far more variable than they are in the wild, and the idea that characters acquired in the course of an individual's lifetime can be inherited (so-called "Lamarckian inheritance").

G. Radick (✉)
University of Leeds, Leeds, UK
e-mail: g.m.radick@leeds.ac.uk

M. Elice Brzezinski Prestes (ed.), *Understanding Evolution in Darwin's "Origin"*,
History, Philosophy and Theory of the Life Sciences 34,
https://doi.org/10.1007/978-3-031-40165-7_12

12.1 Introduction

> [One authority on animal breeding] speaks of the principle of selection as "that which enables the agriculturalist, not only to modify the character of his flock, but to change it altogether. It is the magician's wand, by means of which he may summon into life whatever form and mould he pleases". (Darwin, 1859, p. 31)

Charles Darwin's theory of evolution by natural selection is cosmically thrilling. But the book that introduced it to the world begins in a deliberately un-cosmic, un-thrilling manner. Darwin's ambition is to convince the reader that observations about ordinary processes happening around us right now, and checkable by anyone who takes the trouble to do so, lead ineluctably to the extraordinary conclusions that he defends in the book – conclusions, moreover, that no one can check observationally, however hard they try.[1]

Hence Chap. 1 is set not on the primitive, pre-biotic Earth that no human ever saw but in the humble worlds of the farm, the garden, and the aviary, where humans for millennia have been bringing into being new varieties of animals and plants that our ancestors managed to domesticate. Here Darwin lays the foundation for an analogical argument that he intended the name "natural selection" to keep before the reader's mind, linking what happens in nature to the artful or "artificial selection" engaged in by human breeders.[2] To this end, the chapter mounts a two-part case: first, for how astonishingly variable animals and plants become after domestication (Darwin, 1859, pp. 7–29); second, for how impressively selection by humans has accumulated those domestication-induced variations into new varieties, ever more numerous and ever better adapted to human needs and desires (Darwin, 1859, pp. 29–43).

At the very start of all of his chapters, Darwin provides a list of the topics covered. In what follows, I will take my enumerated subheadings directly from Darwin's list. But I will embed his topics in an organizational structure of my own so that the reader can grasp more fully how each topic fits within the larger scheme of the chapter and, ultimately, the argument.

---

[1]On the "uniformitarian" scientific ideal expressed in Darwin's ordering of the *Origin*, and how it can lead to disappointment for new readers – "they expect fanfare, and they get fantails," as Stephen Jay Gould quipped – see Gould (2002, p. 105).

[2]The most extensive historical and philosophical analysis of the analogical argument to date is in White et al. (2021).

12.2 Inheritable Variation Among Domesticated Animals and Plants

12.2.1 Domesticated Animals and Plants Are Amazingly Variable Due to the Effects of Several Causes

12.2.1.1 Causes of Variability (Darwin, 1859, pp. 7–11)

The opening pages of this opening chapter address something that our textbooks do not recognize as a fact. According to Darwin, if we compare animals and plants from our domesticated varieties with their wild counterparts, we will be struck by how much more *variable* the former are – that is, by how large the individual-to-individual differences are in domesticated organisms, and by how small those differences are in wild organisms, to the point where wild organisms of a particular kind basically appear to be uniform. Having declared that fundamental contrast in the "variability" (Darwin, 1859, p. 7) of domesticated versus wild organisms, Darwin proceeds immediately to the question of what explains it. His lengthy answer is full of fascinating reports, reflections, and qualifications; in brief, he suggests that the main cause is the disruptive effect that a change from natural conditions to human-created ones eventually has on the functioning of the reproductive systems in the domesticated organisms. As to what, exactly, about domestication might be responsible for this variation-making alteration in reproductive functioning, he offers tentative backing for a proposal from the English horticultural improver Thomas Andrew Knight (1759–1838), who credited the excess of food that pampering humans typically provide.

For purposes of following the twists and turns of Darwin's reasoning through his discussion here, it will help, first of all, if you imagine that you need persuading before you accept this explanation; and secondly, if you appreciate that, in trying to overcome your resistance, Darwin is prepared to make all sorts of what he regards as minor concessions. In Darwin's second paragraph, for example, he confronts doubts about whether the disruptive effect of domestic conditions really occurs before conception. Why should we not suppose instead that the act of conception is the key moment, or that the disruption occurs at some time in the long period when the embryo is developing? In reply, Darwin dwells on a familiar frustration of would-be domesticators: the failure of otherwise healthy organisms, apparently thriving after confinement, to produce offspring. From such cases Darwin infers that everything else in newly domesticated organisms can be fine and yet, anomalously, their reproductive systems shut down. He concludes that when, despite this scope for total shutdown, those systems nevertheless function well enough for offspring to be produced, we should be utterly unsurprised to find that the functioning is imperfect, with the result that the offspring are imperfect copies of their parents. Even so, he immediately adds in the third paragraph, sometimes there is no discernible disruption under domestication, and – equally unsurprisingly – the offspring are then nearly as uniform as they would be under natural conditions.

12.2.1.2 Effects of Habit (Darwin, 1859, pp. 11)

The greater variability of domesticated versus wild organisms ceased to be treated as true in the decades around 1900. By no means coincidentally, belief in the inheritance of acquired characters and habits – so-called "Lamarckian" inheritance – declined during the same period, as part of the general embrace of the doctrine, due to the German biologist August Weismann (1834–1914), that what parents experience in their lifetimes has no effect on the hereditary material which they transmit to the next generation. Darwin, however, took permeability between parental experience and offspring inheritance for granted and expected his readers to do so too.[3] In his first paragraph, he reports, with no fuss or fanfare, that after domestication, a few generations of continual exposure to the new conditions are required before animals and plants begin to show the variability that is his major theme in this part of the chapter. In other words, for Darwin, before the disruptive effect of the new conditions on reproductive systems becomes visible, they undergo incremental and invisible weakening, with each new generation inheriting weaker versions and, as the excess food (or whatever) takes its toll, passing on still weaker ones. And when, at last, the weakening reaches the point where the new variability sets in, it becomes permanent – which, Darwin goes on, is why breeders can still develop new varieties from animals and plants whose domestication happened a long time ago.

Although he regards domestication-induced disruption in reproductive systems as the most important cause of variability, he admits other, lesser causes. One is what he calls "the direct action of the conditions of life" (Darwin, 1859, p. 11), as when an organism comes to be one color rather than another depending on whether it is fed on this or that sort of food. Another cause is a new habit acquired under the new conditions, resulting in an organ or limb or faculty becoming enhanced through use or diminished through disuse, and for offspring accordingly to inherit enhanced or diminished versions. Darwin cites several examples in evidence, including his own studies of wing and leg bones in domesticated versus wild ducks. It appears, writes Darwin, that under domestication, ducks have acquired weaker wings and stronger legs – which is exactly what we should expect from domesticated ducks flying less and walking more, generation after generation.

[3] On the long-run history of Lamarckian inheritance, see Gissis and Jablonka (2011).

12.2.1.3 Correlation of Growth (Darwin, 1859, pp. 11–12)

Next, Darwin considers something that is as much a constraint on variability as a cause: the physiological ties within organisms ensuring that whenever some part X, whatever it is, changes in a particular direction, some other part Y will change too, likewise in a particular direction. In illustration, he gives some very odd examples: if a cat has blue eyes, it will also be deaf; if a pigeon has a short beak, it will also have long feet (and vice versa); and so on. Darwin draws two lessons, pointing in opposite directions. One is that breeders do not have a free hand to shape varieties in absolutely every respect since, due to the laws governing such correlations, the targeting of one part for emphasis or de-emphasis may have unavoidable consequences for the emphasis or de-emphasis of some other part, whatever a breeder's wishes. The other is that, under conditions of domestication, organisms become, in every part, variable, as has been impressively documented in the scientific literature on the oldest varieties. As Darwin puts it, "The whole organisation [of these animals and plants] seems to have become plastic, and tends to depart in some small degree from that of the parental type" (Darwin, 1859, p. 12).

12.2.2 Some Associated Topics: Inheritance; Domesticated Varieties as Reverting Back to the Forms of Their Wild Progenitors; the Haziness of the Line Dividing Varieties from Species

12.2.2.1 Inheritance (Darwin, 1859, pp. 12–14)

Inheritance has already come up implicitly in the passages where Darwin presumes that new conditions induce or elicit changes which are then passed on to offspring. However, now he deals with it explicitly. "Any variation which is not inherited", he begins, "is unimportant for us" (Darwin, 1859, p. 12). No matter how valuable to the farmer or gardener commercially or to the organism itself in a state of nature (for escaping predators, for gaining access to food, for attracting mates), if an organism's offspring do not inherit it, then a variation can play no part in the future modification of that kind of organism. For Darwin's purposes, then, a lot hangs on whether the reader comes to believe that inherited variations are common enough to play the role that Darwin wants to assign to them.

Briskly, Darwin sets out his case, citing testimony from two authorities. There are, of course, the breeders whose livelihoods depend on variations generally being inheritable. "No breeder doubts how strong is the tendency to inheritance," writes Darwin: "like produces like is his fundamental belief: doubts have been thrown on this principle by theoretical writers alone" (Darwin, 1859, p. 12). And there are also medical observers of patterns of inheritance in human families, most recently the

French physician Prosper Lucas (1808–1885), whose magisterial two-volume treatise on *l'Heredité Naturelle,* published in 1847–1850, Darwin had studied closely and now praised. As Darwin explains, children who share pretty ordinary features with their parents may have inherited those features or may simply have been exposed to the same, widely distributed environmental causes. But children who, as documented in Lucas' pages, share extraordinary, one-in-several-million features with their parents, although exposed to the same environmental causes as everyone around them, must have inherited those features. And if "strange and rare deviations of structure are truly inherited, less strange and commoner deviations may be freely admitted to be inherited" (Darwin, 1859, p. 13).

So inheritance is the rule and non-inheritance the exception, Darwin concludes. But beyond that baseline, he goes on, much remains mysterious (Olby, 2013). He ends by cataloging some of the mysteries, among them why it is that some individuals end up resembling not their parents but a more distant, and even very distant, ancestor – a phenomenon known as "reversion." With the possibility of reversion affecting individual organisms thus raised, he turns to consider a related topic: reversion affecting domesticated varieties.

12.2.2.2 Character of Domestic Varieties [with a Preview of Natural Selection] (Darwin, 1859, pp. 14–15)

> Having alluded to the subject of reversion, I may here refer to a statement often made by naturalists – namely, that our domestic varieties, when run wild, gradually but certainly revert in character to their aboriginal stocks. Hence it has been argued that no deductions can be drawn from domestic races to species in a state of nature. (Darwin, 1859, p. 14)

From this start, the reader might think that Darwin will now argue against the idea that if a domestic variety goes feral, it will inevitably revert to the character of its wild progenitor species. After all, that idea, as Darwin introduces it here, had been held to undermine the soundness of looking to domesticated varieties or "races" (a casually introduced synonym) for light on the origin of natural species – and Darwin's whole argumentative project is premised on that strategy.

But the argument is in fact headed in a rather different direction. In the first half of the ensuing paragraph, we learn that, although Darwin finds the evidence for the expectation of feral reversion to be surprisingly poor and, for reasons that he explores fascinatingly, difficult to improve upon, he nevertheless thinks it is probably correct. Feral reversion, then, is not the problem. Darwin's real target turns out to be a related but different expectation, which we might call "non-feral reversion." Here is Darwin's epitome, along with his immediate dismissal:

> If it could be shown that our domestic varieties manifested a strong tendency to reversion, – that is, to lose their acquired characters, whilst kept under unchanged conditions, and whilst kept in a considerable body, so that free intercrossing might check, by blending together, any slight deviations of structure, in such case, I grant that we could deduce nothing from

> domestic varieties in regard to species. But there is not a shadow of evidence in favour of this view: to assert that we could not breed our cart and race-horses, long and short-horned cattle, and poultry of various breeds, and esculent [fit for human consumption] vegetables, for an almost infinite number of generations, would be opposed to all experience. (Darwin, 1859, p. 15)

In other words, Darwin concedes that if the distinctive characters of domesticated varieties were so superficially imposed as to be constantly slipping away to reveal the wild-progenitor characters beneath, then nothing could be learned from the farm and the garden about nature. But millennia of experience suggest that the characters which human have bred in to long-established domesticated varieties are more than stable enough to serve Darwin's argumentative purposes.

He ends the paragraph with a short but significant preview of where that argument is headed: "I may add that when under nature the conditions of life do change, variations and reversions of character probably do occur; but natural selection, as will hereafter be explained, will determine how far the new characters thus arising shall be preserved." Here, Darwin spells out crucial preconditions for the analogical argument that he will mount in Chap. 4 between the production of varieties by the actions of human selectors and the production of varieties-unto-species by the selective effects of the struggle for existence. What makes selective modification by humans possible is inheritable variation, brought on by changed conditions. We have reason to believe that, in nature too, changed conditions bring about inheritable variation (though plainly, given the apparent uniformity of wild species, such variation is less conspicuous); and so, in Darwin's view, it follows that we have reason to believe that, provided there is a struggle for existence, selective modification can take place in nature too.

12.2.2.3 Difficulty of Distinguishing Between Varieties and Species (Darwin, 1859, pp. 15–16)

In the analogical argument to come, Darwin will aim to show that natural selection in its effects is different only in degree, not in kind, from artificial selection, and relatedly that the difference between a new species and a new variety is one of degree and not of kind. Here he continues to prepare the ground with some apposite reflections on domesticated varieties and natural species. He suggests that, notwithstanding how much more variable domesticated varieties are when compared with natural species, and how striking, even "somewhat monstrous" (Darwin, 1859, p. 16), are some of the characters to be found only under domestication, domesticated varieties of a single species "differ from each other in the same manner as, only in most cases to a lesser degree than, do closely-allied species of the same genus in a state of nature" (Darwin, 1859, p. 16). And in support of this different-in-degree-not-kind view, he adduces the frequent disagreements among competent naturalists as to whether a particular breed should be judged as a mere variety of a species or as a species in its own right, descended with fidelity from a wild progenitor species. "If any marked distinction existed between domestic races and species," Darwin adds, "this source of doubt could not so perpetually recur" (Darwin, 1859, p. 16).

12.2.3 *Wild Progenitor Species and Their Domesticated Descendants*

12.2.3.1 Origin of Domestic Varieties from One or More Species (Darwin, 1859, pp. 16–20)

Next, Darwin asks whether there is anything in general to be said about how many "parent-species" or "aboriginal stocks" are involved in the begetting of a group of domesticated varieties. His answer is that the number is different for different groups. For some groups, like the domesticated pigeons (as he will argue at length in the next section), the evidence favors descent from a single wild ancestor. For other groups, like the domesticated dogs, he thinks the evidence favors multiple wild ancestors.

Darwin's pigeon answer is justly famous for its intellectual quality as well as for its status as an illustration of the tree-of-life genealogical pattern that we rightly associate with the *Origin*, along with the theory of natural selection. Darwin's dog answer, by contrast, is not at all well known and was and remains controversial. By the lights of current science, it is wrong: all domesticated dog varieties are now believed to descend from ancestral wolves. But what troubled an early reader of the *Origin*, Darwin's geological mentor Sir Charles Lyell (1797–1875), was that, in allowing for multiple-origins stories in principle, and then in backing one such story for dogs, in this chapter as well as in Chap. 7, Darwin had made a mistake that was not just scientific (Lyell backed a single, wolfy origin) but moral and political. For, Lyell intimated to Darwin, people wanting to argue that the different human races had separate origins – a view associated with American slaveholders, in justification of their abhorrent treatment of black men and women – could now hold up the *Origin* in support.

Darwin did not agree, and nor did he change his mind about whether dogs had a single origin or multiple origins, going on to defend the latter at length in the book which is, in effect, a truly staggering scaling up of this chapter: *The Variation of Animals and Plants Under Domestication*, published in two volumes in 1868. But as a passionate and lifelong anti-slaver, Darwin was never going to take lightly the possibility that he had given aid to the enemy. His correspondence with Lyell about dogs and humans took place in the months immediately before the *Origin* was published. More or less the first thing Darwin did after publication was to put together and send out a questionnaire on the expression of the emotions in human races around the world: the start of gathering evidence in support of what Darwin later called "a new argument in favour of the several [human] races being descended from a single parent-stock" (Radick, 2018, p. 141).

12.2.3.2 Domestic Pigeons, Their Differences and Origin [with a Preview of the Tree of Life] (Darwin, 1859, pp. 20–29)

Part of what made Lyell's suggestion plausible to Darwin was that, throughout the 1840s and 1850s, work by naturalists on the single or multiple origins of domesticated varieties had become thoroughly entangled in the slavery question, along just the lines Lyell sketched. When, in the late 1850s, Darwin took up a serious interest in pigeon breeds and their origins, he was not a lone pioneer, but someone joining an up-and-running debate whose implications for the slavery question were sometimes made explicit (Desmond & Moore, 2009, pp. 199–266). What is more, in concluding that all of the domesticated pigeon varieties descend from a single, still extant wild ancestral pigeon species, *Columbia livia*, known in English as the "rock dove" or "rock pigeon" (because it is usually found living on rocky cliffs), he was not being boldly original but endorsing what had become the consensus view among naturalists, as he emphasized:

> Great as the differences are between the breeds of pigeon, I am fully convinced that the common opinion of naturalists is correct, namely, that all have descended from the rock-pigeon (*Columbia livia*), including under this term several geographical races or sub-species, which differ from each other in the most trifling respects. (Darwin, 1859, p. 23)

Even so, Darwin's argument for that conclusion is a masterclass of expression and reasoning. He begins (Darwin, 1859, pp. 21–3) with a virtuoso description of the remarkable diversity of structures and habits across pigeon breeds – the English carrier, the short-faced tumbler, the runt (actually a very large bird), the barb, the pouter, and so on. So extensive are the differences between these breeds, he continues, that if they were wild, an ornithologist would rank them as species. So why think that all of them are descendants of just one wild species, the rock pigeon? Darwin now sets out ten subsidiary arguments in support (Darwin, 1859, pp. 23–28), pausing midway (Darwin, 1859, pp. 26–27) to summarize the first five as follows:

1. "the improbability of man having formerly got seven or eight supposed species of pigeons to breed freely under domestication";
2. "these supposed species being quite unknown in a wild state, and their becoming nowhere feral";
3. "these [hypothetical aboriginal] species having very abnormal characters in certain respects, as compared with all other [wild] Columbidae, though so like in many other respects to the rock-pigeon";
4. "the blue colour and various marks [characteristic of the rock pigeon] occasionally appearing in all the breeds, both when kept pure and when crossed";
5. "the mongrel offspring being perfectly fertile."

He then gives the remaining five:

1. Everywhere that it is found, *Columbia livia* has not only proved domesticable but bears much in common with domesticated pigeon breeds.
2. Even with two maximally divergent breeds, such as the English carrier and the short-faced tumbler, "by comparing the several sub-breeds of these breeds [...] we can make an almost perfect series between the extremes of structure" (Darwin, 1859, p. 27).

3. For whatever features are most distinctive of a particular breed, you will find that, within the breed, there is a lot of variability for those features: a tell-tale sign that the features have been developed cumulatively by selection.
4. Historical evidence suggests that people have been breeding pigeons in a serious way for millennia.
5. Because male and female pigeons form permanent pair bonds, lots of different breeds can be kept together but with no loss of breed distinctiveness.

In closing the section, Darwin tells us (Darwin, 1859, pp. 28–9) that he has set out his reasons at such length because he appreciates intimately how hard it can be to accept that so much diversity traces back to just a single wild progenitor.

> [W]hen I first kept pigeons and watched the several kinds, knowing well how true they bred, I felt fully as much difficulty in believing that they could ever have descended from a common parent, as any naturalist could in coming to a similar conclusion in regard to the many species of finches, or other large groups of birds, in nature. (Darwin, 1859, p. 28)

That sly preview of the farm-to-nature analogical argument to come is the first of two that Darwin now gives. The second comes at the end of a famous passage worth quoting from at length:

> One circumstance has struck me much; namely, that all the breeders of the various domestic animals and the cultivators of plants, with whom I have ever conversed, or whose treatises I have read, are firmly convinced that the several breeds to which each has attended, are descended from so many aboriginally distinct species. Ask, as I have asked, a celebrated raiser of Hereford cattle, whether his cattle might not have descended from long-horns, and he will laugh you to scorn. I have never met a pigeon, or poultry, or duck, or rabbit fancier, who was not fully convinced that each main breed was descended from a distinct species. [...] The explanation, I think, is simple: from long-continued study they are strongly impressed with the differences between the several races; and though they well know that each race varies slightly, yet they ignore all general arguments, and refuse to sum up in their minds slight differences accumulated during many successive generations. May not those naturalists who, knowing far less of the laws of inheritance than does the breeder, and knowing no more than he does of the intermediate links in the long lines of descent, yet admit that many of our domesticated races have descended from the same parents – may they not learn a lesson of caution, when they deride the idea of species in a state of nature being lineal descendants of other species? (Darwin, 1859, pp. 28–29)

12.3 Selection by Human Breeders

12.3.1 Principle of Selection Anciently Followed and its Effects (Darwin, 1859, pp. 29–33)

So, how did humans do it? How did we transform all of that domestication-induced variability into so many distinctive varieties, so well adapted to our sense of the useful and the beautiful? In keeping with English traditions in natural history and natural theology stretching back to the seventeenth century and best represented in the age before Darwin's in the writings of the Reverend William Paley (1743–1805),

Darwin takes this point about well-adaptedness to be of the first importance, for animals and plants in the wild, adapted to their conditions of life, as much as for their human-adapted domesticated descendants.[4] Indeed, Darwin takes adaptedness to be such a pronounced feature of living things that he dismisses out of hand any putative explanatory theory that fails to address it. In the book's Introduction, he says as much in connection with theories of the origin of wild species (Darwin, 1859, pp. 3–4). Now he says it in connection with theories of the origin of domesticated varieties. Yes, in small ways, and maybe even, in rare cases, in large ways, the causes of variability have contributed to the making of new breeds without human intervention. But, in Darwin's view, it is utterly implausible to suppose that natural causes by themselves would have produced breeds as distinctively useful and/or pleasing to humans as, say, heavy horses that pull carts versus light horses that run races. Humans did not just get lucky in finding those suited-to-us varieties as they are now. On the contrary, those varieties are so suited to us now because we made them that way. And the chief instrument in our variety-making toolbox, according to Darwin, is selection: the repeated choosing for mating of those individual animals and plants that vary, however slightly, in the direction that the selecting human favors. As Darwin puts it: "The key is man's power of accumulative selection: nature gives successive variations; man adds them up in certain directions useful to him. In this sense he may be said to make for himself useful breeds" (Darwin, 1859, p. 30).

Darwin then surveys how well documented the use of selection by animal breeders has been, along the way expressing admiration for the combination of talent and skill that the most successful breeders bring to the task of discriminating the best individuals from the rest (often, Darwin reports, the difference is so subtle as to be undetectable to the inexpert). The use of selection among horticulturalists is less well documented, though Darwin says that its signature may be seen in the fact that varieties of a particular kind are typically most diverse in the features of interest to humans and least diverse in the features of no interest to us. He gives as an example an English berry called the gooseberry. In the different varieties of gooseberry, the fruit varies in all sorts of ways – size, color, shape, and so on – but the leaves vary hardly at all.

12.3.2 Methodical and Unconscious Selection (pp. 33–7)

Darwin now addresses an anticipated objection: that selection cannot possibly be responsible for the making of new varieties by humans over millennia because the use of selection for breed improvement has been around for less than a century. In response, Darwin introduces a distinction between what he calls "methodical selection" and "unconscious selection," but which, for the twenty-first-century reader, are

[4] On the distinctively English emphasis on the good design of organisms and Darwin's role in perpetuating it, see Radick (2009, pp. 155–8).

better termed "deliberate selection" and "un-deliberate selection." It is true, Darwin concedes, that fully deliberate selection, in which the breeder aims to improve a variety in a particular direction and so breeds only from individuals judged closest to the ideal, generation after generation, is quite recent. But Roman and Chinese sources show that the general principle was well understood in antiquity. Furthermore, improvement from selection can occur even in the absence of such understanding and deliberation. He gives as an example a case where, in just half a century, two breeders who started out with pure-bred sheep from the same flock, and who each sought only to keep his flock pure by breeding from the best, ended up with flocks so different from each other as to seem to belong to different varieties. All that was needed to produce so much divergence was the common desire to keep up a standard plus slightly different notions of what that standard was.

Indeed, Darwin goes on, selective improvement does not require even that much. It is enough for people looking after domesticated animals and plants simply to have favorites. Here is how Darwin puts the point:

> If there exist savages so barbarous as never to think of the inherited character of the offspring of their domestic animals, yet any one animal particularly useful to them, for any special purpose, would be carefully preserved during famines and other accidents, to which savages are so liable, and such choice animals would thus generally leave more offspring than the inferior ones; so that in this case there would be a kind of unconscious selection going on. We see the value set on animals even by the barbarians of Tierra del Fuego, by their killing and devouring their old women, in times of dearth, as of less value than their dogs. (Darwin, 1859, p. 36)

For Darwin, the Fuegians, encountered on the *Beagle* voyage, were the lowest of the low in the scale of human races. (Like Abraham Lincoln – they were born on the same day – Darwin thought slavery was a moral abomination *and* that each race found its place in a hierarchy that extended from the most highly civilized down to the lowest "savages") (Radick, 2010). Once again, Darwin appeals to what can be observed now – humans at their most savage – to ground a plausible inference about what cannot be observed: the millennia of selective improvement of plants and animals that, in his view, must have taken place long before records began being kept. It is to that long but undocumented period of selection that we owe much of what we find pleasing in our domesticated animals and plants. When Darwin declares unconscious selection to be "more important" (Darwin, 1859, p. 34) than methodical selection, he has in mind the outsized role that the former has played in the making of our varieties.

12.3.3 Unknown Origin of Our Domestic Productions (pp. 37–40)

As Darwin sees it, once we appreciate the scale of this debt to unconscious selection, we cease to wonder at the fact that, often, the origins and history of our domesticated varieties are obscure. He makes this point first in relation to plant varieties, adding that the absence of "a single plant worth culture" (Darwin, 1859, p. 38) in places such as Tierra del Fuego is not due to a weird paucity of cultivatable plants in those places, but to the fact that the cultivatable plants that are there never benefitted from the sustained unconscious selection that improved such plants in more civilized places. Likewise, although the uncivilized places have their share of domesticable animals, the varieties that emerge there, like dogs, are less protected from wild conditions than they would be under civilization, with the result that savage animal breeds are partly the product of unconscious selection and partly the product of natural selection.

> On the view here given of the all-important part which selection by man has played, it becomes at once obvious, how it is that our domestic races show adaptation in their structure and in their habits to man's wants or fancies. (Darwin, 1859, p. 38)

There follows what is, in effect, a fascinating aside on how what attracts unconsciously selecting humans in animals and plants is external novelty, of whatever kind. For Darwin, that quirk of humans explains how the world came to have in it such varietal oddities as tumbling pigeons because, in all probability, no one ever thought, with a sigh of longing: "oh, if only there were pigeons that punctuate their flights with automatic somersaults!" Instead, one day someone noticed a pigeon born with a very slight tendency to in-flight somersaulting, liked what they saw, and gave that pigeon extra love and attention and, crucially, opportunities for mating. It also explains why the outsides of our domesticated plants and animals are endlessly diverse, but their insides are much of a muchness.

Darwin rounds out his discussion by returning to how unsurprising we should find the obscurity of the origins of our varieties, illustrating at length with the example of how, step by unnoticed step, new dog breeds can emerge from existing ones without anyone intending such an outcome, or being able to say afterward when and where, exactly, the new breed made its debut. Better remembered than the illustration itself is a brief but illuminating analogy that he makes in introducing it: "a breed, like a dialect of a language, can hardly be said to have had a definite origin" (Darwin, 1859, p. 40; on Darwin's later development of this analogy, see Radick, 2008).

12.3.4 Circumstances Favorable to Man's Power of Selection (pp. 40–43)

In his final reflections, Darwin turns to the question of the circumstances that promote selective improvement. He picks out four factors. First, among the animals or plants, there should be high variability, "as freely giving the materials for selection to work on" (p. 40). Second, the number of animals or plants kept should be as large as possible, not only to increase the probability of finding interesting variations (though Darwin stresses that the truly skillful breeder can work his magic equally well with uninteresting variations) but to decrease the probability that the best individuals will mate with markedly inferior ones. Third and, in Darwin's view, most importantly, the superintending humans need to be scrupulously attentive, as they are when the animals and plants involved are most valuable. And fourth, there needs to be a way of preventing unwanted, improvement-diluting crossings.

In Darwin's closing paragraph, summarizing the contents as a whole, a theme that has been a minor element in the chapter up to now – the role of crossing as distinct from selection in the improvement of plant and animal varieties – bulks unexpectedly large. It is as though, on reaching the chapter's end, Darwin suddenly realized that he had intended to say more on that theme and so hastily overcompensates. But his final sentence is as expected: a restatement of how "the accumulative action of Selection," now with a capital S, is, when it comes to the adapted-to-human forms of our domesticated animals and plants, "by far the predominant Power" (Darwin, 1859, p. 43).

12.4 Conclusion

To look ahead a little: we now take our leave of variation under domestication and its selection by humans. The scene shifts to the natural counterparts of these two topics, variation under nature (Chap. 2) and the struggle for existence (Chap. 3). In nature too, Darwin aims to persuade us, there is inheritable variation in animals and plants, brought on chiefly by changes in conditions; and there is a process that, by its selective effects, accumulates that inheritable variation in adaptive directions. But, he then argues in Chap. 4, via the analogical argument that he foreshadowed at several points in Chap. 1, the natural selective process can go further than the artificial one, producing not merely new varieties within existing species but new species. The rest of the book amounts to a defense of that claim for the greater, species-making power of natural selection (Chaps. 5–9) and then a demonstration of the remarkable explanatory work it can do when brought to bear on the most disparate patterns (Chaps. 10–13). That, in outline, is the "one long argument" (Darwin, 1859, p. 459) that Darwin boasted about making in his Conclusion (for the argument's conformity to the *vera causa* tradition, in which one proceeds by establishing first a cause's existence, then its adequacy, and then finally its responsibility, see White, et al., 2021, pp. 106–136). But he thought that the argument in outline was unconvincing. To become convinced, people needed not only to read the book but to think through its argument, including all of the arguments-within-arguments composing it, for themselves – to live with it, as Darwin had, mulling it over, with all the many connections kept firmly at the front of their minds. It was, and remains, a dauntingly challenging prospect. I hope that this chapter makes getting started, at least, a little less so.

References

Darwin, C. (1859). *On the origin of species by means of natural selection, or the preservation of favoured races in the struggle for life*. John Murray.

Desmond, A., & Moore, J. (2009). *Darwin's sacred cause: Race, slavery and the quest for human origins*. Penguin.

Gissis, S. B., & Jablonka, E. (Eds.). (2011). *Transformations of Lamarckism: From subtle fluids to molecular biology*. MIT Press.

Gould, S. J. (2002). *The structure of evolutionary theory*. Belknap Press.

Olby, R. (2013). Darwin and heredity. In M. Ruse (Ed.), *The Cambridge encyclopedia of Darwin and evolutionary thought* (pp. 116–123). Cambridge University Press.

Radick, G. (2008). Race and language in the Darwinian tradition (and what Darwin's language-species parallels have to do with it). *Studies in History and Philosophy of Biological and Biomedical Sciences, 39*, 359–370.

Radick, G. (2009). Is the theory of natural selection independent of its history? In J. Hodge & G. Radick (Eds.), *The Cambridge companion to Darwin* (2nd ed., pp. 147–172). Cambridge University Press.

Radick, G. (2010). Did Darwin change his mind about the Fuegians? *Endeavour, 34*, 51–54.

Radick, G. (2018). How and why Darwin got emotional about race. In E. Sera-Shriar (Ed.), *Historicizing humans: Deep time, evolution, and race in nineteenth-century British sciences* (pp. 138–171). University of Pittsburgh Press.

White, R. M., Hodge, M. J. S., & Radick, G. (2021). *Darwin's argument by analogy: From artificial to natural selection*. Cambridge University Press.

Gregory Radick is Professor of History and Philosophy of Science at the University of Leeds. His books include *The Simian Tongue: The Long Debate about Animal Language* (Chicago, 2007), *Disputed Inheritance: The Battle over Mendel and the Future of Biology* (Chicago, 2023), and, as co-editor with Jonathan Hodge, *The Cambridge Companion to Darwin* (Cambridge, 2003; 2nd edition, 2009).

Chapter 13
Origin's Chapter II: Darwin's Ideas on Variation Under the Lens of Current Evolutionary Genetics

Roberto Rozenberg

Abstract The "long argument" presented in the opening chapter of *On the Origin of Species* develops steadily. Darwin's comparison between variation under domestication (Chapter I) and variation under nature (Chapter II) is an argument that approximates artificial and natural selection, rendering the second more intelligible. However, questions remain about whether Darwin conceived the analogy before developing the theory of natural selection or whether its use later fulfilled a didactic strategy. Despite not being part of the book, Darwin's theory of inheritance, Pangenesis, is fundamental for understanding the chapter. Darwin himself recognised that the origin and inheritance of variation were probably the book's most uncertain subject: "The laws that govern inheritance are quite unknown". However, with his ancient concept of inheritance, it becomes clear that in the development of evolutionary thinking, it was the very existence of variation and not its origin that mattered.

13.1 A Quick, Contemporaneous View

Chapter II, "Variation Under Nature," continues the reasoning of Chapter I, "Variation Under Domestication". In those two chapters, Darwin describes phenomena that directly or indirectly refer to current knowledge of genetics and evolution. He does so in such an advanced manner that it takes by surprise any reader not familiar with the history of science. So, before digging into Chapter II, let us look at some of the striking points from the perspective of current evolutionary thinking. By doing so, we are not ignoring the longstanding historiographical debate on presentism and contextualism, which has been ongoing since Herbert Butterfield's criticisms of Whiggism. Instead, we adopt Michael Ruse's suggestion of bringing together the

R. Rozenberg (✉)
Department of Genetics and Evolutionary Biology, Institute of Biosciences, University of São Paulo, São Paulo, Brazil
e-mail: robertor@usp.br

M. Elice Brzezinski Prestes (ed.), *Understanding Evolution in Darwin's "Origin"*, History, Philosophy and Theory of the Life Sciences 34,
https://doi.org/10.1007/978-3-031-40165-7_13

historian and philosopher of science's "looking back" with the scientist's "looking forward" to reveal the potential that these multiple levels of interest and modes of reading can bring to science teaching (Ruse, 2013).

Of course, Darwin did not know molecular biology. But what if we imagine him embarking on a journey to briefly examine our current evolutionary genetics? Take dogs, for instance, one of Darwin's passions. They vary in size more than any other mammal. We now know that ancient dogs, domesticated from wolves in the past 30,000 years, varied in size to some extent. But the current extreme size differences (giant breeds are up to 40 times bigger than the smallest) emerged in the past 200 years when humans established modern breeds. One variant stood out when variation in the region around a gene called *IGF1* was compared to body size in dogs and wild canids. It lies in a stretch of DNA that encodes a molecule known as a "long non-coding RNA", which controls levels of the IGF1 protein, a potent growth hormone (Plassais et al., 2022). Therefore, domestication achieves extreme phenotypes but does not increase variability, as Darwin thought, based on his inheritance theory. On the contrary, it decreases variability as domestic products are more uniform.

Some of Darwin's views on variation in Chapters I and II are unique insights into genetic phenomena that became known many decades or even over a century later. To satisfy the most anxious readers interested in which of Darwin's ideas expressed in the *Origin* are proximate to present knowledge, below are some of them, which are picketed up in Chapter I. Since a detailed discussion of them is not in the scope of this chapter and volume, they are merely quoted and related to what the present knowledge is – even when not wholly consensual (like the so-called "central dogma", for instance).

- The central dogma of molecular biology: "Any variation which is not inherited is unimportant for us" (Darwin, 1859, p. 12). Also: "[...] this shows how unimportant the direct effect of the conditions of life are in comparison with the laws of reproduction, and of growth, and of inheritance" (p. 10).
- Multifactorial inheritance: "When a deviation appears not infrequently, and we see it in the father and child, we cannot tell whether it may not be due to the same original cause acting on both; but when amongst individuals, apparently exposed to the same conditions, any very rare deviation, due to some extraordinary combination of circumstances, appears in the parent – say, once amongst several million individuals – and it reappears in the child, the mere doctrine of chances almost compels us to attribute its reappearance to inheritance" (p. 13).
- Recessive inheritance: "why the child often reverts in certain characters to its grandfather or grandmother or other much more remote ancestor" (p. 13).
- Sex-linked inheritance: "why a peculiarity is often transmitted from one sex to both sexes, or to one sex alone, more commonly but not exclusively to the like sex" (p. 13).
- Genetic expressiveness and anticipation: "[...] at whatever period of life a peculiarity first appears, it tends to appear in the offspring at a corresponding age, though sometimes earlier" (p. 13).

- Mutation: "These remarks are of course confined to the first appearance of the peculiarity, and not to its primary cause, which may have acted on the ovules or male element" (p. 14).
- Isogenic lineages: "It would be quite necessary, in order to prevent the effects of intercrossing, that only a single variety should be turned loose in its new home" (pp. 14–15).
- Coalescence theory and molecular clock: "When we attempt to estimate the amount of structural difference between the domestic races of the same species, we are soon involved in doubt, from not knowing whether they have descended from one or several parent-species" (p. 16).
- Evolution of evolvability: "It has often been assumed that man has chosen for domestication animals and plants having an extraordinary inherited tendency to vary [. . .]" (p. 17).
- Quantitative genetics: "The offspring from the first cross between two pure breeds is tolerable and sometimes (as I have found with pigeons) extremely uniform, and everything seems simple enough; but when these mongrels are crossed one with another for several generations, hardly two of them will be alike [...]" (p. 20).
- Neoteny: "Not a single domestic animal can be named which has not in some country drooping ears" (p. 11).
- Allometry: "Variability is governed by many unknown laws, more especially by that of correlation of growth" (p. 43).
- Evolutionary restrictions: "There are many laws regulating variation, some few of which can be dimly seen, and will be hereafter briefly mentioned. I will here only allude to what may be called correlation of growth" (p. 11).

13.2 Variability: Variation, Selection, and Evolution

Chapter I, "Variation Under Domestication", was a compilation of two chapters from the manuscript of *Natural Selection* that existed before Darwin received Alfred Russel Wallace's letter of June 1858 (Olby, 2009). As Darwin discusses "unconscious selection", meaning domestication, in Chapter I, he goes on to make the key argument of Chapter II, which is nature's similar power to select and thus embrace change and adaption of species to the environment, establishing the "fact" of universal variation in natural populations. This preludes Chapter IV, where natural selection results from variation, heredity, and competition. Darwin recognises that selection is the main evolutionary mechanism:

> Over all these causes of Change I am convinced that the accumulative action of Selection, whether applied methodically and more quickly, or unconsciously and more slowly, but more efficiently, is by far the predominant Power. (Darwin, 1859, p. 43)

The term "natural selection", the title of Chapter IV, appears thrice in the introduction, once in Chapter I, and three times again in Chapter II. The term "evolved" appears only once in the first edition, as the very last word. Darwin instead uses "descent with modification". However, the idea of evolution permeates the entire story, though Darwin himself consciously avoided the term to prevent confusion with the word "evolution" used for embryo development according to theories of preformationism and epigenesis, much discussed from the seventeenth century until his time. The term has also been used for a different theory proposed by his contemporary philosopher Herbert Spencer, 2 years before the publication of the *Origin*. Darwin only appropriated the term "evolution", using it seven times, in the sixth edition of the *Origin*, published in 1872 (Gayon, 2003).

As it derives from the mechanism rational, variation is one of the critical points for evolution, the others being heredity and super fecundity, leading to competition for resources and consequent natural selection of more adapted forms. That is one of the reasons Darwin devoted the book's first two chapters to the variation issue, be it under domestication (Chapter I) or in a state of nature (Chapter II).

13.3 Individual Differences: The Link Between Artificial and Natural Selection

Chapter II begins with an inductive remark about variation under nature:

> Before applying the principles arrived at in the last chapter to organic beings in a state of nature, we must briefly discuss whether these latter are subject to any variation. (Darwin, 1859, p. 44)

Yes, they are! But the "dry facts" should be reserved for future work. In Chapter XIV, the remark will be restated:

> There is no obvious reason why the principles which have acted so efficiently under domestication should not have acted under nature. (Darwin, 1859, p. 467)

The aim here was to build a "bridge" asserting that just as breeders could use selection to change the traits of a particular breed, nature could also select and thus modify a wild species "unconsciously". Darwin's analogy between artificial selection and natural selection is also an analogy between the practised eye of the breeder and nature itself (Largent, 2009).

Does the analogy between variation under domestication and variation under nature precedes, concur, or is it posterior to Darwin's development of the theory of natural selection?

Michael Ruse stated:

> The artificial selection analogy did not have a crucial role in Darwin's discovery of natural selection as a cause of evolutionary change. It was only after he had grasped this later concept that he came to stress the analogy. (Ruse, 1975, p. 344)

Largent dedicated efforts to that issue:

> Was his analysis of artificial selection as significant for the original formulation of his theory of evolution by natural selection [...]? Or did Darwin come to appreciate the analogy between artificial and natural selection *only after* [emphasis added] he had conceived of

> natural selection, which might lead one to believe that the analogy served only as a means to convince the reader of the reality of evolution and natural selection? [...] Analyses over the last several decades suggest that it arose slowly through a reciprocal relationship between his theorizing and his evidence-gathering activities. Darwin and many of the scholars who have described the genesis of his theory have depicted it springing to life between late September and early October of 1838. (Largent, 2009, pp. 14–15)

Largent also quotes Darwin's letters and *Autobiography* to echo such a claim:

> I came to the conclusion that selection was the principle cause of change from the study of domesticated productions; and then, reading Malthus, I saw at once how to apply this principle. (Darwin as cited in Largent, 2009, p. 24)

However, his private notebooks show that Darwin adopted the potential analogy only after conceiving of natural selection. Darwin immersed himself in the world of English pigeon fanciers from 1855 until he began writing the *Origin* in the summer of 1858, suggesting that his reliance on the analogy between artificial and natural selection emerged in concert with, rather than after, his description of the mechanisms of natural selection (Largent, 2009).

13.4 Doubtful Species: Defining Species from Varieties; Polymorphisms Then and Now

Darwin does not discuss but only mentions the various definitions of the term species in Chapter II: "No one definition has yet satisfied all naturalists" (Darwin, 1859, p. 44). Well, *dear Darwin*, over 160 years have passed, and we still do not have one. This definition is crucial in a book about the origin of species. "Yet every naturalist knows vaguely what he means when he speaks of a species" (Darwin, 1859, p. 44).

Darwin also cites the difficulty in defining what variations are "[...], but here community of descent is almost universally implied, though it can rarely be proved" (Darwin, 1859, p. 44). So, there is no question that variation arises from common descent. Blurring the difference between a variety and a species is all it takes to understand transmutation (the idea of unlimited descent with modification originating new species). A few pages later, he will declare unambiguously: "Hence I believe a well-marked variety may be justly called an incipient species" (Darwin, 1859, p. 52). And a couple of paragraphs down: "[...] I look at the term species, as one arbitrarily given for the sake of convenience to a set of individuals closely resembling each other, and that it does not essentially differ from the term variety" (Darwin, 1859, p. 52). And before closing Chapter II: "Finally, then, varieties have the same general characters as species [...]; but the amount of difference considered necessary to give to two forms the rank of species is quite indefinite" (Darwin, 1859, p. 58/59).

Modern molecular genetics has provided satisfactory definitions for variation, although not for speciation. For the past decades, quantifying and qualifying genetic variation has become a significant part of cladistics and evolutionary biology.

A convention has been established among human geneticists where mutations leading to single or a few nucleotides variation in DNA are defined as a *polymorphism* when present at a ≥1% frequency in a given population. Next-generation DNA sequencing strategies, known as NGS, have been immensely influential in detecting and describing variation. It is now possible to detect which species and variants are present in a ground sample without seeing or recognising any traces of organisms or parts of them, but only by extracting total DNA from dirt and sequencing all genomes in it!

A few paragraphs after starting the species discussion, Darwin uses the term "protean" or "polymorphic", referring to genera in which species present a tremendous amount of variation: "[. . .] and hardly two naturalists can agree which forms to rank as species and which as varieties" (Darwin, 1859, p. 46). As stated before, the issue persists despite our current ability to quantify variation.

> Hence, in determining whether a form should be ranked as a species or a variety, the opinion of naturalists having sound judgment and wide experience seems the only guide to follow. (Darwin, 1859, p. 47)

This need for a specialised opinion still holds today. The experience of naturalists is still the primary parameter used to define a species, despite all the analysis and technologies available today to determine population structures, gene flow, microevolutive processes, and so on (Templeton, 2021).

Darwin also states that the "destiny" of a variety can be different according to the results of natural selection acting upon it:

> [. . .] it need not be supposed that all varieties or incipient species necessarily attain the rank of species. They may whilst in this incipient state become extinct, or they may endure as varieties for long periods. (Darwin, 1859, p. 52)

For Darwin, natural selection was probably too slow a process to be recordable in nature, which is a possible reason he never tried to find actual examples (Ruse, 2023, Chap. 15, in this volume), unlike variation, which is extensively recorded in Chapters II and V.

> How fleeting are the wishes and efforts of man! How short his time! And consequently, how poor will his products be, compared with those accumulated by nature during whole geological periods. (Darwin, 1859, p. 84)

Nowadays, new procedures have emerged that enable the quantification of cumulative variation, evolution, and natural selection. The current notion of a molecular clock is defined by measuring evolutionary change. We calculate the amount of genetic difference between two species and then estimate the time they split in the past, considering an average mutation rate for the group studied.

Since the 1990s, extracting DNA from fossils has become progressively possible. This opens up an amazing window of possibilities to calculate the divergence time between the fossil and the most similar species alive, to understand the ancient DNA sequence of the most common ancestor, and even to measure and compare the amount of variation between current species and those from the fossil record, allowing for cladistics and phylogenetic analysis. For example, the group of the Nobel Prize-winning biologist Svante Pääbo sequenced a Neanderthal fossil, 400,000 years old, at the Max Planck Institute of Evolutionary Anthropology in Leipzig. It permits genetic comparisons of current and extinct *Homo sapiens,* Denisovans, and other hominins from the Paleolithic period (2.5 million–12,000 years ago), allowing inferences on specific genes and corresponding phenotypes, such as speech, cognition, nutrition, and others (Krause & Pääbo, 2016).

Although Chapter II concerns variation under nature, Darwin declares to "have been struck with the fact that if any animal or plant in a state of nature is beneficial to man, or from any cause closely attract his attention, varieties of it will almost universally be found recorded" (Darwin, 1859, p. 50). He implies that through domestication, men spread' species and variations would arise in different parts. This makes perfect sense considering current evolutionary theory, as modifications result from species adapting to new environments with new selective pressures. We know today, and Darwin knew back then, that the domestication of plants and most animals started some 12,000–13,000 years ago (during the transition to the Neolithic period). Valuable and domesticable species were spread, conferring adaptive advantage to agropastoral populations, progressively abandoning the hunter-gathering habit that persisted in the hominin lineage for millions of years.

Determining the boundaries of a species is a goal of modern cladistics. Still, as Darwin stated, limits are arbitrary: different groups have unique characteristics, particularities, and variations to be considered by the specialists when deciding on what constitutes an individual difference, a variety, and a distinct species. He declares:

> It should be remembered that systematists are far from pleased at finding variability in important characters, and that there are not many men who will laboriously examine internal and important organs, and compare them in many specimens of the same species. (Darwin, 1859, p. 45)

After returning from the *Beagle* voyage, Darwin found quite a few of these men. Nonetheless, variability was then viewed as an exception rather than a rule. Species fixism was still the norm to be broken by Darwin's most famous book.

But here, Darwin's views on heredity start to show:

> We have also what are called monstrosities; [...] some considerable deviation of structure [...] not generally propagated. Some authors use the term "variation" in a technical sense, as implying a modification directly due to the physical conditions of life; and "variations" in this sense are supposed not to be inherited. (Darwin, 1859, p. 44)

This was initially a striking paragraph that assumed a lack of use of the inheritance mechanism of acquired characteristics for a scientist who believed, like most of his contemporary fellows, in this kind of heredity. However, he goes on:

> [...] but who can say that [...] [some acquired features] would not in some cases be inherited for at least some few generations? and in this case I presume that the form would be called a variety. (Darwin, 1859, pp. 44–45)

Well, there is variation arising in response to the conditions of life, but now we know that, as a rule, these would not form varieties, as their features would not be passed on to the offspring.

Darwin then moved the discussion to what may define a species and what may be considered variation, which refers to the long-held discussion of the species concept. Many definitions have been proposed so far, with parameters varying among modern biology's different academic fields, such as the biological, the morphological, or the phylogenetic species concepts. Reproductive isolation and phenotypic similarity are two key points in several areas. However, it is easy to see that from an evolutionary-time perspective, the concept "dissolves" in time because there are always intermediary variants between extant and the ancestral species that gave rise to them:

> These differences blend into each other in an insensible series; and a series impresses the mind with the idea of an actual passage. (Darwin, 1859, p. 51).

It is an infinite repertoire of slight and successive cumulative variation. Darwin stated that if a naturalist,

> [. . .] confine his attention to one class within one country, he will soon make up his mind how to rank most of the doubtful forms. (Darwin, 1859, p. 50)

This means that a reference has to be considered. In Darwin's view, defining a species depends on the observer's credentials and extensive research but:

> [. . .] at the expense of admitting much variation,–and the truth of this admission will often be disputed by other naturalists. When, moreover, he comes to study allied forms brought from countries not now continuous, in which case he can hardly hope to find the intermediate links between his doubtful forms, he will have to trust almost entirely to analogy, and his difficulties will rise to a climax. (Darwin, 1859, p. 51)

Speciation is currently understood as a process of separation of part of a population that varies and diverges from the original stock until reproductive isolation leads to a new species (Templeton, 2021). It is a difficult concept to define, as stated before, because if evolutionary time is considered, there is always an almost infinite spectrum of successive subpopulations that reproductively connect the different moments of a species and its descendants. Darwin describes what is today understood as *allopatric speciation*:

> The passage of one stage of difference to another and higher stage may be, in some cases, due merely to the long-continued action of different physical conditions in two different regions. (Darwin, 1859, p. 52)

Currently, other modes, such as *sympatric speciation*, are also recognised, where the separation may occur within the territory of a population, structured with different subpopulations.

After the last quote, Darwin goes on:

> "[. . .] but I have not much faith in this view [the Allopatric speciation he just described]; and I attribute the passage of a variety [. . .] to the action of natural selection in accumulating [. . .] differences of structure in certain definite directions". (Darwin, 1859, p. 52)

However, the two are not considered mutually exclusive or alternatives to the phenomena. Instead, natural selection is the consequence of the process after enough variation has been accumulated in allopatric or sympatric speciation.

Here (and everywhere else), using the term "natural selection" as a cause must be understood metaphorically. Natural selection is not a causative agent. Rather, it is the consequence of the descent process with modification and different survival of fittest states that today we may refer to as nucleotides, genomic positions, genes, genotypes, or organisms. The metaphoric use of natural selection as an agent was extensively acknowledged by both Charles Darwin in Chapter IV of the *Origin* and by Alfred Russel Wallace, co-discoverer of the theory.

As the discussion continues, Darwin uses the word *analogy*, but today we would use *homology* to describe the same phenomena he is addressing. Homology refers to a similarity between two organisms that comes from common ancestry rather than a convergent adaptation to the environment. For instance, the paw of a mouse and the bat's wing are homologous, as they both derive from the same ancestral structure. However, the wings of a bat and those of a fly evolved from different ancestral structures that converged during evolution to perform the same activity (flying), an adaptation that arose independently but in an analogous fashion.

13.5 Inheritance (Pangenesis Views)

Variation itself is the subject that received the most standalone coverage in the *Origin* (Olby, 2009). Inheritance is not given a chapter to itself, perhaps as a recognition of how imprecise and hypothetical Darwin's understanding of the subject was, a restriction of which he was very aware: "the laws that govern inheritance are quite unknown" (Darwin, 1859, p. 13) However, the inheritance of variation is briefly discussed in Chapters I, II, and V. In the latter, Darwin summarises: "Our ignorance of the laws of variation is profound" (Darwin, 1859, p. 167).

It is noteworthy that Darwin's "most" favourite hypothesis, Pangenesis, was never formally introduced in the *Origin of Species* but reserved for separate works. One possible reason is Darwin's rush to publish when he faced Alfred Russel Wallace's letter in 1858 (Bizzo, 2008). Darwin's latter books *The Variations of Animals and Plants Under Domestication* (1868) and *The Expression of the Emotions in Man and Animals* (1872), respectively, discuss and applies his hypotheses of inheritance, developed from the Ancient Greece philosophers' ideas of Pangenesis (Castañeda, 1992).

One year after Darwin's death, the German biologist August Weismann rejected the inheritance of acquired traits and, in the next decades, developed his germ plasm theory, describing the separation of cell lines into the *soma* (cells of the body that do not participate in reproduction) and *germ* (gametes and reproductive tissue).

Before him, British biologist George Jackson Mivart (1871) affirmed that different parts of the body specialised in reproduction as opposed to others that participate in the "daily life" activities of the organism. Darwin understood modification by the mechanism of Pangenesis: the idea that gemmules accumulate information from the environment and life history and then are passed to the offspring, which are formed by combining the gemmules of both parents. In this view, the previous generation's phenotypes would directly affect subsequent generations' genotypes. Of course, Darwin did not use the concepts of genotype, phenotype, or even gene, which only appeared later (in 1909, by Danish botanist Wilhelm Ludvig Johannsen). Darwin believed "there can be little doubt" about the inheritance of acquired characteristics, and Pangenesis helped to explain this supposed mechanism.

The term Pangenesis, however, is not introduced in the *Origin*. His hypothesis was probably first formulated in 1841 but only published in 1868 and was an attempt to give a unified account of all kinds of generations. Still, he considered it to be an imperfect solution. He used the term Pangenesis, but it was not a new idea, as mentioned above. He was trying to establish a consilience of inductions. It worked for the concept of natural selection, and Darwin hoped to repeat the strategy with Pangenesis, even though no one had ever detected a gemmule (Endersby, 2003).

Pangenesis made connections between many of Darwin's ideas: the relationships between embryos, adults, and species; the link between sexual and asexual reproduction; the similarity between gametes and asexual buds; and the continuity of reproduction, growth, and healing:

> No other attempt, as far as I am aware, had been made, imperfect as this confessedly is, to connect under one point of view these several grand classes of facts. An organic being is a microcosm – a little universe, form of a host of self-propagating organisms, inconceivably minute and numerous as the stars in heaven. (Darwin as cited in Endersby, 2003, p. 84)

So, among Darwin's three main theories, natural selection, Pangenesis, and sexual selection, the second is the only one that currently does not hold. Some research on historiography has discussed theoretical reasons for his preference for Pangenesis (e.g. Bizzo, 2008; Lorenzano, 2011; Mead et al., 2019). In addition, we can speculate about other reasons, such as Darwin's lack of extensive training in mathematics and statistics (he obtained the proportion of 3:1 by crossing hybrid plants but never crossed the dominants of F2 to obtain 1:2:1 as Mendel did) and a contextual one, such as the absence of techniques at the time to study cytology and meiosis. However, brave Darwin went on, recognising that "the laws of heredity" were "quite unknown". Understanding heredity through Pangenesis is now admittedly wrong.

Interestingly, the origin of variation is not as crucial for the theory of natural selection as the existence of variation itself. For the mechanism to operate, the source of variation is irrelevant if it is somehow linked to inheritance. Hence, this could justify some school proposals to teach evolution before genetics (Bizzo & El-Hani, 2009).

13.6 Variation and Sexual Selection

Darwin recognised the fundamental differences between sexual and asexual reproduction by July 1837. He understood that *mating* was evolutionarily conservative and *maturation* evolutionarily innovative. However, we consider nowadays the opposite to be the case: *mating* implies that before fertilisation, gametes undergo variation due to sexual recombination and crossing over of chromosomes during meiosis 1 of gametogenesis, while somatic *maturation*, by DNA replication and mitosis, is conservative. Even when a mutation occurs in somatic cells, it does not change the sexual reproductive specialised germ cells. However, Darwin understood inheritance as a blending model of gemmules, and the role of crossing (reproduction) in the generation of variability was reduced. In this view, only immature, i.e. maturing organisms, are impressionable by the environmental influence that he refers to as "conditions of life", and this was the only means by which adaptive and heritable variations could be acquired (Endersby, 2003).

As Darwin explains, he is "strongly inclined to suspect" that in both domesticated and wild species, variations originate with changes in the parents' "reproductive elements having been affected prior to the act of conception" (Darwin, 1859, p. 8), "from the indirect and direct action of the external conditions of life, and from use and disuse" (Darwin, 1859, p. 489–490). The first observation is still accepted, but the cause is understood differently. We realise now that mutations constantly arise during DNA replication (and thus gametogenesis) but independently of environmental forces. Darwin's examples of "use and disuse" as a basis for evolution are relatively uncommon. He concludes that direct environmental effects are not potent enough to explain the tremendous range of "indefinite" variation. This was not a "Lamarckian conception", but elsewhere he considers the "direct action of the conditions of life" and the inheritance of "habit" and hence "use and disuse" (Largent, 2009; Olby, 2009).

13.7 Regression Towards the Mean and Inheritance of Acquired Characteristics

Darwin's big problems started with his different explanations for the causes of variation (Pangenesis). He understood that species members do not deviate significantly from average characteristics since parents' characteristics blend out, producing offspring generally intermediate in characters. He did, however, understand a theory of inheritance with a half contribution of each parent, blending the constitutions of two individuals. Nevertheless, this would lead to the problematic fact that the hereditary contribution of each ancestor would halve in each generation:

> After twelve generations, the proportion of blood, to use common expression, of any one ancestor, is only 1 in 2048; and yet, as we see, it is generally believed that a tendency to reversion is retained by this very small proportion of foreign blood [...] (Darwin, 1859, p. 160)

Had Darwin considered an alternative view, such as Mendel's particulate model of inheritance, he could have realised, maybe, that there is no mixing of characteristics in the male and female reproductive elements, what we now call gametes. Different particles (alleles) do not generally mix at a molecular level, remaining as distinct entities, as shown by Mendel's laws. However, Darwin did realise that some information must have been maintained silently (what we call recessive) in an organism to explain reversion, i.e. how a man could transmit characters to his grandson via his daughter (that did not present them). In the latter book *The Variation of Animals and Plants Under Domestication*, Darwin concluded that the ovules and spermatozoa of higher animals "must be crowded with *invisible characters*, proper to both sexes" and to a long line of male and female ancestors separated by hundreds or even thousands of generations from the present time (Darwin, 1868/1875 as cited in Endersby, 2003, p. 77).

Mendel himself, upon closing his famous article "Experiments on Plant Hybridization" from 1865, challenged Darwin's conception of variation, the very idea that variation under domestication was more frequent than variation under nature discussed before:

> No one will seriously maintain that in the open country the development of plants is ruled by other laws than in the garden bed. (Mendel, 1866)

But Mendel's pea characteristics were all strongly marked ones. They were not what Darwin considered to be individual small differences relevant to evolution (Olby, 2009).

Darwin writes about "individual differences", understanding that "no one supposes that all the individuals of the same species are cast in the very same mould" (Darwin, 1859, p. 45). This leads Chapter II to its central idea:

> These individual differences are *highly* important for us, as they afford materials for natural selection to accumulate, in the same manner as man can accumulate in any given direction individual differences in his domesticated productions. (Darwin, 1859, p. 45, emphasis added)

In Chapter I, variation is understood as a condition for artificial selection. It is followed by the same rationale applied in Chapter II, "Variation Under Nature", as a condition for natural selection. Again, the correlation is clear and crystal.

13.8 "Wide Ranging, Much Diffused, and Common Species Vary Most"

On the topic of variation of species, Darwin states that in most polymorphic genera, "variability is independent of the conditions of life" and "of no service or disservice to the species" (Darwin, 1859, p. 46). However, he does not state here the fundamental importance of variation to evolution: when environmental conditions change, only some of the variants in the population will be adjusted to the new conditions and thus survive: "Hence I look at individual differences, though of small interest to the systematics, as of high importance for us" (Darwin, 1859, p. 51). Darwin did not quite understand the fundamental mechanisms we consider today but indeed perceived the importance of individual particularities as the starting point for speciation and evolution. The word "high" in the previous quote was increased to "the highest" in the last two editions of *Origin* (Bordalejo, 2012).

Still, Darwin provoked a profound shift in the understanding of variation. Since the Aristotelian tradition, variation has been seen as an abnormal developmental outcome. However, for evolution, variation is the norm and the first step towards population heterogeneity that may (in ultimate analysis) lead to speciation. He acknowledged variation as a rule and not an exception in the natural world. Cytologists Bruce Alberts and particularly Julian Lewis have brilliantly summarised this view:

> The whole of biology is a counterpoint between the two themes: astonishing variety in individual particularities; astonishing constancy in fundamental mechanisms. (Alberts et al., 2015, p.1)

In Darwin's thinking, doubtful forms are a transitional state, midway between mere varieties and distinct species. Though Darwin was not the first to believe in transformism, his book *On the Origin of Species* is a seminal work investigating this transformation's evidence and *modus operandi*. The idea of evolution by natural selection is relatively simple and can be laid down in only one paragraph. The principle presents three conditions to occur: heritability, variation, and a correlation between them. These conditions meet a scenario of limited resources and competition, leading to natural selection.

To Darwin's surprise, a simple definition of evolution through natural selection had been laid down in 1831, almost three decades before the publication of the *Origin* when Scottish naturalist Patrick Matthew summarised the principle in an appendix of his book *Naval Timber and Arboriculture* (Gould, 1985). This gentleman, however, recognised Darwin's authorship of the idea, as he later considered *On The Origin of Species* a more scientific and complete description of the phenomenon.

13.9 "Species of the Larger Genera in Any Country Vary More Than Species of the Smaller Genera"

Darwin ends the chapter discussing differences in variation found among species of larger genera and species of smaller genera. As he states in the chapter's header:

> [...] many of the species of the larger genera resemble varieties in being very closely but unequally related to each other, and in having restricted ranges. (Darwin, 1859, p. 44)

He conducts the discussion by writing:

> From looking at species as only strongly-marked and well-defined varieties, I was led to anticipate that the species of the larger genera in each country would oftener [larger than

> average] present varieties, than the species of the smaller genera; for wherever many closely related species have been formed, many varieties or incipient species ought, as a general rule, to be now forming. (Darwin, 1859, p. 55)

The rationale was straightforward for Darwin, as he argued for species transformism against the idea of looking "at each species as a special creation". For him, more variability in the past should mean more variation in the present. This is because, as stated before, Darwin believed natural selection to be too slow a process to be detected in a short period, such as a lifetime.

For Darwin, variation and the generation of varieties differ between smaller and larger genera. However, we now know that the mutation rate is similar in the eukaryotes in general and changes in environmental conditions are common and may quickly result in altered selective pressures. So, how do we reconcile this idea with Darwin's assumption that:

> [...] the species of the large genera [...] invariably present a larger average number of varieties than do the species of the smaller genera? (Darwin, 1859, p. 55)

Larger genera, as Darwin understood, evolved to present more variants because selection would still be favourable to variation in that genus. Thus, timing is of the essence for understanding Darwin's argument that "wherever many closed related species have been formed, many varieties or incipient species ought, as a general rule, to be now forming". This justifies Darwin's idea of larger genera presenting varieties with a larger average. Nowadays, natural selection is understood as a much more rapid process, as we have witnessed in the rapid molecular evolution of the coronavirus during the Covid19 pandemic. However, for Darwin, the idea of larger genera presenting more variants was to oppose transmutation to creationism, an argument repeated throughout the book.

He tested the truth of this anticipation with plants from 12 countries. James Costa (2009), in his *Facsimile of the First Edition of On the Origin of Species*, explains that Darwin used a "rule-of-three" proportional approach in comparing 1408 species from Babington's flora of Great Britain, that he divided into large genera (those with five or more species) and small genera (with four or fewer species). The varieties found in 663 species of the larger genera numbered 101. Thus, one would expect proportionally 113 varieties in the 745 species of the smaller genera (663:745:101: 113). However, in fact, only 89 varieties were found among the species of the smaller genera – fewer than the 113 expected. The higher proportion of species with varieties in larger genera was very significant to Darwin: such patterns make no sense if species are specially created and immutable. Today, however, statistical tests render a probability (p-value) of 14% for the deviation in this data, so chance could explain the different results on the groups of plants he studied. But the statistical tools were not available at the time. The definitions of varieties and the size of the genera were also arbitrary. However, Darwin was pushing for the "Divergence of Character" and the transmutation of species. James Costa (2009) also declares that today we would disagree with the statements that genus size correlates with species frequency. He further defines Darwin's botanical arithmetic: "It ties in with Darwin's view of how variation occurs: direct exposure to varying environmental conditions".

So, for Darwin, environmental variation leads to a heritable variation of characteristics. Detecting variation in larger genera downplays the idea of special creation. Therefore, before closing Chapter II, Darwin presents his first anti-creationist argument (others will be repeated throughout the book):

> [...] if we look at each species as a special act of creation, there is no apparent reason why more varieties should occur in a group having many species, than in one having few. (Darwin, 1859, p. 55)

Well, a creationist could think that bigger groups are fonder of god so that he could have increased their variability... This idea Darwin will consider in Chapter V a mockery and a deception. Ironies apart, it seems now interesting to speculate: Could Darwin anticipate that some 160 years later, the creationist antiscientific movement would still be so proficuous, despite all the knowledge evolutionary thinking has grown to accumulate?

References

Alberts, B., Johnson, A., Lewis, J., Morgan, D., & Raff, M. (2015). *Molecular biology of the cell* (6th ed.). Garland Science.

Bizzo, N. (2008). A teoria genética de Charles Darwin e sua oposição ao mendelismo. [Charles Darwin's genetic theory and his opposition to Mendelianism]. *Filosofia e História da Biologia, 3*, 317–333.

Bizzo, N., & El-Hani, C. (2009). O arranjo curricular do ensino de evolução e as relações entre os trabalhos de Charles Darwin e Gregor Mendel [The curricular arrangement of the teaching of evolution and the relationships between the works of Charles Darwin and Gregor Mendel]. *Filosofia e História da Biologia, 4*, 235–257.

Bordalejo, B. (Ed.) (2008–2012). Online variorum of Darwin's *origin of species*. In John van Wyhe (Ed.), *The complete work of Charles Darwin Online*. http://darwin-online.org.uk/Variorum/

Castañeda, L. (1992). *As ideias pré-mendelianas de herança e sua influência na teoria de evolução de Darwin* [Pre-Mendelian ideas of inheritance and their influence on Darwin's theory of Evolution] [Doctoral dissertation, State University of Campinas].

Costa, J. (2009). *The annotated origin: A facsimile of the first edition of on the origin of species*. Belknap Press/Harvard University Press.

Darwin, C. R. (1859). *On the origin of species by means of natural selection, or the preservation of favoured races in the struggle for life*. John Murray.

Endersby, J. (2003). Darwin on generation, pangenesis and sexual selection. In J. Hodge & G. Radick (Eds.), *The Cambridge companion to Darwin* (pp. 73–95). Cambridge University Press.

Gayon, J. (2003). From Darwin to today in evolutionary biology. In J. Hodge & G. Radick (Eds.), *The Cambridge companion to Darwin* (pp. 277–301). Cambridge University Press.

Gould, S. J. (1985). *The Flamingo smile*. Norton.

Krause, J., & Pääbo, S. (2016). Genetic time travel. *Genetics, 203*, 9–12.

Largent, M. A. (2009). Darwins' analogy between artificial and natural selection in the *origin of species*. In R. Richards & M. Ruse (Eds.), *The Cambridge companion to the "origin of species"* (pp. 14–29). Cambridge University Press.

Lorenzano, P. (2011). What would have happened if Darwin had known Mendel (or Mendel's work)? *History and Philosophy of Life Sciences, 33*, 3–48.

Mead, D., Radick, G., & Arney, K. (2019, April 11). Podcast: What would have happened if Darwin and Mendel had been on Twitter? In *Genetic literacy project: Science not ideology*. https://geneticliteracyproject.org/2019/04/11/podcast-what-would-have-happened-if-darwin-and-mendel-had-been-on-twitter/

Mendel, G. (1866). Versuche über Plflanzenhybriden. Verhandlungen des naturforschenden Vereines in Brünn, Bd. IV für das Jahr 1865. *Abhandlungen*, 3–47.

Olby, R. (2009). Variation and inheritance. In R. Richards & M. Ruse (Eds.), *The Cambridge companion to the "origin of species"* (pp. 30–46). Cambridge University Press.

Plassais, J., vonHoldt, B. M., Parker, H. G., Carmagnini, A., Dubos, N., Papa, I., Bevant, K., Derrien, T., Hennelly, L. M., Whitaker, D. T., Harris, A. C., Hogan, A. N., Huson, H. J., Zaibert, V. F., Linderholm, A., Haile, J., Fest, T., Habib, B., Sacks, B. N., et al. (2022). Natural and human-driven selection of a single non-coding body size variant in ancient and modern canids. *Current Biology, 32*(4), 889–897. https://doi.org/10.1016/j.cub.2021.12.036

Ruse, M. (1975). Charles Darwin and artificial selection. *Journal of the History of Ideas, 36*, 339–350.

Ruse, M. (2013). Teaching the classics: The origin of species as a case study. *Science & Education, 22*, 2255–2265.

Templeton, A. R. (2021). *Population genetics and microevolutionary theory* (2nd ed.). Wiley-Blackwell.

Roberto Rozenberg is a biologist. He received his PhD from the University of São Paulo in 2006, researching human genetics. He held a postdoc position in Biochemistry at the Institute of Chemistry of USP. As a molecular biologist, he worked in the fields of human genetics (inborn errors of metabolism) and biotechnology, performing directed evolution of cellulases for biofuels. He taught human evolution to psychology students and human genetics and anthropology at the Institute of Biosciences of the University of São Paulo.

Chapter 14
Origin's Chapter III: The Two Faces of Natural Selection

Robert J. Richards

Abstract Chapter III contains several puzzles and unexpected features. The first puzzle regards the chapter's relationship to Chapter IV: Natural Selection. Both chapters treat of natural selection, so what distinguishes them? Is it that Chapter IV indicates the intelligence behind nature's selections and Chapter III introduces the analog of intelligence? And is it that Chapter III suggests that natural selection performs an eliminative function, while Chapter IV shows the positive impact of selection? In Chapter IV, and in many subsequent chapters, natural selection, Darwin says, operates only for the good of each individual creature, though Chapter III suggests it destroys most individuals. Darwin asserts that he uses the term "struggle" metaphorically, which raises the question of the explanatory weight his metaphors carry, such metaphors as "the face of nature." There are a number of unexpected features of Darwin's analysis displayed in the chapter: for example, his introduction of a new concept of the selective environment and the several experiments he performs to establish his theory, e.g., from mathematical and thought experiments to controlled empirical experiments. Chapter III is a rich though puzzling chapter.

Chapters III ("Struggle for Existence") and IV ("Natural Selection") of Darwin's *Origin of Species* may be regarded as the central chapters of the entire book. If only these 2 chapters out of the 14 total chapters survived some catastrophe, we would still have the essence of Darwin's theory: its main argument, its significant features, and all the perplexities attendant thereto.

In Chapter I ("Variation under Domestication"), Darwin develops the model for natural selection. He draws a crucial distinction in the chapter between "methodical selection" and "unconscious selection." In methodical selection, regarded by Darwin as a more recent technique, a breeder sets out to alter a variety by focusing on a desired trait in a plant or animal and breeding from only those individuals exhibiting

R. J. Richards (✉)
University of Chicago, Chicago, IL, USA
e-mail: rjr6@uchicago.edu

M. Elice Brzezinski Prestes (ed.), *Understanding Evolution in Darwin's "Origin"*, History, Philosophy and Theory of the Life Sciences 34,
https://doi.org/10.1007/978-3-031-40165-7_14

the trait. In unconscious selection, a traditional mode of breeding, the breeder simply picks "the best" plants or animals among his stocks and breeds from them, without intending to alter the breed but actually doing so, gradually over many generations. Crafty Darwin does not simply provide a model for natural selection; "unconscious selection" is natural selection, in which the environment of the breeder's unintentional activity simply selects the best organisms for reproduction, just as nature does. That one of the factors of the selecting environment happens to be the unintentional action of a breeder is irrelevant to the structure and force of that environment. In Chapter IV, Darwin solidifies the identity of the unconscious selections made by man with the unconscious selections of nature (Darwin, 1859a, b, p. 102).

In Chapter II ("Variation under Nature"), Darwin explores the slight variations among individuals, the sliding differences between individuals and varieties, and the arbitrary differences between varieties and species, such that the insensible blending of one into the other "impresses the mind with the idea of an actual passage" (Darwin, 1859a, b, p. 50). So with the first two chapters of the book, Darwin has already, albeit implicitly, made the argument of descent through natural selection. Chapters III and IV make the argument explicit, showing how the changes from individual differences to varieties to species occurs; in other words, how those exquisite adaptations characterizing species arise in nature through natural selection.

At the beginning of Chapter III, Darwin introduces the idea of competition and "struggle for life," as depicted by Lyell and De Condole (Darwin, 1859a, b, p. 62), and immediately draws the conclusion that any variation, if in any degree profitable to the individual, "will tend to the preservation of that individual, and will generally be inherited by its offspring [...]. I have called this principle, by which each slight variation, if useful, is preserved, by the term of Natural Selection" (Darwin, 1859a, b, p. 61). Here, then, arises the first perplexity endemic to Darwin's theory of natural selection: What is principally preserved – the organism, or the trait in the organism, or the trait in the offspring, or the trait in the population? Presumably, those organisms that fare best in the struggle are preserved, and thus the advantageous trait is also preserved, and through inheritance, it is likewise preserved in the offspring, and as the competitive advantage continues to characterize the hereditary line, the trait will be preserved in the population, gradually changing the species. These several modes of preservation are implied but not made explicitly clear in Chapter III.

Two further questions may occur to the careful reader at this juncture: What are the sources of heritable traits? And how often does natural selection operate in a population, constantly or intermittently? Darwin investigates the first question in Chapter V ("The Laws of Variation"), where he allows for the inheritance of the direct effects of a constant climate and the habits of use and disuse – thus, the inheritance of acquired characteristics, a mode of inheritance that Darwin never relinquishes. He will propose another source of traits in Chapter V, which have become more identified with natural selection: traits that arise spontaneously, which he suggests are the result of changes in the environment acting on the sexual organs of parents (Darwin, 1859a, b, pp. 131–132).

As far as the frequency of the operation of natural selection, at the beginning of Chapter IV, Darwin estimates that useful variations in organisms might occur only a few times in thousands of generations. If variations are due to changes in climate, which Darwin initially proposes in his Lyellian fashion, then such variations would keep pace with climate and geological change, that is, occur very slowly and gradually over great lengths of time. Yet, Darwin also asserts in Chapter III that "natural selection ... is a power incessantly ready for action" (Darwin, 1859a, b, p. 61) and in Chapter IV affirms that claim in quasi-Biblical language:

> It may be said that natural selection is daily and hourly scrutinizing, throughout the world, every variation, even the slightest; rejecting that which is bad, preserving and adding all that is good; silently and insensibly working, whenever and wherever opportunity offers, at the improvement of each organic being. (Darwin, 1859a, b, p. 84)

The tension here between natural selection operating only a couple of times in "thousands of generations" and its "daily and hourly scrutinizing every variation" marks both the haste of composition and Darwin's initial commitment to the Lyllian sources of variation. At some point, the different timescales must have hit him since in the fifth edition of his book, the span of "thousands of generations" becomes "many successive generations" (Darwin, 1859a, b, p. 164). Still, even in Chapter III, Darwin suggests that natural selection operates incessantly when he observes that "we behold the face of nature bright with gladness," but "do not see or we forget, that the birds mostly live on insects or seeds, and are thus constantly destroying life" (Darwin, 1859a, b, p. 62). When Darwin assumes the sources of variation and the selecting environment are climate and geography, natural selection must operate slowly; when, however, he recognizes the selecting environment as that of other creatures, the relationships will constantly be altering, and thus natural selection will be daily and hourly scrutinizing variations.

The title of Chapter III, "Struggle for Existence," prompts two more questions: What does "struggle" mean? And with whom or what does an organism struggle? Darwin quickly points out early in the chapter that he means "the term Struggle for Existence in a large and metaphorical sense" (Darwin, 1859a, b, p. 62). He discriminates two fundamental meanings for struggle: a struggle to survive and a struggle to leave progeny. Two dogs, Darwin remarks, are said to struggle with each other over food to live, while a plant could be said to struggle to leave tastier seeds that attract more birds than competitors. So is the struggle properly the former or the latter? The two modes of struggle, of course, are related; but if the emphasis is on the struggle to increase progeny, then the struggle would necessarily be confined to members of the same species. Darwin oscillates between these two possibilities when he distinguishes, in Chapter IV, between natural and sexual selection: sexual selection, he supposes, is "less rigorous," since the result of female choice is "not death to the unsuccessful competitor but few or no offspring" (Darwin, 1859a, b, p. 88).

Darwin nicely calibrates the strength of the struggle occurring each generation because of the geometrical increase in the number of offspring. The struggle will be strongest, he concludes, between individuals of the same species, since they attempt to utilize the same environmental resources. It will be less strong between individuals of different species (e.g., the rabbit and the fox) – yet here is where Darwin is undone by his focus on the principal denotation of the term "struggle": according to the logic of his theory, the struggle is not between two different species but among rabbits to outrun the fox so as to leave more offspring. A comparable case arises when he refers to the struggle between an individual and the environment, as when a plant at the edge of the desert struggles to obtain moisture. But in these two cases (that of the rabbit and the desert plant), Darwin himself struggled with the metaphor and lost the battle. The logic of his analysis indicates that in the examples mentioned, the struggle must be with other individuals of the same species to leave more progeny.

In his Malthusian moment, Darwin observes that the struggle for existence "inevitably follows from the high rate at which all organic beings tend to increase" (Darwin, 1859a, b, p. 63). And to demonstrate to himself and to his reader the power of exponential increase, he calculates the number of descendants of a plant producing just two seeds, with each of its progeny doing likewise; in 20 years, there would be over a million descendants (i.e., $2^{20} = 1{,}048{,}576$). Likewise, with elephants, one pair can produce six caves during 60 years of breeding life; after 500 years, that original pair would have produced 15 million descendants. Since the world is not filled with elephants, the vast majority must be checked by disease, accidents, and predators. In recognition of the checks against over-reproduction, Darwin deploys that startling metaphor: "the face of nature bright with gladness" hides a great deal of death and destruction (Darwin, 1859a, b, p. 62).

It is often assumed that Darwin's long argument in the *Origin* depended only on observation and warranted assumption, certainly not on an experiment. However, as if to confirm his calculations and ground the metaphor of struggle, the empirically oriented Darwin performed two experiments, which he records in Chapter III. In the first, he cleared a plot of ground 3 ×2 feet and counted the number of weeds that initially sprouted and those that ultimately survived. Of the 357 that came up, 295 were destroyed by slugs and insects, thus showing an 83% culling of plants. In another experiment, in which he cleared a three-by-four-foot plot of ground, out of 20 species of plants, nine of those species died off from being crowded out by the other freely growing plants (Darwin, 1859a, b, pp. 67–68).

Darwin subtly shifts from considering the competition among creatures, one on one as it were, to provide examples of more complicated checks on the production of organisms. This important shift alters Darwin's perspective from the selecting environment being climate and geography to the selecting environment being that of other organisms. The most intricate and amusing of these cases is the way cats in a region surprisingly cause the flourishing of red clover and heartsease. Darwin observes that these plants are fertilized by humble-bees, whose ground nests are predated by mice; cats reduce the mouse population, thus facilitating the activities of the humble-bees and the flourishing of the plants they help to pollinate (Darwin, 1859a, b, pp. 73–74).

In recognition of the complexity of the environment, Darwin draws what he takes to be a "corollary of the highest importance," namely, the structure of every organic being is related "in the most essential yet often hidden manner" to all the other organisms with which they compete (Darwin, 1859a, b, p. 77). This is an insight likely derived from Georges Cuvier, whose principle of "conditions of existence" indicates that organisms are integrated into their environment by their several adaptations. Later in Chapter VI, Darwin explicitly credits Cuvier with this insight, which he thinks natural selection neatly explains (Darwin, 1859a, b, 206). The corollary he discriminates in Chapter III suggests that organisms become mirrors of their past selecting environments.

After providing this example of the complex relationship characteristic of the "web of life," he introduces (Darwin, 1859a, b, p. 74) a metaphor, that of an "entangled bank," with which he will also bring the long argument of his book to a close (Darwin, 1859a, b, p. 489). In the third chapter, this metaphor of the entangled bank immediately suggests another feature of the environment, its determinate, and lawful character: throw a handful of feathers in the air, Darwin (1859a, b, p. 75) muses, "all must fall to the ground according to definite laws," even though those laws might be obscure. Finally, at the end of the book, he ties these metaphors together when he concludes that significant laws governing the entangled bank have now been revealed in the light of his theory.

The metaphor of the entangled bank may have come from Milton's *Paradise Lost* (4.ll.172–301), a poem Darwin took with him on the *Beagle* and knew almost by heart: in the poem, as Satan approaches the Garden of Eden, he is momentarily stopped by an entangled bank, over which he flies and lands on the Tree of Life, where he mediates not on true life but "sat devising death." The Christian sentiment of Milton's great poem is that out of death comes life more abundantly – a sentiment sustained in Darwin's book as well: out of the death and destruction wrought by natural selection come abundant varieties of life.

The metaphors Darwin deploys in the third chapter, as well as his exponential calculations, bring him to a startling discovery: the face of nature may be bright with gladness, but that face hides a great deal of death and destruction. He makes that recognition brutally evident when he uses the metaphor in a graphic way: "The face of Nature may be compared to a yielding surface, with ten thousand sharp wedges packed close together and driven inwards by incessant blows, sometimes one wedge being struck, and then another with greater force" (Darwin, 1859a, b, p. 67).

The recognition of the destructive forces of nature is pitted against an attitude that is expressed throughout the *Origin* and epitomized by a remark that appears in the penultimate paragraph of the book: "as natural selection works solely by and for the good of each being, all corporeal and mental endowments will tend to progress towards perfection" (Darwin, 1859a, b, 489). A version of that assessment – that natural selection works for the good of each individual and is an instrument of progress – appears in at least five other places in the book (Darwin, 1859a, b, pp. 83, 84, 149, 194, 201). Against this quite benevolent attitude, the very logic of his conception of a struggle for existence must have struck Darwin with a confusing blow: natural selection does not work for the benefit of each individual; rather, it eliminates most individuals or inhibits their reproduction. In the very last sentence of the third chapter, he attempts to mitigate these lethal consequences of natural selection:

> When we reflect on this struggle, we may console ourselves with the full belief, that the war of nature is not incessant, that no fear is felt, that death is generally prompt, and that the vigorous, the healthy and the happy survive and multiply. (Darwin, 1859a, b, p. 79)

The logic of Darwin's conception of natural selection as an instrument of struggle could not defeat a deeper and more embedded attitude that becomes more fully expressed in the last paragraph of the book, namely, that the "law" of natural selection, as Darwin refers to it, is a surrogate for the hand of the Creator (Darwin, 1859a, b, p. 488). Darwin denied that God directly intervenes to create species as we currently see them, but he does assume that God created the laws by which those species have evolved. In that sense, the laws become "secondary causes," receiving their direction from the primary Cause, the Supreme Being. This is a long-held view of Darwin's, one that he expressed, for example, in Notebook E (1838) of his *Transmutation Notebooks* (Darwin, 1987, p. 409): "Man is one great object, for which the world was brought into present state" (See also Richards, 2013, pp. 39–50). Several Darwin-scholars have rejected the notion that God plays any role in Darwin's science, and they do cartwheels in attempting to deny the plain meaning of Darwin's text. Elliot Sober, for example, dismisses Darwin's language of evolutionary laws as "secondary causes" in respect of God being the primary cause. Sober contends that when Darwin uses that language, he was mounting an "argument for the existence of God," which was a *philosophical* use of his scientific theory; the notion of God as a primary cause didn't penetrate or shape the science itself (Sober, 2011, p. 128). But there is no hint Darwin was trying to prove God's existence; he was assuming it.

When Darwin started writing Chapter III in the spring of 1857, it was going to be entitled "On Natural Selection" (Darwin, 1975, note by Stauffer, p. 172). However, during the course of the writing, he obviously changed his mind and called it "The Struggle for Existence" and added the succeeding chapter, now called "Natural Selection." Why are there then two chapters on natural selection? The answer is not obvious, but I have the following hypothesis.

In his essay of 1842, a 35-page pencil sketch of his theory, and in the extended fair copy of 1844, Darwin moves from briefly considering variations in nature to exploring the work of a superpowerful breeder, an imaginary being that could far outdo man in his methodical selection of animals and plants; this being, unlike man, could see into the interior of creatures and select for imperceptible, advantageous traits. Only after he had characterized this "infinitely sagacious" being he asks if there is anything in nature as the analog to the ideal being. Then, in answer to his own question, he introduces the Malthusian idea of checks to exponential reproduction via the struggle of organisms to survive. Thus, in nature, it's the struggle for existence that does the work of the infinite, ideal selector. But if that were the whole account of the operations of natural selection, it would run against Darwin's abiding conviction that natural selection works for the "good of each individual."

Moreover, it would only be regarded as a destructive force, not a positive, creative force. The ideal selector, as Darwin depicted him in the essays, is such a creative force. Here's the description from the 1844 essay:

> Let us now suppose a Being with penetration sufficient to perceive differences in the outer and innermost organization quite imperceptible to man, and with forethought extending over future centuries to watch with unerring care and select for any object the offspring of an organism produced under the foregoing circumstances; I can see no conceivable reason why he could not form a new race [...] adapted to new ends. As we assume his discrimination, and his forethought, and his steadiness of object to be incomparably greater than those qualities in man, so we may suppose the beauty and complications of the new races and their differences from the original stock to be greater than in the domestic races produced by man's agency. (Darwin, 1909, p. 85)

This Being appears again in Chapter IV of the *Origin*:

> Man can act only on external and visible characters: nature care nothing for appearances, except in so far as they may be useful to any being. She can act on every internal organ, on every shade of constitutional difference, on the whole machinery of life. Man selects only for his own good; Nature only for that of the being which she tends. (Darwin, 1859a, b, p. 83)

In the manuscript of the *Origin*, Darwin adds a line to this passage: "By nature, I mean the laws ordained by God to govern the Universe" (Darwin, 1987, p. 224). In Darwin's conception, then, nature is also a creative force, a surrogate for God, that acts altruistically (acting only for the good of the being which she tends). This means that the creative feature of natural selection is not a mechanistic algorithmic process, as Daniel Dennett (1995, pp. 48–60) proposes, or modeled on the mechanical operations of a Newcomen pump, as Michael Ruse argues (see this volume). The model that Darwin employs for this aspect of natural selection is that of an intelligent, altruistic, and creative mind.

Thus, my hypothesis: Darwin composed two chapters on natural selection to give due weight to both of its aspects, nature as a destructive force and nature as a creative force. The "face of nature bright with gladness" is, consequently, an authentic face, though it sometimes is hidden by its other face, that of death and destruction.

References

Darwin, C. R. (1859a). *On the origin of species*. Murray.

Darwin, C. R. (1859b). In M. Peckham (Ed.), *The origin of species: A variorum text*. University of Philadelphia Press.

Darwin, C. R. (1909). In F. Darwin (Ed.), *Foundations of the origin of species: Two essays written in 1842 and 1844*. Cambridge University Press.

Darwin, C. R. (1975). In R. C. Stauffer (Ed.), *Natural selection: Being the second part of his big species book written from 1856 to 1858*. Cambridge University Press.

Darwin, C. R. (1987). Charles Darwin's notebooks, 1836–1844. In P. H. Barrett, P. J. Gautrey, S. Herbert, D. Kohn, & S. Smith (Eds.), *Charles Darwin's notebooks, 1836–1844*. Cornell University Press.

Dennett, D. (1995). *Darwin's dangerous idea: Evolution and the meaning of life*. Simon and Schuster.

Richards, R. J. (2013). Darwin's theory of natural selection and its moral purpose. In R. J. Richards (Ed.), *Was Hitler a Darwinian? Disputed questions in the history of evolutionary theory* (pp. 13–44). University of Chicago Press.

Sober, E. (2011). *Did Darwin write the origin backwards? Philosophical essays on Darwin's theory*. Prometheus Books.

Robert J. Richards is the Morris Fishbein Distinguished Service Professor of the History of Science at the University of Chicago. His area of research is German Romanticism and the history of evolutionary theory. He is the author of several books, including *The Romantic Conception of Life: Science and Philosophy in the Age of Goethe* (2002); *The Tragic Sense of Life: Ernst Haeckel and the Struggle over Evolutionary Thought* (2008); and, with Michael Ruse, *Debating Darwin* (2016). Currently, he is writing an intellectual biography of Carl Gustav Carus.

Chapter 15
Origin's Chapter IV: The Newton of the Blade of Grass

Michael Ruse

Abstract Charles Darwin, the leading evolutionist, introduces and discusses his key mechanism, natural selection, in Chapter IV of his *On the Origin of Species* (1859). He shows how the mechanism follows from the struggle for existence, together with random variation, and he argues that it not only explains change, but change in the direction of features of adaptive worth. He introduces the secondary mechanism of sexual selection and then, through examples, shows how selection might be expected to work in nature. He argues, based on the economic notion of the division of labor, that this shows how selection can lead to there being different species. He ends the chapter by showing how the overall result is a tree of nature, with the primitive at the bottom and the complex at the top.

15.1 Introduction

Chapter IV of the *Origin of Species*, "Natural Selection," is the key chapter in the whole book. Fully to understand its content and structure, we should think back to Darwin's model, his guide. Under the influence of the leading philosopher-scientists of his day, John F. W. Herschel and William Whewell, Darwin aspired to be what Immanuel Kant, in his *Critique of Judgement* (1790), had declared impossible: "The Newton of the Blade of Grass" (Ruse, 1975a). Darwin's ideal, everybody's ideal, was Isaac Newton's *Philosophiæ Naturalis Principia Mathematica* (Mathematical Principles of Natural Philosophy). In that work, Newton gave a causal explanation – with the force of gravitational attraction at its heart – more formally, with the law of gravity as one of the axioms in what philosophers today call "hypothetico-deductive systems," and Kepler's and Galileo's laws as deductive consequences. He gave a scientific explanation of the Copernican system, the sun at the center, planets, including Earth, in orbits around it. In the *Origin*, most particularly in Chapter IV,

M. Ruse (✉)
Florida State University, Tallahassee, FL, USA
e-mail: mruse@fsu.edu

M. Elice Brzezinski Prestes (ed.), *Understanding Evolution in Darwin's "Origin"*,
History, Philosophy and Theory of the Life Sciences 34,
https://doi.org/10.1007/978-3-031-40165-7_15

Darwin set out to do the same for biology. A cause, natural selection, as an axiom in a hypothetico-deductive system – admittedly a lot less formal than Newton's – explains the consequences, namely, the evolutionary tree of life (Ruse, 1975b). Selection begins the chapter; the tree ends it.

15.2 Natural Selection

Take the discussions of the chapter in order. Having, in previous chapters, given the analogy of selection in the hands of the breeder and established the fact of universal variation in natural populations, going on then to the Malthusian struggle for existence, Darwin was ready to introduce his central cause, the force of natural selection:

> HOW will the struggle for existence, discussed too briefly in the last chapter, act in regard to variation? Can the principle of selection, which we have seen is so potent in the hands of man, apply in nature? I think we shall see that it can act most effectually. Let it be borne in mind in what an endless number of strange peculiarities our domestic productions, and, in a lesser degree, those under nature, vary; and how strong the hereditary tendency is. Under domestication, it may be truly said that the whole organisation becomes in some degree plastic. Let it be borne in mind how infinitely complex and close-fitting are the mutual relations of all organic beings to each other and to their physical conditions of life. Can it, then, be thought improbable, seeing that variations useful to man have undoubtedly occurred, that other variations useful in some way to each being in the great and complex battle of life, should sometimes occur in the course of thousands of generations? If such do occur, can we doubt (remembering that many more individuals are born than can possibly survive) that individuals having any advantage, however slight, over others, would have the best chance of surviving and of procreating their kind? On the other hand, we may feel sure that any variation in the least degree injurious would be rigidly destroyed. This preservation of favourable variations and the rejection of injurious variations, I call Natural Selection. (Darwin, 1859, pp. 81–82)

Change, but change of a particular kind, in the direction of "adaptive advantage." In Chapter III, Darwin had asked:

> How have all those exquisite adaptations of one part of the organisation to another part, and to the conditions of life, and of one distinct organic being to another being, been perfected? We see these beautiful co-adaptations most plainly in the woodpecker and missletoe; and only a little less plainly in the humblest parasite which clings to the hairs of a quadruped or feathers of a bird; in the structure of the beetle which dives through the water; in the plumed seed which is wafted by the gentlest breeze; in short, we see beautiful adaptations everywhere and in every part of the organic world. (Darwin, 1859, pp. 60–61)

Now he had his answer. Natural selection!

At once, the criticisms started. The codiscoverer of natural selection, Alfred Russel Wallace, opined that the term "natural selection" was unduly anthropomorphic and that it would be better to talk about "survival of the fittest." In a letter dated July 2, 1866, he refers to a critic who:

> [. . .] by an extract from which I see that he considers your weak point to be, that you do not see that "thought & direction are essential to the action of 'Nat. Selection'." The same objection has been made a score of times by your chief opponents, & I have heard it as often stated myself in conversation.

Continuing:

> Now I think this arises almost entirely from your choice of the term "Nat. Selection" & so constantly comparing it in its effects, to Man's selection, and also to your so frequently personifying Nature as "selecting" as "preferring" as "seeking only the good of the species" &c. &c. To the few, this is as clear as daylight, & beautifully suggestive, but to many it is evidently a stumbling block. I wish therefore to suggest to you the possibility of entirely avoiding this source of misconception in your great work, (if not now too late) & also in any future editions of the "Origin," and I think it may be done without difficulty & very effectually by adopting Spencer's term (which he generally uses in preference to Nat. Selection) viz. "Survival of the fittest." (DCP-LETT-5140)

Although, in later editions of the *Origin*, Darwin did start using the term "survival of the fittest," already in the third edition of the *Origin* (1861) he had given his answer. Answering a list of criticisms of his work, Darwin continued:

> Others have objected that the term selection implies conscious choice in the animals which become modified; and it has even been urged that as plants have no volition, natural selection is not applicable to them! In the literal sense of the word, no doubt, natural selection is a misnomer; but who ever objected to chemists speaking of the elective affinities of the various elements?—and yet an acid cannot strictly be said to elect the base with which it will in preference combine. It has been said that I speak of natural selection as an active power or Deity; but who objects to an author speaking of the attraction of gravity as ruling the movements of the planets? Every one knows what is meant and is implied by such metaphorical expressions; and they are almost necessary for brevity. So again it is difficult to avoid personifying the word Nature; but I mean by Nature, only the aggregate action and product of many natural laws, and by laws the sequence of events as ascertained by us. With a little familiarity such superficial objections will be forgotten. (Darwin, 1861, pp. 84–85)

Notwithstanding the fact that versions of this criticism keep occurring even unto the present, Darwin's point was well taken. Science uses metaphor all the time: attraction, repulsion, adaptive landscape, arms race, and Oedipus complex. And as students of metaphor have long realized, metaphors are not just shorthand for ideas that can be expressed literally; they bring added or extra understanding of the situation. Above all, they are heuristic (Hesse, 1966; Ruse, 2022a). To speak of the interaction between two species as an arms race encourages you to think in terms of what human arms races involve – not just brute force but intelligence, computers, and the like. Do we find something similar in the nonhuman world? In the case of natural selection, it drives us to look for the features being selected and to ask why they are selected – namely, they help in the struggle for existence. We are encouraged to think teleologically, in terms of "final cause." A feature is selected. Why? What end or purpose will it serve?

Kant, in the *Third Critique*, although explicitly rejecting evolution, realized this. You cannot think about organisms without thinking about the final cause. What purpose or end does the eye, the hand, the heart serve? However, so taken was Kant with the Newtonian rejection of final causes, he concluded that final-cause thinking in biology, although necessary, is in a sense a sign of weakness. He spoke of regulative principles, with heuristic value, rather than acceptable constitutive principles. This is why he was led to such a demeaning conclusion about biology as a science. Darwin's genius was to speak to this. He did not deny final cause. He thought it was as important as did Kant. The eye does exist in order to see, and that is as true of the newborn baby who has not yet opened its eyes as it is for us adults. However, whereas in the past, final cause meant stepping outside mechanism – blind law operating endlessly – and adopting either the Platonic solution of a mind creating and directing things or the Aristotelian solution of special forward-looking forces, Bergson's *élans vitaux*, he gave an explanation withing the machine metaphor paradigm, mechanism. We think the eye of the newborn exists in order to see because generations before have used their eyes for exactly that reason. We could be wrong. The world might be cast into darkness where no one can see. This is a risk we take, although, of course, both Platonist and Aristotelian face the same nonfunctioning prospect. Darwin completed the Scientific Revolution. He showed how to explain organisms in a neo-Newtonian fashion – teleology without tears (Ruse, 2017, 2021).

15.3 Sexual Selection

Moving on through Chapter IV, we come to Darwin's secondary mechanism, sexual selection:

> This depends, not on a struggle for existence, but on a struggle between the males for possession of the females; the result is not death to the unsuccessful competitor, but few or no offspring. Generally, the most vigorous males, those which are best fitted for their places in nature, will leave most progeny. But in many cases, victory will depend not on general vigour, but on having special weapons, confined to the male sex. A hornless stag or spurless cock would have a poor chance of leaving offspring.

Adding:

> Amongst birds, the contest is often of a more peaceful character. All those who have attended to the subject, believe that there is the severest rivalry between the males of many species to attract by singing the females. The rock-thrush of Guiana, birds of Paradise, and some others, congregate; and successive males display their gorgeous plumage and perform strange antics before the females, which standing by as spectators, at last choose the most attractive partner. (Darwin, 1859, pp. 88–89)

Starting with Wallace, many have been less than impressed by sexual selection. In the same vein as his criticism of natural selection, Wallace felt it was altogether too anthropomorphic. Generally, sexual selection is (in line with the quotations above) divided into "male combat" and "female choice," and Wallace argued that sexual selection through female choice supposed, illicitly, that animals would have the same standards of beauty as humans. These kinds of worries meant that, for a 100 years after the *Origin*, professional scientists downplayed the mechanism, at least as something distinct from natural selection (Milam, 2011). Inasmuch as it worked, it was natural selection, and inasmuch as it didn't, it wasn't. In the last 50 years or so, however, with the rise of evolutionary studies of social behavior – "sociobiology" – sexual selection has made a strong comeback, and your average professional evolutionist today is quite comfortable with the concept.

There are a couple of points of interest about sexual selection. First, note that it occurs within the species. It is, therefore, a matter of struggle between individuals and not groups – "individual selection" rather than "group selection" (Ruse, 2022b). As Darwin said in Chapter III:

> [. . .] as more individuals are produced than can possibly survive, there must in every case be a struggle for existence, either one individual with another of the same species, or with the individuals of distinct species, or with the physical conditions of life. (Darwin, 1859, p. 63)

Note, therefore, for Darwin, groups competing are not necessarily manifesting group selection in the sense of the group, as a whole, competing, having individuals within the group altruistically giving for the good of the group despite no return. Generally, groups competing are truly engaged in individual selection because it is the individuals in groups who compete and win or lose. Later in the *Origin* (Chapter VII), Darwin was careful to emphasize that by individual, he included interrelated social groups. There is no genuine disinterested altruism here because non-reproducers reproduce by proxy, as it were, through the success of the reproducers. Today's evolutionists refer to this as "kin selection":

> [S]election may be applied to the family, as well as to the individual, and may thus gain the desired end. Thus, a well-flavoured vegetable is cooked, and the individual is destroyed; but the horticulturist sows seeds of the same stock, and confidently expects to get nearly the same variety; breeders of cattle wish the flesh and fat to be well marbled together; the animal has been slaughtered, but the breeder goes with confidence to the same family. I have such faith in the powers of selection, that I do not doubt that a breed of cattle, always yielding oxen with extraordinarily long horns, could be slowly formed by carefully watching which individual bulls and cows, when matched, produced oxen with the longest horns; and yet no one ox could ever have propagated its kind. (Darwin, 1859, pp. 237–238)

This meant, incidentally, that the sterility of hybrids like the mule could not have been caused by selection, specifically for the benefit of the parent species (horse and donkey), who would not have second-rate, reproducing offspring. Instead, it must be an accident of two different reproductive systems not working harmoniously together.

> The importance of the fact that hybrids are very generally sterile, has, I think, been much underrated by some late writers. On the theory of natural selection the case is especially important, inasmuch as the sterility of hybrids could not possibly be of any advantage to them, and therefore could not have been acquired by the continued preservation of successive profitable degrees of sterility. I hope, however, to be able to show that sterility is not a specially acquired or endowed quality, but is incidental on other acquired differences. (Darwin, 1859, p. 245)

The reason why Darwin was so adamant about individual over group selection (modern terms and not used by him or others back then) was clearly, as today, chiefly a function of not being able to see how group selection could work. Pure altruists would be quickly eliminated by those who took advantage of their labors while offering nothing in return. It would be idle to deny, however, that social factors may have played a role. Charles Darwin came from an upper-middle-class family. His father was a physician, Silicon Valley rich because Darwin's maternal grandfather – also, incidentally, the grandfather of Emma, Darwin's first cousin and wife – was Josiah Wedgwood, of pottery fame. The philosophy of such industrialists was very much that of the Scottish political economist Adam Smith:

> It is not from the benevolence of the butcher, the brewer, or the baker that we expect our dinner, but from their regard to their own self-interest. We address ourselves not to their humanity but to their self-love, and never talk to them of our own necessities, but of their advantages. (Smith, 1777, book 1, chapter 2)

Belief in the primacy of individual selection follows at once.

Wallace, of a social standing far lower, barely in the middle classes, was – thanks to hearing a lecture by Robert Owen – a socialist from early years. Group selection, for him, was as natural as individual selection for Darwin. Not that Wallace's position cut any ice with Darwin:

> Let me first say that no man could have more earnestly wished for the success of N. selection in regard to sterility, than I did; & when I considered a general statement, (as in your last note) I always felt sure it could be worked out, but always failed in detail. The cause being as I believe, that natural selection cannot effect what is not good for the individual, including in this term a social community. (Darwin, 1985–, Vol. 16, p. 374; letter to Wallace April 6, 1868)

Wallace rarely, if ever, cut any ice with Darwin. Where they really fell out was over human evolution, with Wallace arguing that here we must appeal to spiritual forces. Wallace claimed that:

> [. . .] the mental requirements of savages, and the faculties actually exercised by them, are very little above those of animals. The higher feelings of pure morality and refined emotion, and the power of abstract reasoning and ideal conception, are useless to them, are rarely if ever manifested, and have no important relations to their habits, wants, desires, or well-being. They possess a mental organ beyond their needs.

Adding, in explanation:

> Natural Selection could only have endowed savage man with a brain a little superior to that of an ape, whereas he actually possesses one very little inferior to that of a philosopher. (Wallace, 1870, pp. 355–356).

Countering this, in the *Descent*, Darwin tackled the evolution of morality, making clear that not only is it natural selection at work, but it is individual selection at work:

> It must not be forgotten that although a high standard of morality gives but a slight or no advantage to each individual man and his children over the other men of the same tribe, yet that an advancement in the standard of morality and an increase in the number of well-endowed men will certainly give an immense advantage to one tribe over another.

> There can be no doubt that a tribe including many members who, from possessing in a high degree the spirit of patriotism, fidelity, obedience, courage, and sympathy, were always ready to give aid to each other and to sacrifice themselves for the common good, would be victorious over most other tribes; and this would be natural selection. (Darwin, 1871, Vol. 1, p. 166)

"Victorious over most other tribes"? Surely this is an appeal to group selection? Not at all! Immediately before this passage, Darwin implies that (what today is known as) "reciprocal altruism" is a major causal factor. You scratch my back, and I will scratch yours:

> [...] as the reasoning powers and foresight of the members [of a tribe] became improved, each man would soon learn from experience that if he aided his fellow-men, he would commonly receive aid in return. (Darwin, 1871, Vol. 1, p. 163).

This is not the disinterested altruism of group selection. The individual alone is benefiting: individual selection. (There is also a veiled appeal to kin selection. The members of a tribe are interrelated, or think they are, and so help for others is – or is thought to be – help for oneself.)

The second point of interest about sexual selection is that, although professional scientists of the day were not that impressed by it – generally they were not that impressed by natural selection, being more interested in morphology and paleontology, two subjects with little use for mechanisms – the general public, especially through novelists and poets, took it up with enthusiasm. George Eliot's *Daniel Deronda* (1876) is all about choosing appropriate spouses, or not, as the case might be. It was not just the posh literary folk who turned to Darwin. At the end of *Tarzan of the Apes*, published in 1912, Darwin enthusiast Edgar Rice Burroughs has Jane caught in the throes of sexual selection. Sensibly, she suppresses "the psychological appeal of the primeval man to the primeval woman in her nature" and, following her Darwinian nature, makes the wise decision to marry the apparent Lord Greystoke (William Cecil Clayton) instead of the (unacknowledged) true Lord Greystoke (Tarzan).

> Did not her best judgment point to this young English nobleman, whose love she knew to be of the sort a civilized woman should crave, as the logical mate for such as herself?
>
> Could she love Clayton? She could see no reason why she could not. Jane was not coldly calculating by nature, but training, environment and heredity had all combined to teach her to reason even in matters of the heart. (Burroughs, 1912/ 2008, p. 340)

Fortunately, in the next novel in the series, Jane sees reason to revise her decision and we are all set for 25 sequels.

In the same mode, turn to Constance Naden and her lightweight frolic, written about 1885. Titled "Natural Selection," it is really about sexual selection:

> I HAD found out a gift for my fair, I had found where the cave men were laid: Skulls, femur and pelvis were there, And spears that of silex they made.
>
> But he ne'er could be true, she averred, Who would dig up an ancestor's grave— And I loved her the more when I heard Such foolish regard for the cave.
>
> My shelves they are furnished with stones, All sorted and labelled with care; And a splendid collection of bones, Each one of them ancient and rare;

One would think she might like to retire To my study— she calls it a "hole"! Not a fossil I heard her admire But I begged it, or borrowed, or stole.

But there comes an idealess lad, With a strut and a stare and a smirk; And I watch, scientific, though sad, The Law of Selection at work.

Of Science he had not a trace, He seeks not the How and the Why, But he sings with an amateur's grace, And he dances much better than I.

And we know the more dandified males By dance and by song win their wives— 'Tis a law that with avis prevails, And ever in Homo survives.

Shall I rage as they whirl in the valse? Shall I sneer as they carol and coo? Ah no! for since Chloe is false I'm certain that Darwin is true. (Naden, 1999, pp. 207–208)

15.4 Selection at Work

Moving on through Chapter IV, next we have Darwin offering some pretend cases of selection at work to illustrate the plausibility of his mechanism:

> Let us take the case of a wolf, which preys on various animals, securing some by craft, some by strength, and some by fleetness; and let us suppose that the fleetest prey, a deer for instance, had from any change in the country increased in numbers, or that other prey had decreased in numbers, during that season of the year when the wolf is hardest pressed for food. I can under such circumstances see no reason to doubt that the swiftest and slimmest wolves would have the best chance of surviving, and so be preserved or selected,—provided always that they retained strength to master their prey at this or at some other period of the year, when they might be compelled to prey on other animals. I can see no more reason to doubt this, than that man can improve the fleetness of his greyhounds by careful and methodical selection, or by that unconscious selection which results from each man trying to keep the best dogs without any thought of modifying the breed. (Darwin, 1859, pp. 90–91)

It is interesting and surely significant that Darwin never really tried to find actual examples, leading to the strong suspicion that he always thought the action of natural selection was too slow to be recorded in nature. Adding to the suspicion is the fact that, after the *Origin,* Darwin went off to look at orchids and climbing plants and worms. Never, as was possible for a man of his means, did he set up or sponsor attempts to find or simulate the working of natural selection in observable time. And, confirming the supposition, surely, is the fact that, when Wallace's collecting partner in Brazil, Henry Walter Bates, came up with his theory of mimicry thanks to bird predation on butterflies, although Darwin was excited and grateful, to the extent of getting Bates a job as Secretary to the Royal Geographical Society, only toward the end of later editions of the *Origin* did he introduce Bates's findings. They had no place in Chapter IV.

Truly remarkable is that, late in Darwin's life, an enthusiastic butterfly collector, a civil servant, not a professional scientist, sent the great naturalist a letter documenting the action of natural selection in observable time:

> My dear Sir,
>
> The belief that I am about to relate something which may be of interest to you, must be my excuse for troubling you with a letter.

> Perhaps among the whole of the British Lepidoptera, no species varies more, according to the locality in which it is found, than does that Geometer, Gnophos obscurata. They are almost black on the New Forest peat; grey on limestone; almost white on the chalk near Lewes; and brown on clay, and on the red soil of Herefordshire.
>
> Do these variations point to the "survival of the fittest"? I think so. It was, therefore, with some surprise that I took specimens as dark as any of those in the New Forest on a chalk slope; and I have pondered for a solution. Can this be it?
>
> It is a curious fact, in connexion with these dark specimens, that for the last quarter of a century the chalk slope, on which they occur, has been swept by volumes of black smoke from some lime-kilns situated at the bottom: the herbage, although growing luxuriantly, is blackened by it.
>
> I am told, too, that the very light specimens are now much less common at Lewes than formerly, and that, for some few years, lime-kilns have been in use there.
>
> These are the facts I desire to bring to your notice.
>
> I am, Dear Sir, Yours very faithfully,
>
> A. B. Farn. (Darwin, 1985–, Vol. 26, p. 440; letter from Albert Brydges Farn on November 18, 1878)

One might have expected Darwin to bring out a new edition of the *Origin* with this letter printed opposite the title page. But, as far as is known, he never responded.

Part of what is going on here is that Darwin's definitive professional biological work was as a morphologist, studying the nature of barnacles. You can then use morphology to ferret out the past and its paths, giving flesh to studies of fossils and paleontology. The important point about morphology, however, is that natural selection is very much not needed because, generally, the morphologist is trying to ignore adaptation and get to the underlying shared body plans, "archetypes." Note that this very much does not imply that overall natural selection is unimportant. In Chapter (VI), Darwin makes it clear that he thinks archetypes, giving rise to patterns of similarity, what Richard Own called "homologies," are explained ultimately by selection:

> It is generally acknowledged that all organic beings have been formed on two great laws—Unity of Type, and the Conditions of Existence. By unity of type is meant that fundamental agreement in structure, which we see in organic beings of the same class, and which is quite independent of their habits of life. On my theory, unity of type is explained by unity of descent. The expression of conditions of existence, so often insisted on by the illustrious Cuvier, is fully embraced by the principle of natural selection. For natural selection acts by either now adapting the varying parts of each being to its organic and inorganic conditions of life; or by having adapted them during long-past periods of time: the adaptations being aided in some cases by use and disuse, being slightly affected by the direct action of the external conditions of life, and being in all cases subjected to the several laws of growth. Hence, in fact, the law of the Conditions of Existence is the higher law; as it includes, through the inheritance of former adaptations, that of Unity of Type. (Darwin, 1859, pp. 206–207)

We can see how, paradoxically, as a professional scientist – as a morphologist – natural selection was not of vital importance to Darwin. We can also see why, at the immediate level, Darwin did not think that every characteristic had to be adaptive. This said, as the passage just quoted shows well, in the end, everything comes down to natural selection.

15.5 Division of Labor

Moving on briskly through Chapter IV, Darwin introduces Adam Smith's notion of the "division of labor." Things work much more efficiently if, rather than everyone aspiring to be a jack of all trades, they take up individual tasks, being part of a whole. First, Darwin speculates on a physiological division of labor. Talking of plants:

> No naturalist doubts the advantage of what has been called the "physiological division of labour;" hence we may believe that it would be advantageous to a plant to produce stamens alone in one flower or on one whole plant, and pistils alone in another flower or on another plant. In plants under culture and placed under new conditions of life, sometimes the male organs and sometimes the female organs become more or less impotent; now if we suppose this to occur in ever so slight a degree under nature, then as pollen is already carried regularly from flower to flower, and as a more complete separation of the sexes of our plant would be advantageous on the principle of the division of labour, individuals with this tendency more and more increased, would be continually favoured or selected, until at last a complete separation of the sexes would be effected. (Darwin, 1859, pp. 93–94)

Then, after some discussion of sundry topics like the most favorable conditions for natural selection to be effective and why and how extinction occurs, Darwin comes to the major topic of "divergence of character." Why do organisms divide themselves up into different species? Turning, as always, to the domestic world for guidance, we find that divergence comes from different interests – one person breeds for stronger horses, another for fleeter horses, and so two varieties appear and the differences get ever more marked. Do we see a like force working in the natural world?

> I believe it can and does apply most efficiently, from the simple circumstance that the more diversified the descendants from any one species become in structure, constitution, and habits, by so much will they be better enabled to seize on many and widely diversified places in the polity of nature, and so be enabled to increase in numbers. We can clearly see this in the case of animals with simple habits. Take the case of a carnivorous quadruped, of which the number that can be supported in any country has long ago arrived at its full average. If its natural powers of increase be allowed to act, it can succeed in increasing (the country not undergoing any change in its conditions) only by its varying descendants seizing on places at present occupied by other animals: some of them, for instance, being enabled to feed on new kinds of prey, either dead or alive; some inhabiting new stations, climbing trees, frequenting water, and some perhaps becoming less carnivorous. The more diversified in habits and structure the descendants of our carnivorous animal became, the more places they would be enabled to occupy. What applies to one animal will apply throughout all time to all animals—that is, if they vary—for otherwise natural selection can do nothing. So it will be with plants. (Darwin, 1859, pp. 112–113)

With good reason, Darwin concludes:

> The advantage of diversification in the inhabitants of the same region is, in fact, the same as that of the physiological division of labour in the organs of the same individual body (Darwin, 1859, p. 115).

15.6 The Tree of Life

We are now prepared for the final part of Chapter IV. The only diagram given in the *Origin* is here, and it is of a tree (Fig. 15.1). But it is not intended in any way to be representative of what has happened in the past. As can be seen, Darwin introduces it to illustrate the discussion of divergence and of how species break into two. However, this segues naturally and easily into discussion of what did actually happen:

> The affinities of all the beings of the same class have sometimes been represented by a great tree. I believe this simile largely speaks the truth. The green and budding twigs may represent existing species; and those produced during each former year may represent the long succession of extinct species. At each period of growth all the growing twigs have tried to branch out on all sides, and to overtop and kill the surrounding twigs and branches, in the same manner as species and groups of species have tried to overmaster other species in the great battle for life. The limbs divided into great branches, and these into lesser and lesser branches, were themselves once, when the tree was small, budding twigs; and this connexion of the former and present buds by ramifying branches may well represent the classification of all extinct and living species in groups subordinate to groups. Of the many twigs which flourished when the tree was a mere bush, only two or three, now grown into great branches, yet survive and bear all the other branches; so with the species which lived during long-past geological periods, very few now have living and modified descendants. From the first growth of the tree, many a limb and branch has decayed and dropped off; and these lost branches of various sizes may represent those whole orders, families, and genera which have now no living representatives, and which are known to us only from having been found in a fossil state. (Darwin, 1859, pp. 129–130)

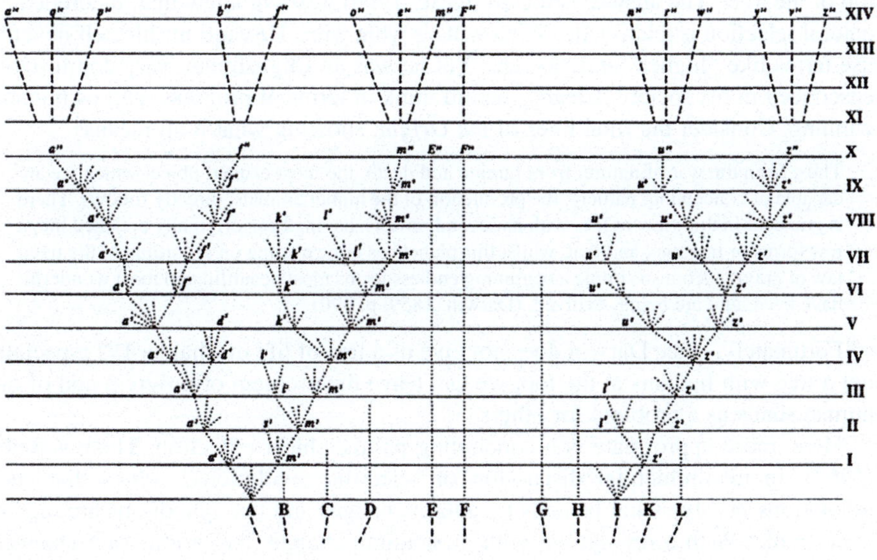

Fig. 15.1 *Tree diagram in the* Origin. (Charles Darwin, *The Origin of Species,* p. 116)

Darwin was not a paleontologist, so he had little immediate interest in the actual course of evolution, the "phylogenies" traced by evolving organisms. There is no illustration of an actual tree, as Darwin's followers like the German Ernst Haeckel were eager to provide (Fig. 15.2). This means also that, in the *Origin,* Darwin sidesteps the question of human evolution and whether we won, meaning we are at the top of the tree because we are the best. Picking up on this point, note that here, as so much earlier in the chapter – the Malthusian struggle, natural selection, adaptation, and division of labor – Darwin is drawing on tropes and metaphors from the culture of his society. Malthus (1826) had sounded the trumpet on population growth, a major issue as Britain moved from an agricultural to an industrial society. Natural selection from artificial selection, so vital to feeding so many in urban areas from the labors of so few left in rural areas (Sebright, 1809). Adaptation, straight out of Archdeacon William Paley's (1802) turn-of-the-century discussions of the argument from design, when so much is made of the teleological, functioning nature of organic characteristics. The eye is like a telescope, indeed. Division of labor, straight out of Adam Smith (1766) explaining the ways of industrialism. No longer do we have the same farmer, ploughing, planting, reaping the harvest. Now each person to their individual job. And finally, of course, the tree of life straight out of Genesis.

> 8 And the Lord God planted a garden eastward in Eden; and there he put the man whom he had formed. 9 And out of the ground made the Lord God to grow every tree that is pleasant to the sight, and good for food; the tree of life also in the midst of the garden, and the tree of knowledge of good and evil. (Genesis, 8, 9)

Does this mean that Darwin would naturally have picked up the prejudice of his age, particularly the British Victorian age, that humans are naturally superior, at the top of the tree? The answer is mixed (Ruse, 1996). Darwin knew that, theoretically, natural selection is relativistic. What wins is what wins. He cautions himself never to use terms like "higher" and "lower." Yet he was an Englishman, very comfortably ensconced in his society. Clearly, he did think in terms of progress with us humans winning. Consider the final lines of the *Origin*, showing what it all means:

> Thus, from the war of nature, from famine and death, the most exalted object which we are capable of conceiving, namely, the production of the higher animals, directly follows. There is grandeur in this view of life, with its several powers, having been originally breathed into a few forms or into one; and that, whilst this planet has gone cycling on according to the fixed law of gravity, from so simple a beginning endless forms most beautiful and most wonderful have been, and are being, evolved. (Darwin, 1859, p. 490)

Fortunately, since Darwin does not give us a tree of life in Chapter IV, especially not a tree with humans at the top, we can leave the problem of progress and of our human status as a problem for others.

Here, more appropriate is a concluding philosophical reflection. Thomas Kuhn (1962), in his influential discussion of scientific revolutions, argues that such revolutions involve what he calls paradigm changes and that paradigms are incommensurable. With a revolution, with a paradigm change, the world itself changes, literally. The Old World and the New World are different dimensions of reality.

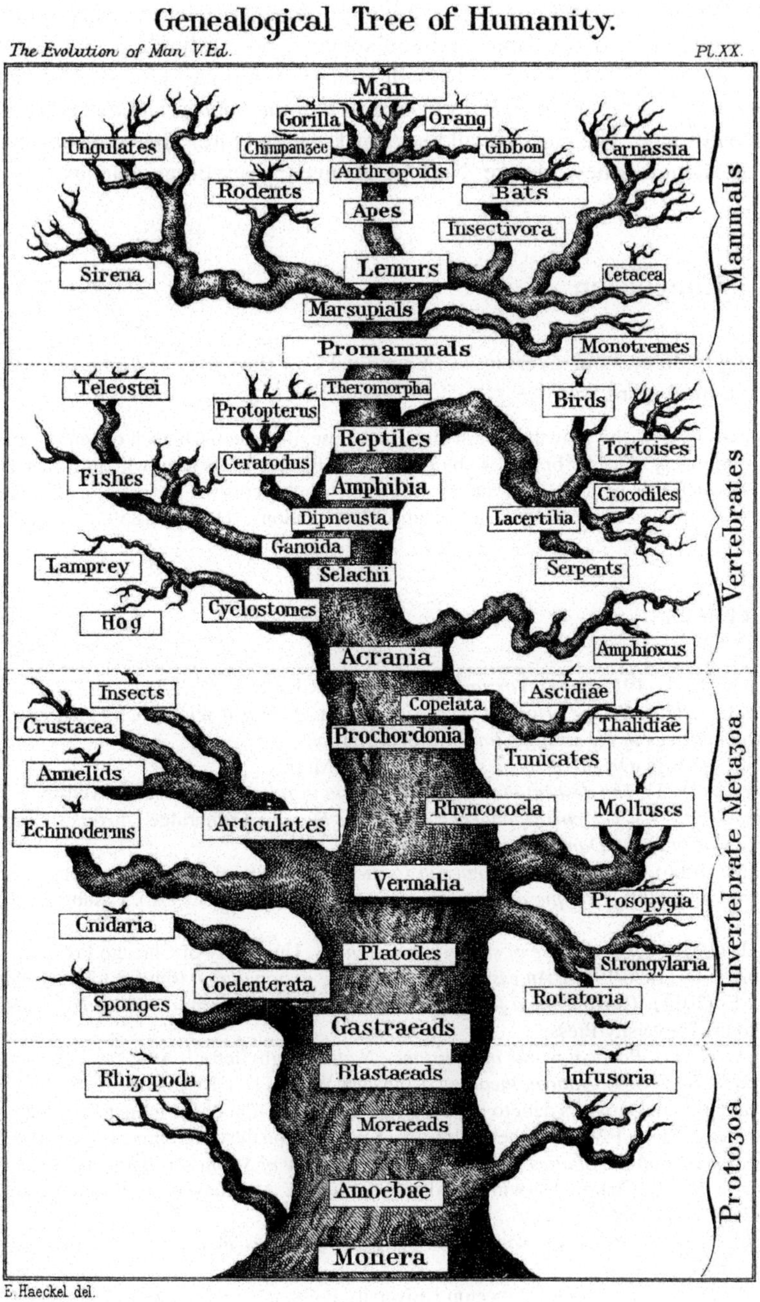

Fig. 15.2 *Tree diagram in* The Evolution of Man. (Ernest Haeckel, *The Evolution of Man*, London: Watts and Co., 1909, Plate XX)

Darwin's theory of evolution through natural selection tells us that, simply, this is not true. Darwin's theory came out of his society, and, as shown by our references to literature – novels and poetry – Darwin's theory returned to his society. This does not minimize its importance. It does suggest that the appropriate metaphor is that of the kaleidoscope. The parts are all there, making a picture. They are shaken, and we have an altogether new picture. Nothing changes. Everything changes.

15.7 Conclusion

Bringing to an end our overview of Chapter IV of the *Origin of Species*, let Darwin have the final word:

> As buds give rise by growth to fresh buds, and these, if vigorous, branch out and overtop on all sides many a feebler branch, so by generation I believe it has been with the great Tree of Life, which fills with its dead and broken branches the crust of the earth, and covers the surface with its ever branching and beautiful ramifications. (Darwin, 1859, p. 130)

References

Burroughs, E. R. (1912/1914). *Tarzan of the apes*. McClurg.

Darwin, C. (1859). *On the origin of species by means of natural selection, or the preservation of favoured races in the struggle for life*. John Murray.

Darwin, C. (1861). *Origin of species* (3rd ed.). John Murray.

Darwin, C. (1871). *The descent of man, and selection in relation to sex*. John Murray.

Darwin, C. (1985–). *The correspondence of Charles Darwin*. Cambridge University Press.

Eliot, G. (1876/1967). *Daniel Deronda*. Penguin.

Hesse, M. (1966). *Models and analogies in science*. University of Notre Dame Press.

Kant, I. (1790/2000). *Critique of the power of judgment* (P. Guyer., Ed.). Cambridge University Press.

Kuhn, T. (1962). *The structure of scientific revolutions*. University of Chicago Press.

Malthus, T. R. (1826/1914). *An essay on the principle of population* (6th Ed.). Everyman.

Milam, E. (2011). *Looking for a few good males: Female choice in evolutionary biology*. Johns Hopkins University Press.

Naden, C. (1999). *Poetical works of Constance Naden*. High Sierra Books.

Paley, W. (1802/1819). *Natural theology* (Collected Works: IV). Rivington.

Ruse, M. (1975a). Darwin's debt to philosophy: An examination of the influence of the philosophical ideas of John F.W. Herschel and William Whewell on the development of Charles Darwin's theory of evolution. *Studies in History and Philosophy of Science, 6*, 159–181.

Ruse, M. (1975b). Charles Darwin's theory of evolution: An analysis. *Journal of the History of Biology, 8*, 219–241.

Ruse, M. (1996). *Monad to man: The concept of progress in evolutionary biology*. Harvard University Press.

Ruse, M. (2017). *On purpose*. Princeton University Press.

Ruse, M. (2021). The scientific revolution. In S. Bullivant & M. Ruse (Eds.), *The Cambridge history of atheism* (Vol. I, pp. 258–277). Cambridge University Press.

Ruse, M. (2022a). *Understanding natural selection*. Cambridge University Press.

Ruse, M. (2022b). *Why we hate: Understanding the roots of human conflict*. Oxford University Press.
Sebright, J. (1809). *The art of improving the breeds of domestic animals in a letter addressed to the Right Hon. Sir Joseph Banks, K.B.* Privately published.
Smith, A. (1766/1937). *The wealth of nations*. Modern Library.
Smith, A. (1777). *The wealth of nations*. Strahan, W., & Cadell, T.
Wallace, A. R. (1870). *Contributions to the theory of natural selection*. Macmillan.

Michael Ruse, the retired Lucyle T. Werkmeister Professor of Philosophy at Florida State University, is a philosopher and historian of science, with a special interest in Charles Darwin's theory of evolution through natural selection. The founding editor of the journal *Biology and Philosophy*, he is the author or editor of over 60 books. He is a Fellow of the Royal Society of Canada and the recipient of four honorary degrees.

Chapter 16
Origin's Chapter V: How "Random" Is Evolutionary Change?

Sander Gliboff

Abstract Darwin's fifth chapter, "The Laws of Variation," may stand in the shadow of the first four that climax with his presentation of "Natural Selection," but its importance should not be underestimated. It deals with philosophical and methodological issues in the study of variation that would be hotly debated for decades after the publication of the book, many of which are still unsettled today. As the chapter title suggests, Darwin felt that a proper scientific study of variation had to discover the laws of nature that governed it. He could not simply let it be random. He looked for laws in the patterns of co-occurrence of changes in different parts of the body or the same part in different species and also in the apparent consistencies in the effects of environment and habit. There is no one main line of argument in this chapter, but rather an exploration of multiple possible conceptions, patterns, and laws of variation—all ways in which variation might not be entirely random. For example: To what extent was variation responsive to environmental changes or to the organism's needs and habits? And if Darwin admitted such responses, then how was his theory any different from Lamarck's? Do all parts of the organism vary freely and independently or are there hidden connections, correlations, or trade-offs between varying parts of the body? Why do related species seem often to vary in similar ways? Many of these questions are still with us, in modern evolutionary developmental biology ("evo-devo") and studies of epigenetics.

16.1 Introduction

Darwin's fifth chapter is the anticlimactic one. The case for natural selection as the mechanism of evolutionary change has just been made in the fourth, and a reader could be forgiven for the urge to skip ahead to the discussion of the "difficulties" with it in the sixth and on to the triumphant problem-solving of the later chapters.

S. Gliboff (✉)
Indiana University, Bloomington, IN, USA
e-mail: sgliboff@indiana.edu

M. Elice Brzezinski Prestes (ed.), *Understanding Evolution in Darwin's "Origin"*, History, Philosophy and Theory of the Life Sciences 34,
https://doi.org/10.1007/978-3-031-40165-7_16

Besides, this is already the third chapter on variation. Do we really need to go into that again?

Yes, we do! Variation is essential to Darwin's theory because there can be no selection without variations to select from. The nature and causes of variation determine what selection can and cannot accomplish. Darwin anticipates criticism of the element of randomness in his system and does not want to leave the reader with the impression that variation is a mystery or uncaused or not governed by laws of nature. As the chapter title implies, Darwin ideally would like to identify some of the laws or at least some of the regular patterns of variation.

The chapter also anticipates some of the questions later to be raised by Darwin's critics and rivals, ongoing philosophical discussions of the meaning of "randomness" in evolution, and even some aspects of recent evolutionary developmental biology, or "evo-devo."

Darwin looks in many different directions for laws and patterns of variation. There is no one line of investigation or argument in this chapter, but rather a series of interesting cases, observations, and generalizations. I will introduce a few of the main ones and then analyze the chapter in more detail.

From the late nineteenth century on, competing theories of evolution have differed with Darwin precisely on the nature and causes of variation: Were hereditary variations inborn or acquired during an individual's lifetime? Various schools of "Lamarckism" emphasized the acquisition of new and beneficial characteristics as a result of environmental effects or the effects of habit or use and disuse of body parts. Some versions of Lamarckism relied on the influence of the mind or perceived needs of the organism to generate adaptive changes. On these accounts, there would not be much for natural selection to do, since there would be few bad variations to weed out. Its role as the exclusive explanation of adaptation would be diminished or eliminated. Such ideas about the adaptiveness of variation were long rejected, but there are increasing calls for reconsidering the possibilities, in light of new developments in epigenetics (Jablonka et al., 1998, pp. 206–210, see also Balter, 2000).

Which side was Darwin on? Was he secretly a Lamarckian? Here in Chap. 5, he was more than open to environmental effects and especially the effects of habit, just not to the extent that they could replace natural selection as the cause of adaptation. Such effects indeed seemed to produce variations that were mostly beneficial, but they were not finished adaptations, and they did not solve complex problems such as interactions with other evolving organisms. In short, such beneficial effects of the environment were no substitute for natural selection but could complement it in special situations.

Other Darwinians, such as Ernst Haeckel in Germany or George Romanes in England, continued to see no opposition between Lamarckism and Darwinism. The opposition only set in after Darwin's death in 1882. Neo-Darwinians, led by August Weismann in the 1880s and 1890s, sought to rule out any inheritance of acquired characteristics. And Haeckel was labeled "Lamarckian" for retaining Darwin's original account of the causes of variation (Gliboff, 2011).

The chapter also discusses the mechanics by which the body produces anatomical variations and whether there are laws, patterns, and limitations to what changes can occur. Did all parts of the body vary freely and independently or was each change necessarily correlated with, or limited by, changes elsewhere? Such interconnections between body parts could diminish the effectiveness of natural selection, in cases where a beneficial variation in one part came with useless or even maladaptive baggage elsewhere or where certain combinations of adaptive traits were not feasible. Modern critiques of "adaptationism" and "constraints" (Gould & Lewontin, 1979) on variation have raised such questions anew, though using different language from Darwin's (without all the talk of "laws"). Several more modern ideas about development also have their roots in this chapter. One is that there are trade-offs between growth in one part of the body and growth in other parts because the developing body has only finite resources (Moczek & Nijhout, 2004). Another is that the ancestry of a species can affect its ability to vary (what we would now call phyletic constraints). And yet another is that the process of development has options. It can proceed in different ways in different environments (what we would now call phenotypic plasticity).

The chapter also touches upon the philosophical and theological questions of how variation could be "random," or what randomness even means in the context of biological variation. Darwin develops his conception of randomness further in later works but seems already to have it in mind in 1859 (Harman, 2009).

In his correspondence with the American botanist Asa Gray after the publication of *The Origin*, Darwin thinks of randomness not in a statistical sense, or in the sense of being uncaused by the laws of nature, but rather as *unintended* by the Creator or any purposeful force of nature. Gray was a theistic evolutionist. He accepted the general mechanism of variation, struggle, and selection but wanted the Creator to be able to intervene, undetected, to make sure that favorable variations always occurred when they were needed. Darwin argued against this idea using the example of his own ugly nose.

The nose had made a poor first impression on Captain FitzRoy and almost cost Darwin his opportunity to join the voyage of the *Beagle*. FitzRoy believed in the pseudoscience of physiognomy, which taught that facial features reveal personality traits, and he thought that a tough, determined sailing companion would come with a better-looking nose. Anyway, Darwin challenged Gray, as he had also done with the theistically inclined geologist Charles Lyell, to explain whether Darwin's particular nose shape had been purposefully designed or was just a mindless and purposeless variation on the general form of the human nose (Darwin Correspondence Project, letter no. 3256). The implication was that variations were random in the sense of not being designed or intended for any purpose. Natural selection had the job of choosing the variations that suited a purpose.

In a later book, Darwin used the analogy of stones used to build a house. Would anyone think that nature or God intentionally shaped every stone to fit in its future place? No, the builder had lots of stones to choose from and could always find what he needed for his own purposes:

> Let an architect be compelled to build an edifice with uncut stones, fallen from a precipice. The shape of each fragment may be called accidental; yet the shape of each has been determined by the force of gravity, the nature of the rock, and the slope of the precipice,

> —events and circumstances, all of which depend on natural laws; but there is no relation between these laws and the purpose for which each fragment is used by the builder. In the same manner the variations of each creature are determined by fixed and immutable laws; but these bear no relation to the living structure which is slowly built up through the power of selection. (Darwin, 1868 pp 248–249)

This is essentially the modern idea of a random variation: not uncaused or inexplicable, not necessarily even random in the statistical sense, but shaped independently of its usefulness as an adaptation.

Although Darwin clearly did not accept the theistic account of variation, he does not make an issue of it in *The Origin*. He does, however, explicitly reject divine intervention in the design and special creation of species. He takes several shots at special creation in this chapter, which will be discussed below.

Finally, there is the methodological problem of what counts as a variation or how variations are to be detected. Darwin uses the term "variation" in two different senses, referring to two different points in the cycle of variation-struggle-selection-new variation. One sense refers to the variation among all the offspring of a species at birth (or hatching or germinating), regardless of whether it is beneficial or not. These variations are the raw material on which natural selection acts, and many of them might not persist very long in nature if they are not beneficial. That makes some of them difficult to observe except under domestication, as in Chap. 1.

The variations that we do commonly observe in nature and were the subject of Chap. 2 are the ones that have survived after natural selection has done at least some of its work. That is the second sense of "variation." It refers to the emerging directions and patterns of evolutionary change, as individual differences are molded into new varieties and, over the longer term, distinct subspecies or species.

We will be most concerned with variation in the sense of raw material for evolution, as indeed is most of Darwin's Chap. 5.

16.2 Effects of External Conditions

The chapter begins with philosophical questions about the role of chance in the production of variation. Though he has said before that variations were due to chance, Darwin now emphasizes that he did not mean to say that they were not caused in accordance with regular laws of nature: "This [i.e., 'chance'], of course, is a wholly incorrect expression, but it serves to acknowledge plainly our ignorance of the cause of each particular variation" (Darwin, 1859, p. 131). He now wants to explore as much as possible what the causes of variation could be and what laws or at least patterns they follow. He begins by considering how environmental effects, or as Darwin calls them, "external conditions" or "conditions of life" can cause variations and supply the raw materials for natural selection.

They can do so in several ways. The main one is by somehow interfering with the reproductive system and making it fail to preserve the full resemblance of offspring to parents:

> The reproductive system is eminently susceptible to changes in the conditions of life; and to this system being functionally disturbed in the parents, I chiefly attribute the varying or plastic condition of the offspring. (Darwin, 1859, pp. 131–132)

In such cases, it should be noted, the environmental effects are not of the Lamarckian sort, not necessarily adaptive responses to environmental change. These environmentally induced disturbances produce variations in all possible directions and give natural selection much raw material to choose from. They are Darwin's main cause of variation.

Darwin does allow that the Lamarckian sort of adaptive responses to the environment can occur but denies that they are of much importance in evolution. They are only responses to the physical environment, and therefore:

> Cannot have produced the many striking and complex co-adaptations of structure between one organic being and another, which we see everywhere throughout nature. (Darwin, 1859, p. 132)

As he made clear in his Introduction (Darwin, 1859, p. 3) and as is implicit in the concept of struggle in Chap. 3, Darwin considered biological interactions within and between species to be more important than interactions with the physical environment, and they were complex and everchanging. Competition with an evolving competitor, predation on an evolving prey, and especially mutualism that has to be continually negotiated over evolutionary time, could not be understood as simple responses to an environmental stimulus.

Darwin does acknowledge some general trends in observable variation with respect to the environment. For example, seashells in warmer, southern waters are more brightly colored than in the north. Even shells of a single species are more brightly colored at the southern end of its range than the northern. But for Darwin, such observations do not undermine natural selection. On the contrary, they argue against the design and special creation of each species. Was the species created with bright colors? If so, it would have to change in order to lose them as it expanded to the north, or if not, it would have to acquire them as it expanded to the south. The design proponent could not avoid at least some variation and evolution:

> He who believes in the creation of each species, will have to say that this shell, for instance, was created with bright colours for a warm sea; but that this other shell became bright-coloured by variation when it ranged into warmer or shallower waters. (Darwin, 1859, p. 133)

One last thought on environmental effects concerns the methodological problem of how to distinguish the Lamarckian sort from the effects of natural selection. Consider the effects of cold weather on mammalian fur:

> Thus, it is well known to furriers that animals of the same species have thicker and better fur the more severe the climate is under which they have lived; but who can tell how much of this difference may be due to the warmest-clad individuals having been favoured and preserved during many generations, and how much to the direct action of the severe climate? (Darwin, 1859, p. 133)

All that can be concluded is that there is a correlation between cold climate and warm fur. Such observations and generalizations do not reveal the causal connection between the two and should not be taken as evidence against natural selection.

Darwin concludes this section with the point that it is mostly natural selection that creates adaptations, but that the environment, or conditions of life, creates the variations on which selection acts. It does so by affecting the reproductive system in a mostly undirected manner.

16.3 Effects of Use and Disuse

Another form of variation associated with Lamarck and inheritance of acquired characteristics results from the use or disuse of body parts, depending on habits or behaviors. This one Darwin takes more seriously than environmental effects as a cause of adaptation, or at least a contributing factor. He refers back to examples from domesticated animals in Chap. 1 and adds a few from animals in nature. An interesting one is the ostrich, which he assumes is descended from something like a bustard, a very large and long-legged bird that could still fly: "As natural selection increased in successive generations the size and weight of its body, its legs were used more, and its wings less, until they became incapable of flight" (Darwin, 1859, p. 135). Here Darwin envisions habit or use and disuse working together with natural selection and coordinating the needed changes to the different parts of the body. Again, as in many other examples, a "Lamarckian" mechanism of adaptive change is inadequate by itself and is no substitute for natural selection.

Only in the case of blind cave animals does Darwin accept disuse alone as the cause of an evolutionary change— but not of an adaptation: "As it is difficult to imagine that eyes, though useless, could be in any way injurious to animals living in darkness, I attribute their loss wholly to disuse" (Darwin, 1859, p. 137). The loss of eyesight is of no advantage but also of no disadvantage, so natural selection takes no action.

Lest anyone think the example of cave animals undermines the theory of natural selection, Darwin quickly turns it against the idea of special creation. Cave environments being very similar no matter where the cave is located, one might expect a wise and benevolent Creator to design just one set of perfectly adapted cave species and populate every cave with it. But on the contrary, animals from widely separated caves do not resemble one another but do resemble animals found near their caves. They cannot all be of the best possible design for the cave environment. Why would God care to make them resemble their open-air neighbors?

Such a distribution pattern only makes sense in Darwinian terms:

> On my view we must suppose that American animals, having ordinary powers of vision, slowly migrated by successive generations from the outer world into the deeper and deeper recesses of the Kentucky caves, as did European animals into the caves of Europe. . . . By the time that an animal had reached, after numberless generations, the deepest recesses, disuse

> will on this view have more or less perfectly obliterated its eyes, and natural selection will often have effected other changes, such as an increase in the length of the antennae or palpi, as a compensation for blindness. Notwithstanding such modifications, we might expect still to see in the cave-animals of America, affinities to the other inhabitants of that continent, and in those of Europe, to the inhabitants of the European continent. And this is the case.... (Darwin, 1859, p. 148)

Note also that in the end, the effects of disuse on the eyes do not by themselves suffice to adapt the animal to cave life. This is another example of use and disuse complementing natural selection and coordinating multiple changes in the body.

16.4 Acclimatization

By "acclimatization," Darwin means the ability of a species to survive and even thrive when moved to a new climate or environment, or when its native climate or environment changes. The matter had been studied in domesticated animals, in cases where a breed is exported to new places, as well as in species imported to Europe from other parts of the world. Darwin believes that species are not always narrowly specialized and capable of thriving in just one environment or climate zone. If they are confined to one particular region, it must be because they have competitors that prevent them from expanding into new territories:

> We may infer this from our frequent inability to predict whether or not an imported plant will endure our climate, and from the number of plants and animals brought from warmer countries which here enjoy good health. We have reason to believe that species in a state of nature are limited in their ranges by the competition of other organic beings quite as much as, or more than, by adaptation to particular climates. (Darwin, 1859, p. 140)

Here again, we see the primacy, for Darwin, of biological interactions over the effects of the physical environment.

But the main reason why Darwin brings up the problem of acclimatization is the question of how it occurs and again the interplay of natural selection with the supposedly Lamarckian mechanisms of change: changed habits, modification by the new environment, or some kind of innate flexibility of the organism's "constitution." In cases of adaptation to a new climate, Darwin argues that natural selection favors inborn variations in the internal constitution that make the individual better adapted to the new environment. As in the example above of the evolution of thicker fur in cold climates, Darwin finds that the evidence is not conclusive, but it is at least consistent with natural selection as a necessary cause of adaptive change: "I can see no reason to doubt that natural selection will continually tend to preserve those individuals which are born with constitutions best adapted to their native countries" (Darwin, 1859, p. 142). Darwin concludes this section by reinforcing his previous point that all the other modifying forces only complement natural selection and do not by themselves bring about adaptation or, in the present case, acclimatization to new environments:

> On the whole, I think we may conclude that habit, use, and disuse, have, in some cases, played a considerable part in the modification of the constitution, and of the structure of various organs; but that the effects of use and disuse have often been largely combined with, and sometimes overmastered by, the natural selection of innate differences. (Darwin, 1859, pp. 142–143)

To this point, the chapter has served to clarify Darwin's views on the causes of variation and change and the extent to which he can admit environmental effects and the effects of use and disuse. Although he makes use of the latter two effects, he uses them differently from Lamarck and later Lamarckians. He does not allow them to create fully formed adaptations, but only variations that either provide the raw material for selection to work on, or to complement and coordinate complex adaptive changes as selection gradually brings them about.

The rest of the chapter turns to observations on patterns of variation and the rules or perhaps laws they seemed to follow.

16.5 Correlation of Growth

By "correlation of growth," Darwin means changes in one part of the body that tend to occur together with changes elsewhere. As he expresses it, "the whole organization [of the organism] is so tied together during its growth and development, that when slight variations in any one part occur, and are accumulated through natural selection, other parts become modified" (Darwin, 1859, p. 143).

Darwin offers various explanations for such correlations. Sometimes, he says, a change in an early stage of development has ramifications later, since the later stages build upon the earlier:

> The most obvious case is, that modifications accumulated solely for the good of the young or larva, will, it may safely be concluded, affect the structure of the adult; in the same manner as any malconformation affecting the early embryo, seriously affects the whole organisation of the adult. (Darwin, 1859, p. 143)

Homologous structures, which have similar origins in the early embryo, tend to change together:

> The several parts of the body which are homologous, and which, at an early embryonic period, are alike, seem liable to vary in an allied manner: we see this in the right and left sides of the body varying in the same manner; in the front and hind legs, and even in the jaws and limbs, varying together, for the lower jaw is believed to be homologous with the limbs. (Darwin, 1859, p. 143)

Some other sorts of correlations are more mysterious and perhaps could limit the efficacy of natural selection. For example:

> What can be more singular than the relation between blue eyes and deafness in cats, and the tortoise-shell colour with the female sex; the feathered feet and skin between the outer toes in pigeons, and the presence of more or less down on the young birds when first hatched, with the future colour of their plumage; or, again, the relation between the hair and teeth in the naked Turkish dog, though here probably homology comes into play? With respect to this

> latter case of correlation, I think it can hardly be accidental, that if we pick out the two orders of mammalia which are most abnormal in their dermal covering, viz. Cetacea (whales) and Edentata (armadilloes, scaly anteaters, &c.), that these are likewise the most abnormal in their teeth. (Darwin, 1859 p. 144)

Darwin was evidently mistaken about the extent of the correlation between blue eyes and deafness in cats and can be seen backtracking in later revisions of *The Origin*, changing it in the fourth and fifth editions to a correlation, "between complete whiteness with blue eyes and deafness" before deleting the example entirely in the sixth.

The tortoiseshell coloring of cats, on the other hand, is really associated with the female sex, but for genetic reasons, rather than developmental ones. The cat's mix of orange and black requires two different X chromosomes, one with the gene for each color. The correlation of hairlessness with tooth abnormalities was probably due to the fact that hair and teeth (as well as nails, sweat glands, and mammary glands) all derive from the embryonic ectoderm, by means of the same signaling pathway, controlled by a small number of genes. A mutation in any one of them can affect more than one adult characteristic.

Some correlations are more apparent than real. The modifications only go together because natural selection brought them together in a common ancestor:

> We may often falsely attribute to correlation of growth, structures which are common to whole groups of species, and which in truth are simply due to inheritance; for an ancient progenitor may have acquired through natural selection some one modification in structure, and, after thousands of generations, some other and independent modification; and these two modifications, having been transmitted to a whole group of descendants with diverse habits, would naturally be thought to be correlated in some necessary manner. (Darwin, 1859, p. 146)

Regardless of the reason for correlations, they raise questions about the efficacy of natural selection, because they make it possible, or even likely for parts of the body to become modified in ways that are neither useful nor the result of environmental effects. Correlated changes occur "independently of utility and, therefore, of natural selection" (Darwin, 1859, p. 144). But in some cases, natural selection can gradually decouple the changes and make one occur without the other. This can occur most readily in the first two sorts of correlations, where a later developmental stage is affected by a change at an earlier stage or where there is a homology: "These tendencies, I do not doubt, may be mastered more or less completely by natural selection" (Darwin, 1859, p. 143).

16.6 Compensation and Economy of Growth

These cases are the reverse of correlations, where growth in one place seems to inhibit growth elsewhere, presumably because of some sort of trade-off in the utilization of material or nourishment. Darwin quotes the German poet-scientist Johann Wolfgang von Goethe for the explanation that "in order to spend on one side, nature is forced to economize on the other side" (Darwin, 1859, p. 147). Darwin has many examples from agricultural plants and animals, which cannot be bred for every possible purpose at the same time: "Thus it is difficult to get a cow to give much milk and to fatten readily. The same varieties of the cabbage do not yield abundant and nutritious foliage and a copious supply of oil-bearing seeds" (Darwin, 1859, p. 147).

Darwin does not think such trade-offs diminish the efficacy of natural selection. In fact, he thinks natural selection would have the same effects as any internal, developmental trade-offs and that the two are not distinguishable in practice:

> I see hardly any way of distinguishing between the effects, on the one hand, of a part being largely developed through natural selection and another and adjoining part being reduced by this same process or by disuse, and, on the other hand, the actual withdrawal of nutriment from one part owing to the excess of growth in another and adjoining part. (Darwin, 1859, p. 147)

In other words, natural selection can be expected to favor the efficiency of resource use and to make the necessary trade-offs itself:

> Natural selection is continually trying to economise in every part of the organisation. If under changed conditions of life a structure before useful becomes less useful, any diminution, however slight, in its development, will be seized on by natural selection, for it will profit the individual not to have its nutriment wasted in building up an useless structure. (Darwin, 1859, pp. 147–148)

16.7 Other Patterns of Variation

Next, Darwin considers patterns of variation observable in nature, which perhaps is not a measure of how much variation can be produced but of how many variants can survive well. He notices that snakes, which have a large number of vertebrae, are also highly variable in the number of vertebrae. One bone more or less evidently makes little difference for survival if there are many with the same function.

Unspecialized parts, he says, vary more than specialized, because, "In the same way that a knife which has to cut all sorts of things may be of almost any shape; whilst a tool for some particular object had better be of some particular shape" (Darwin, 1859, p. 149).

Rudimentary (vestigial) parts are highly variable, "owing to their uselessness, and therefore to natural selection having no power to check deviations in their structure" (Darwin, 1859, pp. 149–150). In the opposite case, when a part is developed to an unusually high degree, such as the long arms of the orangutan, it is also highly variable, despite the fact that it must be very important to the species and continually acted on by natural selection. Darwin's explanation is that such parts could not have developed if they were not variable in the first place and that they are still undergoing variation and selection:

> An extraordinary amount of modification implies an unusually large and long-continued amount of variability, which has continually been accumulated by natural selection for the

> benefit of the species. But as the variability of the extraordinarily developed part or organ has been so great and long continued within a period not excessively remote, we might, as a general rule, expect still to find more variability in such parts than in other parts of the organisation, which have remained for a much longer period nearly constant. (Darwin, 1859, p. 153)

The same reasoning applies to characteristics of only one species in a genus. The species-specific characteristics have also developed more recently and are therefore more likely still to be varying than the older, better-established characteristics of the whole genus.

Variations in secondary sexual characteristics are commonly observed in nature because they are not crucial for survival:

> We can see why these characters should not have been rendered as constant and uniform as other parts of the organisation; for secondary sexual characters have been accumulated by sexual selection, which is less rigid in its action than ordinary selection, as it does not entail death, but only gives fewer offspring to the less favoured males. (Darwin, 1859, p. 156–157)

Another interesting tendency is that related species often have subspecies or varieties that vary in parallel: "Distinct species present analogous variations; and a variety of one species often assumes some of the characters of an allied species" (Darwin, 1859, p. 159). Darwin illustrates this tendency with examples from Chap. 1 on domesticated plants and animals. Pigeon fanciers working independently in different countries, starting with different stocks, have come up with some of the same modifications: "The most distinct breeds of pigeons, in countries most widely apart, present subvarieties with reversed feathers on the head and feathers on the feet,—characters not possessed by the aboriginal rock-pigeon" (Darwin, 1859, p. 159). Such cases make no sense under the assumption of special creation, which would have to posit: "three separate yet closely related acts of creation" (Darwin, 1859, p. 159), one for the main species and one for each subspecies or variety.

One last example is atavism, or the tendency for ancestral characteristics to reappear. Again, Darwin draws on the pigeons and notes:

> the occasional appearance in all the breeds, of slaty-blue birds with two black bars on the wings, a white rump, a bar at the end of the tail, with the outer feathers externally edged near their bases with white. As all these marks are characteristic of the parent rock-pigeon, I presume that no one will doubt that this is a case of reversion, and not of a new yet analogous variation appearing in the several breeds. (Darwin, 1859, pp. 159–160)

Darwin also discusses at length the occurrence of stripes in quaggas, asses, the hemonius (better known as the onager), and horses, and he argues that they are reversions to an ancestral color pattern, still retained by the zebra.

Darwin draws two important conclusions from such examples. One is that they give us a glimpse of the ancestor of the group. This is one of the very few places in the book where Darwin ventures to reconstruct a specific line of descent:

> For myself, I venture confidently to look back thousands on thousands of generations, and I see an animal striped like a zebra, but perhaps otherwise very differently constructed, the common parent of our domestic horse..., of the ass, the hemionus, quagga, and zebra.

The other conclusion is, once again, that the evolutionary view makes more sense than special creation of each species:

> He who believes that each equine species was independently created, will, I presume, assert that each species has been created with a tendency to vary, both under nature and under domestication, in this particular manner, so as often to become striped like other species of the genus; and that each has been created with a strong tendency, when crossed with species inhabiting distant quarters of the world, to produce hybrids resembling in their stripes, not their own parents, but other species of the genus. (Darwin, 1859, p. 167)

But why would a purposeful and reasonable Creator do such a thing? Just to mislead Darwin into thinking that the living equines had a striped common ancestor? "To admit this view is, as it seems to me, to reject a real for an unreal, or at least for an unknown, cause. It makes the works of God a mere mockery and deception" (Darwin, 1859, p. 167).

16.8 Summary

The chapter is difficult to summarize because it makes so many different points about variation, but the following are its main recurring themes:

1. That variation only appears to be produced by chance because we are usually ignorant of its causes. The chapter identifies patterns and regularities in the occurrence of variation that at least suggest that there are underlying causes and laws of variation to be discovered.
2. That the causes of evolutionary change that we now think of as "Lamarckian," i.e., the direct effects of the environment and the effects of habit or use and disuse, are still acceptable to Darwin and compatible with natural selection. He only wants to limit their ability to create fully formed, complex adaptations on their own. They are partners to natural selection, not replacements.
3. That the element of chance, in the sense of purposelessness, in the production of variations gives Darwin's theory an advantage over theories of special creation and divine design. The patterns of variation we observe in nature are often illogical and have no discernable purpose.

References

Balter, M. (2000). Was Lamarck just a little bit right? *Science, 288*(5463), 38–38.

Darwin, C. R. (1859). *On the origin of species by means of natural selection: Or the preservation of favored races in the struggle for life*. John Murray.

Darwin, C. R. (1868). *The variation of animals and plants under domestication*. John Murray.

Darwin Correspondence Project, "letter no. 3256". https://www.darwinproject.ac.uk/letter/DCP-LETT-3256.xml

Gliboff, S. (2011). The golden age of Lamarckism, 1866–1926. In S. B. Gissis & E. Jablonka (Eds.), *Transformations of Lamarckism: From subtle fluids to molecular biology* (pp. 45–55). MIT Press.

Gould, S. J., & Lewontin, R. C. (1979). The spandrels of San Marco and the Panglossian paradigm: A critique of the adaptationist programme. *Proceedings of the Royal Society of London Series B: Biological Sciences, 205*(1161), 581–598.

Harman, O. S. (2009). Neo-Darwinism and random variation: Or why it's important to read Darwin carefully. In J. Seckbach & R. Gordon (Eds.), *Divine action and natural selection: Science, faith, and evolution* (pp. 139–152). World Scientific.

Jablonka, E., Lamb, M. J., & Avital, E. (1998). 'Lamarckian' mechanisms in Darwinian evolution. *Trends in Ecology and Evolution, 13*(5), 206–210.

Moczek, A. P., & Nijhout, H. F. (2004). Trade-offs during the development of primary and secondary sexual traits in a horned beetle. *American Naturalist, 163*(2), 184–191.

Sander Gliboff's is associate professor and director of graduate studies at Indiana University. His research and teaching interests are in the history of biology, especially evolution and genetics, and science in modern Germany and Austria. He is the author of *H.G. Bronn, Ernst Haeckel, and the Origins of German Darwinism: A Study in Translation and Transformation* (Cambridge: MIT Press, 2008).

Chapter 17
Origin's Chapter VI: The Initial Difficulties of Darwin's Theory

Martins R. A.

Abstract The sixth chapter of Charles Darwin's *Origin of Species* is called "Difficulties of the Theory." In that part of his work, Darwin reveals some possible objections to his theory and attempts to provide an answer to all of them. Such a chapter was part of the first edition of the *Origin of Species*; therefore, the difficulties described were not reactions derived from the publication of the book. Instead, friends like Charles Lyell have probably presented some of them. Others can be found in works published before Darwin's work, such as *Natural Theology* by William Paley, which argued for creationism and introduced arguments against any natural explanation of the origin of species. Many other problems dealt with in the sixth chapter were born, however, from Darwin's own internal dialogue, his own initial doubts concerning the theory, and his anticipation of criticism, as one can gather from his manuscripts. This paper describes the main difficulties shown in the sixth chapter of the *Origin of Species*, providing a more detailed exposition of a few topics, and analyzing Darwin's defense, in those cases. It also discusses some of the weak points in Darwin's line of reasoning from a diachronic point of view.

17.1 Introduction

The title of the sixth chapter of Charles Darwin's *On the Origin of Species* is "Difficulties of the Theory." In that part of his work, Darwin revealed some possible objections to his theory and attempted to provide an answer to all of them. As this chapter was already part of the first edition of *On the Origin of Species*, the difficulties described were not reactions derived from the publication of the book. Some of them have probably been presented by friends, such as Charles Lyell. Others can be found in works published before Darwin's work, such as *Natural*

Martins R. A. (✉)
Federal University of São Paulo, São Paulo, Brazil

Group of History, Theory and Science Teaching, University of São Paulo, São Paulo, Brazil

M. Elice Brzezinski Prestes (ed.), *Understanding Evolution in Darwin's "Origin"*, History, Philosophy and Theory of the Life Sciences 34,
https://doi.org/10.1007/978-3-031-40165-7_17

Theology by William Paley, which argued for creationism and introduced arguments against any natural explanation of the origin of species. Criticism presented against former evolutionary proposals – such as those of Lamarck and Chambers – was also taken into account by Darwin (Fisher, 1954).

Many other problems dealt with in the sixth chapter were born, however, from Darwin's internal dialogue, his own initial doubts concerning the theory, and his anticipation of criticism, as one can gather from his manuscripts. For instance, in 1841, he annotated in one of his notebooks: "Bats are a great difficulty not only are no animals known with an intermediate structure, but it is not possible to imagine what habits an animal could have had with such structure" (Barrett et al., 1987, p. 493). That is exactly one of the difficulties he addressed in the sixth chapter of *On the Origin of Species*.

This paper reviews the main difficulties presented by Darwin in the sixth chapter of the first edition of *On the Origin of Species*, providing a more detailed exposition of a few topics, and analyzing Darwin's defense, in those specific cases. It also discusses some of the weak points in Darwin's line of reasoning from a diachronic point of view. No attempt will be made to deal with the chapter's changes in the succeeding editions of the book. Those interested in the variations may consult the *Online Variorum of Darwin's Origin* of *Species*, edited by Barbara Bordalejo (2008–2012), which superseded Morse Peckham's variorum edition (Peckham, 1959). It is well known that St. George Jackson Mivart (1827–1900) was one important critic of the *Origin of Species* (Gruber, 1960; Brooke, 2009, pp. 271–272). Darwin presented a detailed discussion of his objections only in the sixth edition of his book, and for that reason, they will not be dealt with here.

17.2 The Main Difficulties

The opening of the sixth chapter acknowledges that there are many difficulties to the acceptance of his theory:

> Long before having arrived at this part of my work, a crowd of difficulties will have occurred to the reader. Some of them are so grave that to this day I can never reflect on them without being staggered; but, to the best of my judgment, the greater number are only apparent, and those that are real are not, I think, fatal to my theory. (Darwin, 1859, p. 171)

At this particular point, one perceives Darwin's use of a rhetoric technique: he invites the reader to identify himself with the author since both have felt the same kind of difficulties; once the emotional identification is made, the reader will tend to follow and accept Darwin's arguments.

The difficulties are first divided into two broad types (Lustig, 2009):

> These difficulties and objections may be classed under the following heads: Firstly, why, if species have descended from other species by insensibly fine gradations, do we not everywhere see innumerable transitional forms? Why is not all nature in confusion instead of the species being, as we see them, well defined?

> Secondly, is it possible that an animal having, for instance, the structure and habits of a bat, could have been formed by the modification of some animal with wholly different habits? Can we believe that natural selection could produce, on the one hand, organs of trifling importance, such as the tail of a giraffe, which serves as a fly-flapper, and, on the other hand, organs of such wonderful structure, as the eye, of which we hardly as yet fully understand the inimitable perfection? (Darwin, 1859, p. 172)

The extreme perfection and complexity of the eye had been used by William Paley (1743–1805) in his work Natural Theology or evidences of the existence and attributes of the Deity, collected from the appearances of nature (1802) as strong evidence that God himself could only produce such an organ.

Darwin also pointed out two other topics discussed in the following chapters of the *Origin* and which, therefore, will not be dealt with here:

Acquisition and modification of instincts by natural selection
Fertility of the offspring of crossings between varieties and sterility of the crossings between distinct species

17.3 Absence or Rarity of Transitional Forms

The first broad difficulty can be subdivided into two different situations, although Darwin did not clearly distinguish them:

Transitional forms in time, between extinct species and those existing nowadays
Transitional forms in space, at places where two different species coexist

The first problem, then, can be described as follows: if the species that exist today came from different species that existed in the past, shouldn't we find now, or in the geological record, all the gradations between ancient and modern forms? Why don't we see this?

Again, the difficulty can be divided into two: one related to the actual existence of all the links between the old and the new species at present and the other referring to the preservation of links in the geological record. Darwin's answer to the first one is:

> As natural selection acts solely by the preservation of profitable modifications, each new form will tend in a fully-stocked country to take the place of, and finally to exterminate, its own less improved parent or other less-favoured forms with which it comes into competition. Thus extinction and natural selection will, as we have seen, go hand in hand. Hence, if we look at each species as descended from some other unknown form, both the parent and all the transitional varieties will generally have been exterminated by the very process of formation and perfection of the new form. (Darwin, 1859, p. 172)

That is a good answer! If one supposes that the gradual transformation was *linear*, without ramifications. A more complex argument would be required in the case when, from a single parent species, several different variations arise, each of them with some peculiar advantage, different from the vantages of the other ones.

The second part of the difficulty concerns the gaps in the geological record: even admitting that the successive forms superseded and extinguished the former ones, we would expect to find the sequential links preserved as fossils. However, this is not the case. Darwin's reply to this difficulty is the imperfection of the geological record: preservation of old forms as fossils is the exception because only rarely the conditions for preservation are met (Darwin, 1859, pp. 172–173). The detailed argument appears in the ninth chapter of the first edition of the *Origin of Species*, so it will not be dealt with here.

Let us now consider the second part of the difficulty concerning simultaneous transitional forms in space:

> But it may be urged that when several closely-allied species inhabit the same territory we surely ought to find at the present time many transitional forms. Let us take a simple case: in travelling from north to south over a continent, we generally meet at successive intervals with closely allied or representative species, evidently filling nearly the same place in the natural economy of the land. These representative species often meet and interlock; and as the one becomes rarer and rarer, the other becomes more and more frequent, till the one replaces the other. But if we compare these species where they intermingle, they are generally as absolutely distinct from each other in every detail of structure as are specimens taken from the metropolis inhabited by each. (Darwin, 1859, p. 173)

When Darwin took as his "simple case" someone traveling from north to south over a continent, he was probably benefitting from his personal experience in South America during the travel of the Beagle (Eldredge, 2009, pp. 41–42). Hence, I will illustrate his argument using a map of this continent (Fig. 17.1).

> By my theory these allied species have descended from a common parent; and during the process of modification, each has become adapted to the conditions of life of its own region, and has supplanted and exterminated its original parent and all the transitional varieties between its past and present states. Hence we ought not to expect at the present time to meet with numerous transitional varieties in each region, though they must have existed there, and may be embedded there in a fossil condition. But in the intermediate region, having intermediate conditions of life, why do we not now find closely-linking intermediate varieties? This difficulty for a long time quite confounded me. But I think it can be in large part explained. (Darwin, 1859, pp. 173–174)

The difficulty addressed at this point is what nowadays we call the mechanism of speciation – a process of divergence and branching (Kohn, 2009, pp. 95–101). That was one of the most problematic points in Darwin's original theory. Species present a discontinuity in several ways: there are well-defined qualitative differences between similar species, and they are mutually sterile. But, on the other hand, natural selection operates upon slight changes that Darwin supposed to have no discontinuous character – except in the rare cases of "sports" (sudden changes) that were not particularly relevant to his theory.[1] If new varieties – and, afterwards, new species – are produced by the selection of continuously variable changes, and if the environment does also vary gradually (continuously) from one place to another, how can a discontinuity arise (see Waters, 2009, p. 134)? At that time, there was no theory that could conciliate continuous causes with discontinuous effects.

[1] Thomas Huxley did believe that species could arise suddenly, but that was not Darwin's idea (Blitz, 2013, p. 36).

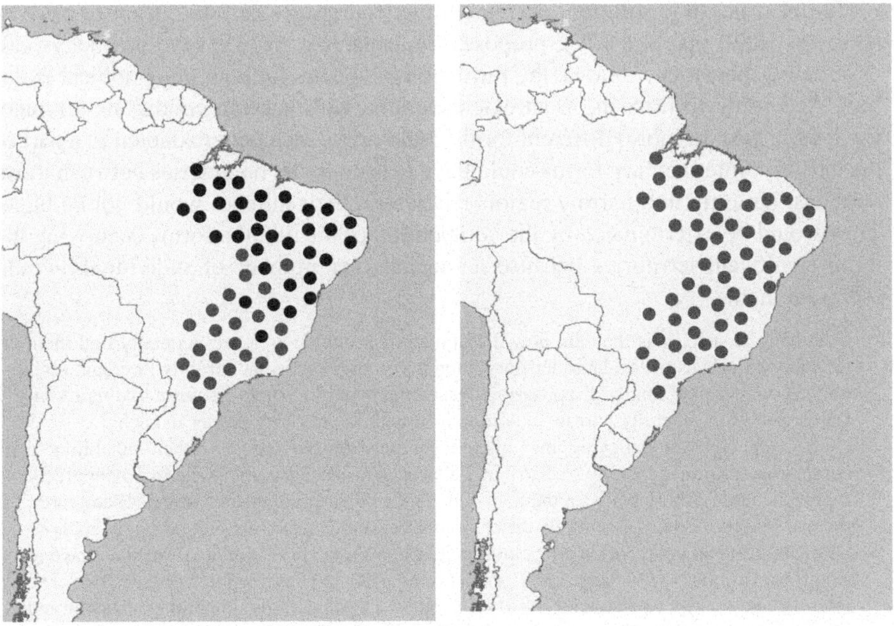

Fig. 17.1 South America map representing species in continental Brazil. *Note*. Left: According to Darwin, two closely allied species can be found scattered in a large piece of land, each being distinct even in the places where they coexist. Right: Darwin attempted to explain this situation supposing that both species have descended from a common parent, different from both, which was supplanted and exterminated by them

In chapter six, he proposed two ways out of the difficulty. The first one is temporary geographical isolation, which would allow the development of two different species in separated areas, and the later disappearance of the barrier, by geological changes (Darwin, 1859, p. 174). A geographical discontinuity could give rise to a biological discontinuity – a very simple solution.

> [...] areas now continuous must often have existed within the recent period in isolated portions, in which many forms, more especially amongst the classes which unite for each birth and wander much, may have separately been rendered sufficiently distinct to rank as representative species. In this case, intermediate varieties between the several representative species and their common parent, must formerly have existed in each broken portion of the land. Still, these links will have been supplanted and exterminated during the process of natural selection, so that they will no longer exist in a living state. (Darwin, 1859, p. 178)

However, he did not regard this mechanism as the most important one:

> But I will pass over this way of escaping from the difficulty; for I believe that many perfectly defined species have been formed on strictly continuous areas; though I do not doubt that the formerly broken condition of areas now continuous has played an important part in the formation of new species, more especially with freely-crossing and wandering animals. (Darwin, 1859, p. 174)

The next possible solution, presented over four pages of *The Origin of Species* (Darwin, 1859, pp. 174–177), proposes a mechanism for what we nowadays call "sympatric speciation," that is, the formation of species without geographical isolation. According to Darwin, in an extensive area with varying conditions (climate, food, etc.), two (or more) different forms could arise, each better adapted to a part of the territory. Intermediary forms could have existed in the boundaries between those domains, but only in a narrow region. Therefore, their number would not be huge. They would be exterminated by the competition of the definite forms occupying the more significant territories because a much larger number of individuals would represent them.

> In looking at species as they are now distributed over a wide area, we generally find them tolerably numerous over a large territory, then becoming somewhat abruptly rarer and rarer on the confines, and finally disappearing. Hence the neutral territory between two representative species is generally narrow in comparison with the territory proper to each. [. . .]
>
> If I am right in believing that allied or representative species, when inhabiting a continuous area, are generally so distributed that each has a wide range, with a comparatively narrow neutral territory between them, in which they become rather suddenly rarer and rarer; then, as varieties do not essentially differ from species, the same rule will probably apply to both; and if we in imagination adapt a varying species to a very large area, we shall have to adapt two varieties to two large areas, and a third variety to a narrow intermediate zone. The intermediate variety, consequently, will exist in lesser numbers from inhabiting a narrow and lesser area; and practically, as far as I can make out, this rule holds good with varieties in a state of nature. [...]
>
> Now, if we may trust these facts and inferences, and therefore conclude that varieties linking two other varieties together have generally existed in lesser numbers than the forms which they connect, then, I think, we can understand why intermediate varieties should not endure for very long periods; why as a general rule they should be exterminated and disappear, sooner than the forms which they originally linked together.
>
> For any form existing in lesser numbers would, as already remarked, run a greater chance of being exterminated than one existing in large numbers; and in this particular case the intermediate form would be eminently liable to the inroads of closely allied forms existing on both sides of it. But a far more important consideration, as I believe, is that, during the process of further modification, by which two varieties are supposed on my theory to be converted and perfected into two distinct species, the two which exist in larger numbers from inhabiting larger areas, will have a great advantage over the intermediate variety, which exists in smaller numbers in a narrow and intermediate zone. For forms existing in larger numbers will always have a better chance, within any given period, of presenting further favourable variations for natural selection to seize on, than will the rarer forms which exist in lesser numbers. Hence, the more common forms, in the race for life, will tend to beat and supplant the less common forms, for these will be more slowly modified and improved. (Darwin, 1859, pp. 174–177)

There are some flaws in Darwin's argument. First, he tried explaining the discontinuity between species (an effect) without assuming any discontinuous cause.

New species do not arise by jumps but by gradual changes. Therefore, the different forms inhabiting adjoining regions will be initially different varieties, not distinct species – and, consequently, they could cross and produce a mixed descendance in the intermediate region. Darwin accepted that heredity worked by a continuous combination of characters – something called "blending inheritance" by later authors – and therefore, there would be a continuous production of intermediate forms at the boundaries between the two regions. The process would continue indefinitely, and the intermediate forms would always be found – and that is not the case.

Some kind of discontinuity must be introduced. George John Romanes (1848–1894), Darwin's friend and collaborator, did not think that the explanation presented in *On the Origin of Species* was adequate, and he suggested a new mechanism: reproductive isolation, or "physiological selection" as he named it (Martins, 2017). He supposed that the new varieties could only become new species if they could not cross with the parental species. Hence, he introduced a discontinuous cause (sudden reproductive isolation) to account for the discontinuity of species.

Ronald Aylmer Fisher (1890–1962) analyzed Darwin's argument, claiming that the only way out of this difficulty is the acceptance of discontinuous, Mendelian heredity – a mechanism not contemplated in *On the Origin of Species* (see especially the first chapter of Fisher, 1930). Ernst Mayr and several other authors also claimed that Darwin did not provide a sound explanation of speciation (Mallet, 2008). Unfortunately, it is impossible to review all the arguments that have been presented for and against the mechanism proposed in the book's sixth chapter. However, the reader must be aware that Darwin's explanation is not generally accepted.

17.4 On the Origin and Transitions of Organic Beings with Peculiar Habits and Structure

The second leading group of difficulties pointed out by Darwin concerned the origin of very peculiar organs and habits, such as terrestrial animals that originate animals with aquatic habits, and flying mammals, such as bats.

Darwin's strategy is the same in all cases: he attempts to find intermediary instances to show that there was no discontinuous and sudden change, and he argues that the intermediary cases present some adaptative advantage, leading to increased development of the differences under examination. Let us see his first example:

> It has been asked by the opponents of such views as I hold, how, for instance, a land carnivorous animal could have been converted into one with aquatic habits; for how could the animal in its transitional state have subsisted? It would be easy to show that within the same group carnivorous animals exist having every intermediate grade between truly aquatic and strictly terrestrial habits; and as each exists by a struggle for life, it is clear that each is well adapted in its habits to its place in nature. Look at the Mustela vison of North America, which has webbed feet and which resembles an otter in its fur, short legs, and form of tail; during summer this animal dives for and preys on fish, but during the long winter it leaves the frozen waters, and preys like other polecats on mice and land animals. (Darwin, 1859, pp. 179–180)

Darwin pointed out a species with an intermediary habit in this particular instance. However, he admitted that, in other cases, no intermediary species is known:

> If a different case had been taken, and it had been asked how an insectivorous quadruped could possibly have been converted into a flying bat, the question would have been far more difficult, and I could have given no answer. Yet I think such difficulties have very little weight. (Darwin, 1859, p. 180)

Darwin's strategy was simple and effective: he acknowledged that sometimes no intermediary case is known. However, in other parallel cases, the transitional situation is known. Therefore, *it is not impossible* that such transitional forms existed in the other subjects. Moreover, in the particular case of the bat, Darwin made use of a parallel analysis: that of flying squirrels.

> Look at the family of squirrels; here we have the finest gradation from animals with their tails only slightly flattened, and from others, as Sir J. Richardson has remarked, with the posterior part of their bodies rather wide and with the skin on their flanks rather full, to the so-called flying squirrels; and flying squirrels have their limbs and even the base of the tail united by a broad expanse of skin, which serves as a parachute and allows them to glide through the air to an astonishing distance from tree to tree. We cannot doubt that each structure is of use to each kind of squirrel in its own country, by enabling it to escape birds or beasts of prey, or to collect food more quickly, or, as there is reason to believe, by lessening the danger from occasional falls. [...] Therefore, I can see no difficulty, more especially under changing conditions of life, in the continued preservation of individuals with fuller and fuller flank-membranes, each modification being useful, each being propagated, until by the accumulated effects of this process of natural selection, a perfect so-called flying squirrel was produced. (Darwin, 1859, pp. 180–181)

Then, Darwin introduced the remarkable instance of the so-called flying lemur (Fig. 17.2), which he describes as much closer to a bat:

> Now look at the Galeopithecus or flying lemur, which formerly was falsely ranked amongst bats. It has an extremely wide flank-membrane, stretching from the corners of the jaw to the tail, and including the limbs and the elongated fingers: the flank membrane is, also, furnished with an extensor muscle. Although no graduated links of structure, fitted for gliding through the air, now connect the Galeopithecus with the other Lemuridæ, yet I can see no difficulty in supposing that such links formerly existed, and that each had been formed by the same steps as in the case of the less perfectly gliding squirrels; and that each grade of structure had been useful to its possessor. Nor can I see any insuperable difficulty in further believing it possible that the membrane-connected fingers and fore-arm of the Galeopithecus might be greatly lengthened by natural selection; and this, as far as the organs of flight are concerned, would convert it into a bat. In bats which have the wing-membrane extended from the top of the shoulder to the tail, including the hind-legs, we perhaps see traces of an apparatus originally constructed for gliding through the air rather than for flight. (Darwin, 1859, p. 181)

Darwin found both the flying squirrels and the flying lemur cases in Lamarck's *Philosophie Zoologique* and took notes about them in his 1838–1839 *Notebook E* (Barrett et al., 1987, p. 445).

Notice that Darwin did not describe how bats originated. He did not attempt to identify any nonflying animal with similar habits that could have been related to bats. However, by describing the gliding squirrels and the flying lemur, he conveyed the belief that *it would be possible* that bats had originated by similar intermediate steps.

After examining this case, Darwin turned to wings and flight in other animals. He suggested that the so-called "flying fishes" could have given rise to animals that could fly: "it is conceivable that flying-fish, which now glide far through the air,

Fig. 17.2 Galeopithecus rufus (Audebert, 1797, plate facing p. 34). (Public domain. Biodiversity Heritage Library (https://doi.org/10.5962/bhl.title.163531))

slightly rising and turning by the aid of their fluttering fins, might have been modified into perfectly winged animals" (Darwin, 1859, p. 182). Although he did not explicitly state it, at this point, he was thinking about the origin of birds, and he immediately replied to the possible objection that fossils intermediate between fishes and birds had never been found:

> When we see any structure highly perfected for any particular habit, as the wings of a bird for flight, we should bear in mind that animals displaying early transitional grades of the structure will seldom continue to exist to the present day, for they will have been supplanted

by the very process of perfection through natural selection. Furthermore, we may conclude that transitional grades between structures fitted for very different habits of life will rarely have been developed at an early period in great numbers and under many subordinate forms. Thus, to return to our imaginary illustration of the flying-fish, it does not seem probable that fishes capable of true flight would have been developed under many subordinate forms, for taking prey of many kinds in many ways, on the land and in the water, until their organs of flight had come to a high stage of perfection, so as to have given them a decided advantage over other animals in the battle for life. Hence the chance of discovering species with transitional grades of structure in a fossil condition will always be less, from their having existed in lesser numbers, than in the case of species with fully developed structures. (Darwin, 1859, pp. 182–183)

Next, Darwin presented several instances of animals with diversified habits and structures, but at that point, he was not trying to answer any objection to the theory.

17.5 Organs of Extreme Perfection and Complication

The extreme perfection and complexity of some organs of the animals had been used as an argument for their creation by God. Darwin discussed some of those examples in this chapter section, beginning with the eye.

To suppose that the eye, with all its inimitable contrivances for adjusting the focus to different distances, for admitting different amounts of light, and for the correction of spherical and chromatic aberration, could have been formed by natural selection, seems, I freely confess, absurd in the highest possible degree. (Darwin, 1859, p. 186)

The eye had been presented as a very strong argument for God's design by William Paley (1743–1805) in his famous 1802 essay on *Natural Theology*. He devoted most of the third chapter of his book to a detailed analysis of the eye, initially comparing it to the telescope but afterward showing that it is much superior to any contrivance made by humankind (Paley, 2006, pp. 16–28). Immediately after admitting the difficulty of understanding the formation of the eye, Darwin sketches his argument for natural selection. I will decompose it into topics for easier understanding:

Yet reason tells me, that

1. if numerous gradations from a perfect and complex eye to one very imperfect and simple, each grade being useful to its possessor, can be shown to exist;
2. if further, the eye does vary ever so slightly, and the variations be inherited, which is certainly the case;
3. and if any variation or modification in the organ be ever useful to an animal under changing conditions of life,
4. then the difficulty of believing that a perfect and complex eye could be formed by natural selection, though insuperable by our imagination, can hardly be considered real. (Darwin, 1859, pp. 186–187).

Darwin did not attempt to show that there are inherited variations in the eyes of animals (2); he simply admitted that those variations could be useful (3). He just tried to show that there are gradations between eyes of different degrees of perfection.

Of course, the eyes of extinct animals cannot be known. Hence, when Darwin referred to the gradations in perfection (1), he could not be talking about development in time. Indeed, he attempted to show a gradation *comparing different groups of animals*:

> In looking for the gradations by which an organ in any species has been perfected, we ought to look exclusively to its lineal ancestors; but this is scarcely ever possible, and we are forced in each case to look to species of the same group, that is to the collateral descendants from the same original parent-form, in order to see what gradations are possible, and for the chance of some gradations having been transmitted from the earlier stages of descent, in an unaltered or little altered condition. (Darwin, 1859, p. 187)

Moreover, if one limits the comparison to the vertebrates, it will be impossible to find "very imperfect and simple" eyes:

> Amongst existing Vertebrata, we find but a small amount of gradation in the structure of the eye, and from fossil species we can learn nothing on this head. In this great class we should probably have to descend far beneath the lowest known fossiliferous stratum to discover the earlier stages, by which the eye has been perfected. (Darwin, 1859, p. 187)

Therefore, Darwin had to compare the eyes of higher animals to those of invertebrates. In this case, he was able to show a remarkable gradation, from very simple sensitive nerve terminations to a reasonably complex eye in crustacea:

> In the Articulata we can commence a series with an optic nerve merely coated with pigment, and without any other mechanism; and from this low stage, numerous gradations of structure, branching off in two fundamentally different lines, can be shown to exist, until we reach a moderately high stage of perfection. In certain crustaceans, for instance, there is a double cornea, the inner one divided into facets, within each of which there is a lens-shaped swelling. In other crustaceans the transparent cones which are coated by pigment, and which properly act only by excluding lateral pencils of light, are convex at their upper ends and must act by convergence; and at their lower ends there seems to be an imperfect vitreous substance. (Darwin, 1859, pp. 187–188)

Darwin did not attempt to exhibit intermediates between the eyes of crustacea and those of fishes and other vertebrates. Neither did he present a detailed comparison between the eyes of the vertebrates. Indeed, he simply acknowledged that in many cases, no gradations are known, and he asked the reader to *admit* that it could be produced by natural selection:

> He who will go thus far, if he find [*sic*] on finishing this treatise that large bodies of facts, otherwise inexplicable, can be explained by the theory of descent, ought not to hesitate to go further, and to admit that a structure even as perfect as the eye of an eagle might be formed by natural selection, although in this case he does not know any of the transitional grades. His reason ought to conquer his imagination; though I have felt the difficulty far too keenly to be surprised at any degree of hesitation in extending the principle of natural selection to such startling lengths. (Darwin, 1859, p. 188)

In the final part of his argument, Darwin took Pailey's analogy between the eye and the telescope, and he attempted to invalidate it with a theological argument:

> It is scarcely possible to avoid comparing the eye to a telescope. We know that this instrument has been perfected by the long-continued efforts of the highest human intellects; and we naturally infer that the eye has been formed by a somewhat analogous process. But may not this inference be presumptuous? Have we any right to assume that the Creator works by intellectual powers like those of man? (Darwin, 1859, p. 188)

Afterward, he presented a hypothetical chronological development of the eye under the assumption of the existence of variations and the action of natural selection:

> If we must compare the eye to an optical instrument, we ought in imagination to take a thick layer of transparent tissue, with a nerve sensitive to light beneath, and then suppose every part of this layer to be continually changing slowly in density, so as to separate into layers of different densities and thicknesses, placed at different distances from each other, and with the surfaces of each layer slowly changing in form. Further we must suppose that there is a power always intently watching each slight accidental alteration in the transparent layers; and carefully selecting each alteration which, under varied circumstances, may in any way, or in any degree, tend to produce a distincter image. We must suppose each new state of the instrument to be multiplied by the million; and each to be preserved till a better be produced, and then the old ones to be destroyed. In living bodies, variation will cause the slight alterations, generation will multiply them almost infinitely, and natural selection will pick out with unerring skill each improvement. Let this process go on for millions on millions of years; and during each year on millions of individuals of many kinds; and may we not believe that a living optical instrument might thus be formed as superior to one of glass, as the works of the Creator are to those of man? (Darwin, 1859, pp. 188–189)

Notice that, in the above quotation, Darwin replaced God with a personified natural selection (Depew, 2009, pp. 248–249).

I have dealt with the eye argument at some length because it was a well-known difficulty of any evolutionary theory. That was the central instance discussed by Darwin in this section. Other examples of complex organs he dealt with were not so important as the development of the electric organ of some fishes (Darwin, 1859, pp. 192–193).

While discussing the formation of complex organs, Darwin also dealt with the possibility of *change of function* of preexisting structures. One of the instances he presented was that of the transformation of the swim bladder of fishes into the lungs of higher vertebrate animals:

> The illustration of the swimbladder in fishes is a good one, because it shows us clearly the highly important fact that an organ originally constructed for one purpose, namely flotation, may be converted into one for a wholly different purpose, namely respiration. [...] All physiologists admit that the swimbladder is homologous, or "ideally similar," in position and structure with the lungs of the higher vertebrate animals: hence there seems to me to be no great difficulty in believing that natural selection has actually converted a swimbladder into a lung, or organ used exclusively for respiration. (Darwin, 1859, pp. 190–191)

However, Darwin did not present a detailed description of how this transformation could have arisen, nor did he describe intermediate stages between the two functions.

17.6 Organs of Little Apparent Importance

The next difficulty discussed by Darwin was the existence of organs with minor importance for their owners. Natural selection can only explain *adaptations*, that is, changes that improve the ability to survive and produce descendants. If a given organ has no use or minor importance, how could natural selection have made it?

> As natural selection acts by life and death, – by the preservation of individuals with any favourable variation, and by the destruction of those with any unfavourable deviation of structure,– I have sometimes felt much difficulty in understanding the origin of simple parts, of which the importance does not seem sufficient to cause the preservation of successively varying individuals. I have sometimes felt as much difficulty, though of a very different kind, on this head, as in the case of an organ as perfect and complex as the eye. (Darwin, 1859, pp. 194–195)

The tail of the giraffe and other mammals was the first example discussed by Darwin:

> The tail of the giraffe looks like an artificially constructed fly-flapper; and it seems at first incredible that this could have been adapted for its present purpose by successive slight modifications, each better and better, for so trifling an object as driving away flies; yet we should pause before being too positive even in this case, for we know that the distribution and existence of cattle and other animals in South America absolutely depends on their power of resisting the attacks of insects: so that individuals which could by any means defend themselves from these small enemies, would be able to range into new pastures and thus gain a great advantage. It is not that the larger quadrupeds are actually destroyed (except in some rare cases) by the flies, but they are incessantly harassed and their strength reduced, so that they are more subject to disease, or not so well enabled in a coming dearth to search for food, or to escape from beasts of prey. (Darwin, 1859, p. 195)

Notice that, at that time, it was not known that insects could transmit diseases – this was only established in the late nineteenth century. So, what Darwin actually claimed was that flies could harass and reduce the strength of the animals, and they could, for that reason, become easier prey to diseases or death. And he presented the specific instance of cattle and other animals in South America that can only survive if they can resist the attack of insects. Therefore, organs that seem useless (or of little importance) might be very important, indeed.

Besides that, Darwin also stated that organs of little importance could have been highly important to the ancestors of that animal. For example, the tail is a fundamental structure in many aquatic animals; assuming that land animals came from aquatic animals, they would have inherited their tails, and they could afterward have undergone a change of function:

> Organs now of trifling importance have probably in some cases been of high importance to an early progenitor, and, after having been slowly perfected at a former period, have been transmitted in nearly the same state, although now become of very slight use; and any actually injurious deviations in their structure will always have been checked by natural selection. Seeing how important an organ of locomotion the tail is in most aquatic animals, its general presence and use for many purposes in so many land animals, which in their lungs or modified swimbladders betray their aquatic origin, may perhaps be thus accounted for. A well-developed tail having been formed in an aquatic animal, it might subsequently come to

> be worked in for all sorts of purposes, as a fly-flapper, an organ of prehension, or as an aid in turning, as with the dog, though the aid must be slight, for the hare, with hardly any tail, can double quickly enough. (Darwin, 1859, pp. 195–196)

Darwin also called the attention of the reader to the possibility of *other causes* – besides natural selection – that could produce organs of little importance:

> In the second place, we may sometimes attribute importance to characters which are really of very little importance, and which have originated from quite secondary causes, independently of natural selection. We should remember that climate, food, &c., probably have some little direct influence on the organisation; that characters reappear from the law of reversion; that correlation of growth will have had a most important influence in modifying various structures; and finally, that sexual selection will often have largely modified the external characters of animals having a will, to give one male an advantage in fighting with another or in charming the females. Moreover when a modification of structure has primarily arisen from the above or other unknown causes, it may at first have been of no advantage to the species, but may subsequently have been taken advantage of by the descendants of the species under new conditions of life and with newly acquired habits. (Darwin, 1859, p. 196)

17.7 Final Comments

I have shown here some crucial features of the chapter addressing the difficulties of Darwin's theory. There are, however, many other relevant topics that one should study by reading *The Origin of Species*.

Darwin's main strategy is to show that the difficulties are not fatal to the theory of natural selection. His arguments do not *prove* that all the organs and features of the existing species were produced by natural selection – and, indeed, he always contemplated other causes, as shown at the end of the last section. However, in the case of useful characteristics, he attempted to show that there are plausible ways to develop those features by natural selection.

The weightiest difficulty dealt with by Darwin in this chapter was the mechanism of speciation. His proposal was not generally accepted, even by some of his friends, and it was rejected in later times. However, in the case of the other difficulties he addressed, evolutionists generally accepted his answers.

References

Audebert, J. B. (an VII [1797]). *Histoire naturelle des singes et des makis*. Desray.

Barrett, P. H., Gautrey, P. J., Herbert, S., Kohn, D., & Smith, S. (Eds.). (1987). *Darwin's notebooks, 1836–1844: Geology, transmutation of species, metaphysical enquiries*. Cambridge University Press.

Blitz, D. (2013). *Emergent evolution: Qualitative novelty and the levels of reality*. Springer.

Bordalejo, B. (Ed.) (2008–2012). Online Variorum of Darwin's *Origin of Species*. In John van Wyhe (Ed.), *The complete work of Charles Darwin online*. http://darwin-online.org.uk/Variorum/

Brooke, J. H. (2009). "Laws impressed on matter by the creator"? The origin and the question of religion. In M. Ruse & R. J. Richards (Eds.), *The Cambridge companion to the "origin of species"* (pp. 256–275). Cambridge University Press.
Darwin, C. (1859). *On the origin of species by means of natural selection, or the preservation of favoured races in the struggle for life*. John Murray.
Depew, D. J. (2009). The rhetoric of the *origin of species*. In M. Ruse & R. J. Richards (Eds.), *The Cambridge companion to the "origin of species"* (pp. 237–255). Cambridge University Press.
Eldredge, N. (2009). Experimenting with transmutation: Darwin, the beagle, and evolution. *Evolution: Education and Outreach, 2*, 35–54.
Fisher, R. A. (1930). *The genetical theory of natural selection*. Oxford University Press.
Fisher, R. A. (1954). Retrospect of the criticisms of the theory of natural selection. In J. Huxley, A. Hardy, & E. B. Ford (Eds.), *Evolution as a process* (pp. 84–98). Allen & Unwin.
Gruber, J. W. (1960). *A conscience in conflict: The life of St. George Jackson Mivart*. Columbia University Press.
Kohn, D. (2009). Darwin's keystone: The principle of divergence. In M. Ruse & R. J. Richards (Eds.), *The Cambridge companion to the "origin of species"* (pp. 87–108). Cambridge University Press.
Lustig, A. J. (2009). Darwin's difficulties. In M. Ruse & R. J. Richards (Eds.), *The Cambridge companion to the "origin of species"* (pp. 109–119). Cambridge University Press.
Mallet, J. (2008). Mayr's view of Darwin: Was Darwin wrong about speciation? *Biological Journal of the Linnean Society, 95*(1), 3–16.
Martins, R. A. (2017). *George John Romanes and the struggle for Darwin's mantle*. Privatus Typus.
Paley, W. (2006). *Natural theology or evidence of the existence and attributes of the Deity, collected from the appearances of nature*. (M. D. Eddy & D. Knight, Edition, Intr., and Notes). Oxford University Press.
Peckham, M. (1959). *The origin of species by Charles Darwin: A variorum text*. Oxford University Press.
Waters, C. K. (2009). The arguments in the origin of species. In J. Hodge & G. Radick (Eds.), *The Cambridge companion to Darwin* (2nd ed., pp. 120–146). Cambridge University Press.

Martins R. A. is a Brazilian scholar. His main areas of research are history and philosophy of science and science education. He is a retired professor of the State University of Campinas (UNICAMP) and a visiting professor of PECMA, Federal University of São Paulo (UNIFESP). He has been President of the Brazilian Society for History of Science (SBHC) and the South Cone Association for Philosophy and History of Science (AFHIC).

Chapter 18
Origin's Chapter VII. Darwin and the Instinct: Why Study Collective Behaviors Performed Without Knowledge of Their Purposes?

Nelio Bizzo and Lucas Marino Vivot

Abstract *Origin of Species* brings a chapter that deals with instinct, summarizing a series of reflections by Charles Darwin, started in 1837, in his Notebook B, which is interpreted here as a dialogue with the Natural Theology and the Moral Philosophy of the Anglican elite of that time. Darwin discussed crucial aspects that had been taught in his university course at Cambridge, but which he saw manifest themselves quite differently in nature. He had been taught that moral philosophy should be a mirror of the divine will, from which benevolence and compassion would derive. However, in this chapter, Darwin shows that struggle in nature is a law that explains morally repulsive natural behaviors such as ant slavery, the infanticide of certain wasps, and the matricide of honeybees. Does instinct, as "natural behavior," have an intelligent agent that continually encourages the opposite of benevolence and compassion? How would the "instinct to make slaves," matricide and infanticide relate to the benevolence of nature? Would these disgusting behaviors have been designed and would continue to be encouraged by an omnipotent Creator? Instead of a myriad of phenomena that required continuous intervention by an intelligent agent, nature could be seen as a number of mechanisms following fixed laws, in the Newtonian tradition, which implacably leads to the "advancement of all organic beings, namely, multiply, vary, let the strongest live and the weakest die." This new Darwinian interpretation led to the conclusion that the "intelligent agent" could not be cruel enough to encourage repugnant behavior in nature. Thus, Darwin somewhat anticipated his defense of the need for new theoretical frameworks for Anglican Theology, from which he did not expect any greetings, even more after the first critical

N. Bizzo (✉)
Faculty of Education, University of São Paulo; Institute of Environmental Science, Chemical & Pharmaceutical Sciences, Federal University of São Paulo, Diadema, SP, Brazil
e-mail: bizzo@usp.br; bizzo@unifesp.br

L. M. Vivot
UniBTA University Center – Brazilian Institute of Advanced Technology, Education and History of Science at the Federal University of ABC (UFABC), Santo André, SP, Brazil

M. Elice Brzezinski Prestes (ed.), *Understanding Evolution in Darwin's "Origin"*,
History, Philosophy and Theory of the Life Sciences 34,
https://doi.org/10.1007/978-3-031-40165-7_18

reviews of his book in the fall of 1859. In fact, one of the few modifications Darwin in the second edition of *Origin* was the insertion of an epigraph of one of the books he had to read at the university, written by Joseph Butler (1692–1752), with the traditional Anglican Moral Philosophy. These reflections may have important consequences for the teaching and learning of biological evolution today, discussing, from the historical point of view, how the moral and religious aspects were turned compatible with strictly scientific questions.

18.1 Introduction

In this text we deal with instinct, synthesizing a series of reflections by Charles Darwin, started in 1837, in his Notebook B, which is here interpreted as a dialogue with the Natural Theology and Moral Philosophy of the Anglican elite of the time. Darwin discussed crucial aspects that had been taught in his university course at Cambridge, but which he saw manifested quite differently in nature. Let us analyze, in detail, a part of the *Origin of Species*, more precisely Chap. VII, on instinct. We consider this chapter more important than it has been acknowledged, and it does not seem to have been the part of the book that has most attracted the attention of historians and biologists. His ideas on the evolution of morality were referenced mainly by his writings in *Descent of Man* (1871), understanding that morality is linked to the will, to the judgment of present and past actions, recognizably human characteristics.

We do not intend to claim that the chapter on instinct has been neglected by specialists, as one can easily remember the emergence of Ethology, in the well-known works of Konrad Lorenz (1903–1989) and, soon after, Nikolaas Tinbergen (1907–1988), which accompanied the emergence of the so-called New Synthesis, since the early 1930s. However, let us not forget that the theory of heredity in which Darwin operated was entirely different from the one Neo-Darwinism was based on.

Therefore, we present evidence in support of our thesis, namely, that the chapter on instinct in the first edition of *Origin of Species* played a key role for argumentation regarding morality, which Darwin would develop in later years within an evolutionary framework. We want to demonstrate how the chapter on instinct had strategic importance for Charles Darwin in the debate with Cambridge professors, clerics of the religious elite of the time, as it extended the discussion beyond material causes, reaching the field of moral philosophy and Natural Theology. After all, he was well aware of the theological implications of his ideas, as he had explicitly written in letters to many correspondents, as the American Botanist Asa Gray (1810–1888) and the Anglican divine Charles Kingsley (1819–1875).

Darwin had studied the mandatory bibliography of theology taught at "Oxbridge," including writings on ethics and morals, but had evidenced contradictions between behaviors supposedly based on values held by Christians, such as love of neighbors, compassion, and solidarity, and those reported as instinctive in scientific treatises, describing how the "young cuckoo ejects its foster-brothers," the enslavement of certain ants, and how "the larvae of ichneumonidae feeds within the live bodies of caterpillars."

Darwin ends the chapter by stating that such instinctive behaviors, totally different from what could be called "natural virtues," could never be attributed to a "special creation," or the result of repeated divine interventions, but the result of a "general law" which was not limited to natural selection, but went far beyond it, as it included the reproductive processes and those that gave rise to variation. The chapter ends with the compromising sentence about the "general law" of nature: "multiply, vary, let the strongest live and the weakest die" (Darwin, 1859/2018, p. 253).

So we ask: how could Chap. VII of the *Origin of Species* have been written as simple additional evidence of natural selection, if it were just the last part of that great "general law" which, after all, would lead to a continual "advancement" of every living being?

18.2 "Natural" Behaviors and "Natural" Government

Starting from this provocative question, we intend to present evidence that the chapter on instinct has clear allusions to the moral philosophy of Anglican theology studied by Darwin, such as the writings of Bishop Joseph Butler (1692–1752) in his *Analogy of Religion* (Fig. 18.1). In the same final paragraph of the chapter, Darwin speaks of the "advancement of all living beings" coinciding with what he had learned about the development of virtue.

Initially, it is worth remembering that the first edition of *Origin of Species* has two epigraphs, but all other editions gained an additional one, whose author was Joseph Butler. The Anglican bishop's "un-deistic quotation" was "inserted on the verso of the title page at the instigation of Gray" (Moore, 1979, p. 324), at the same time that explicit allusions to the "Creator" were added, including the supposed original miraculous act of human creation, stating that life was originally "breathed by the Creator." Thus, the Bishop's entry on the scene revealed the final causes of some linguistic choices in the first edition. The verb "to breathe," referring to the creation of life, was a clear allusion to the biblical passage from Genesis 2:7: "and breathed into his nostrils the breath of life." By making the subject explicit in the sentence, inserting "Creator" with a capital letter, Darwin left no doubt about the origin of the reference, which would certainly be noticed by the clerics of "Oxbridge." However, in the following edition, he introduced some changes aiming at eliminating "the possibility that his God might play an active role in the world. His break from Paley's theology was unequivocally complete" (Moore, 1979, p. 325).

Darwin had studied moral philosophy, listening to what we should and should not do, what would be the "natural state of humanity," which would somehow have retained what would have been infused in the "original breath." This would be the "nature of virtue," the name of one of the parts of the book highlighted by Darwin. Maternal love, for example, would be a manifestation of this natural state, since

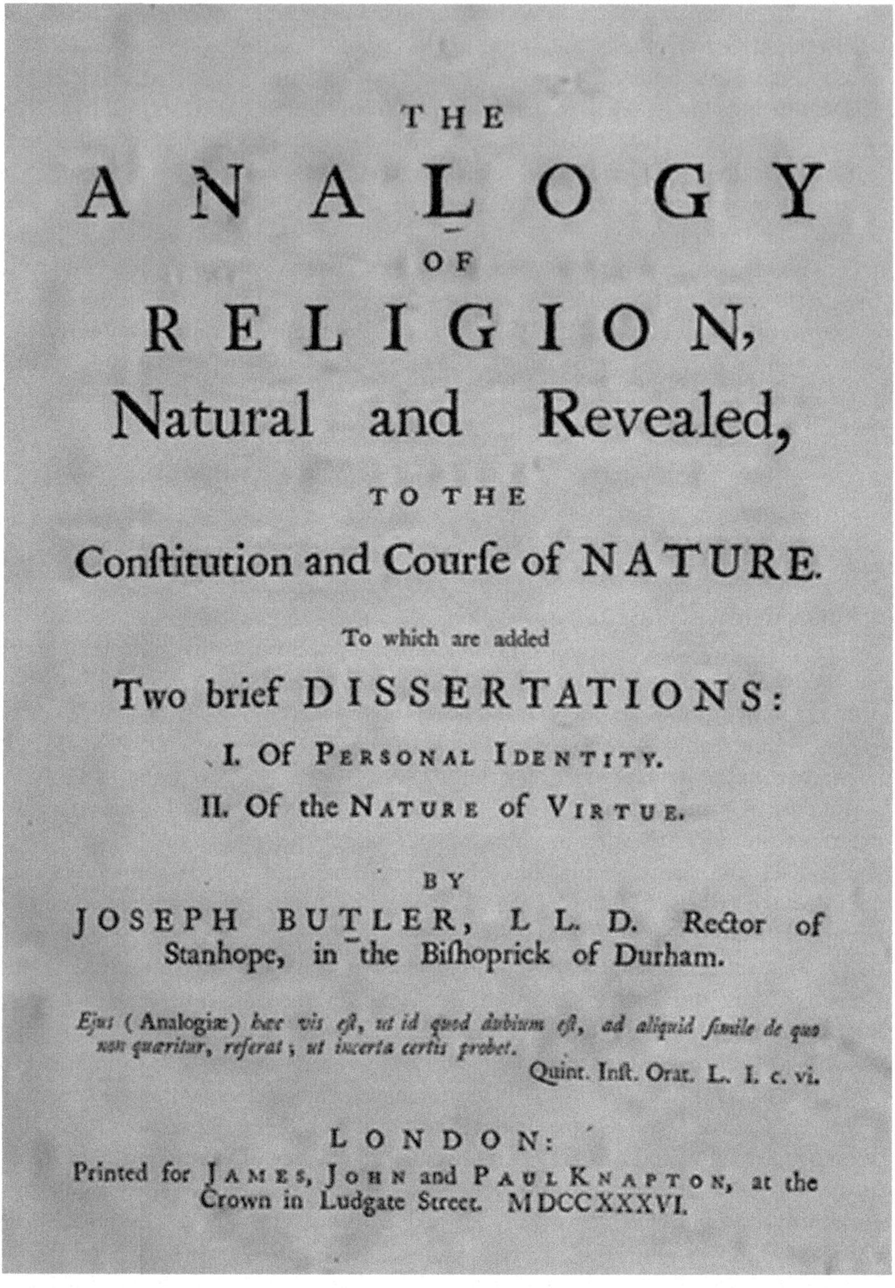

THE

ANALOGY

OF

RELIGION,

Natural and Revealed,

TO THE

Conſtitution and Courſe of NATURE.

To which are added

Two brief DISSERTATIONS:

I. Of Personal Identity.

II. Of the Nature of Virtue.

BY

JOSEPH BUTLER, LL.D. Rector of Stanhope, in the Biſhoprick of Durham.

Ejus (Analogiæ) *hæc vis eſt, ut id quod dubium eſt, ad aliquid ſimile de quo non quæritur, referat; ut incerta certis probet.*

Quint. Inſt. Orat. L. I. c. vi.

LONDON:

Printed for James, John and Paul Knapton, at the Crown in Ludgate Street. MDCCXXXVI.

Fig. 18.1 *Joseph Butler's book mentioned by Darwin.* (Fonte: The Open Library. License: Public Domain Mark. https://archive.org/details/analogyofreligio02butl)

mothers have instinctive and involuntary behaviors to care for others, especially young children. Mothers do not judge their child after birth to decide whether the new child deserves their love and attention from their older siblings. His love, in each birth, would be instinctive or "natural." In Butler's theology, however, what was "natural" did not always start from an unambiguous understanding, for life, as we know it from evidence, is "natural." However, the afterlife, which we do not know from hard evidence, would also be "natural." He argues that it is very comfortable to use the word "natural" to refer to both processes, but only very short reasoning would allow us to believe only what we can see before our eyes, for the immortality of the soul and the possible path afterward death, is "upon the evidence of reason." And here he establishes a precise definition of what is natural, in the sense of something that can be "stated, fixed or settled", as being the direct result of an "intelligent agent," who would instigate, at every moment, or intervals, the behaviors so-called natural, which presupposes a continuous, daily, that is, non-miraculous action (Butler, 1736, p. 29).

The continual administration, repeated promptings, careful and meticulous government, scrutiny of every act, however, would not have been necessary for what would have been miraculous. Miracles would not be ordinary phenomena, and their effects would be permanent. Soon after the passage highlighted by Darwin to appear as an epigraph in all subsequent editions, we can find the existence of a project, which serves a purpose, a final cause of the constitution of nature, and this active agent is an intelligent designer, who acts by constantly sending alerts, rewards, and punishments.

In the next chapter (Chap. II "Of the Government of God by Rewards and Punishments; and particularly of the latter"), he will treat the administration of rewards and punishments, in this life and the future one, as something predictable, since we would be under the control of a meticulous government. After all, all people would naturally act knowing the consequences of avoiding evil or obtaining good. Our "natural faculties of knowledge and experience" derive from his recommendations, and our presentiments of the consequences would be warnings given by him at every moment (Butler, 1736, p. 36).

In Chap. III ("Of the Moral Government of God"), Butler will detail how this uninterrupted "natural government" of the "Intelligent Mind" intermittently distributing "pleasure and pain" takes place:

> As the manifold appearances of design and of final causes, in the constitution of the world, prove it to be the work of an intelligent mind, so the particular final causes of pleasure and pain, distributed amongst his creatures, prove that they are under his government; what can be called his natural government of creatures endued with sense and reason. This, however, implies somewhat more than seems usually attended to, when we speak of God's natural government of the world. It implies government of the very same kind that which a master exercises over his servants, or a civil magistrate over his subjects. These latter instances of final causes as really prove an intelligent governor of the world (...) (Butler, 1736, p. 44).

The exercise of this "natural government" maintained master-servant relationships, or similar to those of the civil magistrate over his subjects, that is, a highly hierarchical relationship of evident and inevitable submission. Judging by the number of editions of Joseph Butler's books on both sides of the Atlantic, there is no doubt that this was a dominant, truly inescapable perspective that Darwin knew he could not dodge. The inclusion of a third epigraph in subsequent editions of *Origin of Species* could, in fact, only be intended to clarify its alignment with the Anglican mainstream.

18.3 The Intelligent Agent and Natural Theology

At the same time that Butler was writing his sermons on the scrutiny of every act performed in the concrete world by the "Intelligent Mind," Carl von Linné (1701–1778) was also certain that living beings had been created with marks of lessons to be learned. He wrote: "The calyx is the conjugal bed, the corolla is the curtain, the filaments are the spermatic vessels, the anthers are the testes, the pollen is the fertilizing principle" (von Linné, 1749, p. 373).

Linné is describing a flower based on human anatomy and behavior, with different morals designating men and women of the time. Men would be "naturally" promiscuous, while women would be "naturally" demure, with floral anatomy reflecting a supposed final cause of the anatomy of male and female gamete-producing organs. But Linné did not speak of superficial analogies, stating categorically a functional anatomical correspondence, saying that "the stigma is the vulva," "the style is the vagina," "the germ is the ovary," "the pericarp is the fertilized ovary," "the seed is the ovum" (von Linné, 1749, p. 373). This was the corrected version of the previous small published work, *Sponsalia Plantarum* (Linné and Wahlbom, 1746) where they discussed how the plants "nuptias celebrant," explaining the paralleled anatomy.[1] Thus, he traced a perfect anatomical correspondence between animals and plants, and even went so far as finding the floral correspondence of the "Tubæ Falopianæ"! Nevertheless, far more than that, a similarity between the sexual behavior of animals and plants, the result of the "natural government" of the material and tangible world.

But the question of universal sexuality in plants and animals was so perplexing that the Imperial Academy in St. Petersburg instituted a prize in 1759 for the best essay on fertilization, which ended up winning none other than Carl von Linné himself. Moreover, in that essay, he attributed the existence of hybrids to the promiscuity between the male and female parts, which could explain the emergence of new species. From a moral point of view, in the Puritan perspective, botany should be a subject reserved for men, which would explain the fact that Erasmus Darwin (1731–1802) anonymously published his verses about sex in vegetables in

[1] In the copy of *Sponsalia Plantarum* available online on the Linnean Society of London website, it is possible to see the correction made by hand, as the style was identified as being both the vulva and the vagina, which was then modified to be the stigma (p. 53).

1789, exploring a form of eroticism potentially scandalous in that moral atmosphere (Browne, 1989).

The times and context of his grandson, Charles Darwin, gave Botany a Victorian air, reviving the myth of Proserpina and her double character of innocence and sexuality, which would have been revived by the somewhat libertine verses of Erasmus Darwin and the botanical poetry of Catherine Maria Fanshawe (1765–1834). Victorian romantic literature valued the connection of flowers to the feminine character (Catsikis, 2009).

When Charles Darwin makes his reference to Joseph Butler's Anglican moral philosophy explicit, the unanswered question is: did Darwin change his mind about the origin of species? When analyzing an excerpt from the letter Darwin wrote in May 1861 to Sir J. F. W. Herschel, he says:

> [...] we cannot look at all living productions and the Universe without believing that it has been intelligently designed, but when I look at each individual organism, I cannot see any evidence of it (DCP-LETT-3154; letter from Darwin to Sir J.F.W. Herschel, May 23, 1861).

Here Darwin makes it clear, writing to his greatest idol of the day, that he has not changed his conception of the intelligent agent. Joseph Butler, in 1736, in his famous sermons, explains that human benevolence, such as love of neighbor and charity – which Anglicans extolled as praiseworthy behaviors – can only be exercised because of the project that created us made it possible. In 1744, while Butler was reading his sermons from English pulpits, the British Crown was paying no less than 50 pounds sterling for the scalp of an Indian woman or child from the Massachusetts colony. For an adult male Indian scalp, the reward was doubled. Shortly thereafter, in 1757, the murder of adult men was rewarded at triple the value, which made Indian genocide a very profitable business (Seybolt, 1930).

By that time, the Merseyside shipowners grew rich from the trade in enslaved Africans, including British colonies in the New World. An estimated 12.5 million enslaved people were transported from Africa to the Americas. At least 1.8 million would have died crossing the Atlantic, with about half being trafficked by English slave ships (Gomes, 2019). The abolition of slaves by British law would not occur until 1833, and even then, it would persist in some colonies, such as Jamaica, for some more years. The Church of England itself owned slaves in the region, in Barbados, in the *Society for the Propagation of the Gospel in Foreign Parts*. Anglicans recognized that the act of abolition brought pecuniary reward to them, as they owned 665 slaves, but no aid to the enslaved people (BBC News, 2006). In Angola alone, the Jesuits had become the largest slaveholders in the mid-seventeenth century, reaching some 10,000 people; most of them toiled in farms near Luanda (Ferreira, 2012, p 93).

Therefore, the Church of England, directly subordinate to the Crown, preached a "natural government," which followed everything, rewarded every act performed in the tangible world, filing benevolent deeds for just reward in the hereafter. However, the Church not only owned slaves but also did not disapprove of the genocidal policy practiced in the American colonies. And Butler said that everyone would be being watched for the practice of "benevolence." It is interesting that in Butler's book key terms such as "slavery," "slave," and "enslaved" do not appear at all.

Another very important author who continued Butler's work was William Paley (1743–1805) who wrote the book *Natural Theology* (Paley, 1802/1829), mandatory for more than 100 years in Cambridge courses. This famous bibliography continues this argument about the "author of nature," this deity, who would have created the world in his likeness. The "bad things" would be human "perversions," acts to be restrained and punished. At the beginning of the book, he praises this clever project, showing, for example, how the shape of the human hand and fingers of different lengths are perfect for picking up and manipulating objects, as when we bend our hand we have a better grip, which would be a clear demonstration of the nature's perfection.

At the end of the book, he stresses the question of morality, of how we should behave, and extols the great "benevolence" of this deity and how it manifests itself in our behavior. He says, in Chap. XXVI ("Goodness of the Deity"):

> Assuming the necessity of food for the support of animal life; it is requisite that the animal be provided with organ, fitted for the procuring, receiving and digesting of its food. It may be also necessary that the animal be impelled by its sensations to exert its organs, But the pain of hunger would do all this. Why add pleasure to the act of eating, sweetness and relish to food? Why a new and appropriate sense for the perception of pleasure? Why should – the juice of a peach, applied to the palate, affect the part so differently from what it does when rubbed upon the palm of the hand? This is a constitution which, so far as appears to me, can be resolved into nothing but the pure benevolence of the Creator. Eating is necessary; but the pleasure attending it is not necessary: and that this pleasure depends, not only upon our being in possession of the sense of taste, which is different from every other, but upon a particular state the organ in which it resides, a felicitous adaptation of the organ to the object, will be confessed by any one who may happen to have experienced that vitiation of taste which frequently occurs in fevers, when every taste is irregular, and every one bad (Paley, 1802/1829, p. 270).

For Paley, there is no way to discredit the extreme benevolence of the Creator, since he makes us eat, rather than because of hunger, but because of the pleasant flavors he instilled in food. At the same time, he designed organs suitable for transmitting the sensation of pleasure that leads us to perform the acts that explain our existence and subsistence. It would not do any good for the peach to be so sweet if we did not have taste buds to perceive the sweetness. For him, the intelligent design is evident, almost definitively proven, because by putting the sugar in the fruit, at the same time the sensitivity of the sugar was created in our mouth and not in our hands, for instance. If we were to capture the sweetness in our hands, the pleasure we get from taking a peach to our mouths would not be generated. This extreme benevolence of the Creator would also be evident when verifying the distortions caused by unhealthy states: in fevers, our taste remains distorted, and our desire to eat is restrained, as this would do us harm, and not good.

The author goes on to say that, as this benevolence of the Creator is extreme, the human being must abstain from pleasures, because, if we do not have any kind of "brake," we will exaggerate in the enjoyment of this benevolence, of these pleasures. Therefore, we would lose ourselves and move away from the idea of Anglican morals. Looking at nature, we would be seeing intelligent design, designed around what this great designer wants to be done and imitated. Moreover, this "natural ruler," supreme sovereign, would constantly remain watching each animated being, human or not, instigating, stimulating, and warning about the consequences of pleasure and pain of each behavior to be exercised: "The works of nature just want to be contemplated. When contemplated, they have everything that can astonish by their greatness." In addition, that magnitude can be of several orders, from the immensity of Saturn's rings to the particularities of the hummingbird's tongue. Furthermore, he stated, "we have evidence, not only of both these works proceeding from an intelligent agent, but of their proceeding from the same agent." Therefore, he concluded: "one mind hath planned, or at least hath prescribed a general plan for all these productions. One Being has been concerned in all." (Paley, 1802/1829, p. 298).

Darwin had read these words meticulously as a student. There would be evidence that everything we find in nature comes from a single intelligent mind, and it is only up to us to admire these grandiose productions. Nevertheless, Darwin found ways to question this evidence and was aware that this would not please the Anglican elite of his day.

18.4 The Creator's Benevolence in His Works

In Chap. VII of the *Origin of Species*, Darwin states that he will not put forward a clear definition of instinct, but he states a very objective functional definition, even if he claims that it is not universal, saying that an action is instinctive "when performed by an animal, more especially by a very young one, without any experience, and when performed by many individuals in the same way, without their knowing for what purpose it is performed" (Darwin, 1859/2018, p. 218). Here we see a clear break with Natural Theology, by affirming the independence between the exercise of behavior and the knowledge of its purpose, sublimating the final cause that would prove the intelligent project, according to Anglican theologians such as Butler and Paley. The latter presented a definition of instinct, saying that it was "a propensity prior to experience, and independent of instruction" opening the chapter on the subject in his famous book. However, he has restricted himself to discussing examples of behaviors such as suckling and oviposition in this chapter on instinct (XVIII).

Paley wonders if the hen loves her eggs or knows what is inside them to take such good care of the nest. He seems to find the answer by citing the cuckoo as an example that would settle the question: the female, who never knows her young, carefully lays her eggs in other birds' nests, not in any hollow place. He then quotes the salmon, on their journey upriver to spawn in complete ignorance of what they are doing, as none of these fish will ever know their offspring.

The Jamaican Violet Crab is mentioned next as it makes a long march up the slopes of mountains to spawn on the beach, returning shortly afterward with no news of its development. Moths and butterflies are other examples of mothers who have no idea about their children. The argument is persuasive to the point of exploring examples that can counter this idea that instinctive behaviors are unmotivated, including appealing to an unusual resource among defenders of the superiority of the final cause, as Paley says that the exceptions are few and that could be explained by the "tolerable probability," that is, chance could exert some influence depending on the circumstances of some cases.

It is almost impossible not to compare what Darwin wrote about instinct in the *Origin of Species* with the chapter on the same subject by the theologian he studied closely in his university course. In addition, here a cunning strategy insinuates, as Darwin makes use of recent discoveries, which were not known when Paley wrote his book.

One of the most important is perhaps the mention of the work of Pierre Huber (1777–1840), who observed in June 1804, in Geneva, where he lived, and reported in the book *Recherches sur les Moeurs des Formis Indigenes* (1810), translated into English in 1820 under the title Natural History of Ants (Huber, 1820). In this book (Fig. 18.2), he noted what has come to be known as the "enslavement" of ants, even though he does not use the corresponding terms "enslavement" or "slaves" for dark gray ants (noir cendré), referring to them as "associates" and "assistants." Nor does he use the terms corresponding to "ladies" or "masters" for *F. rufescens* (which means "reddish" in Latin), already recognized as "warriors" and called "Amazons" and "legionaries."

However, the English translator of the book, J. R. Johnson, decided to replace the expression "dark gray ant" with "black" and "black ant," due to the "coloring" and "situation maintained in the colony" (of subservience), a clear reference to the condition of enslaved. The translator recorded the "freedom" of the translation in a footnote, on page 252 (Fig. 1), where he recorded:

> To avoid the too frequent repetition of the same word, I shall take the liberty, when speaking of the dark ash-colored Ant, occasionally to use the appellation of Negro, or Negro Ant; a term not inapplicable, when we consider the dark color of this species, and the situation it holds in the colony, of providing for and administering to, the wants, &c. of the Amazons. – T. (Huber, 1820, p. 252).

However, not even in the English version are the terms related to slavery such as "slave," "lord," or "master." Darwin had a copy of the original in French, but he has used some commentator terms since the first review of the book was published in English in 1812. Even so, he does not use the term black ant.

Pierre Huber was heavily criticized for "naturalizing" slavery as if his description was just a figment of his imagination, but he proved to have been a very strict observer. But he spoke of "dark gray" ants kidnapped before developing their instinct, being raised by the Amazons, "divide with them the fruit of their industry," and stating that "we here trace neither servitude nor oppression," because "they live under the same roof in brotherly and sisterly union" and "no aversion is excited in the breasts of those who witness their birth" (Huber, 1820, pp. 308-9).

It is not unreasonable to point out that the period is extremely delicate in the British context, due to the discussion about the slave trade and, mainly, the English

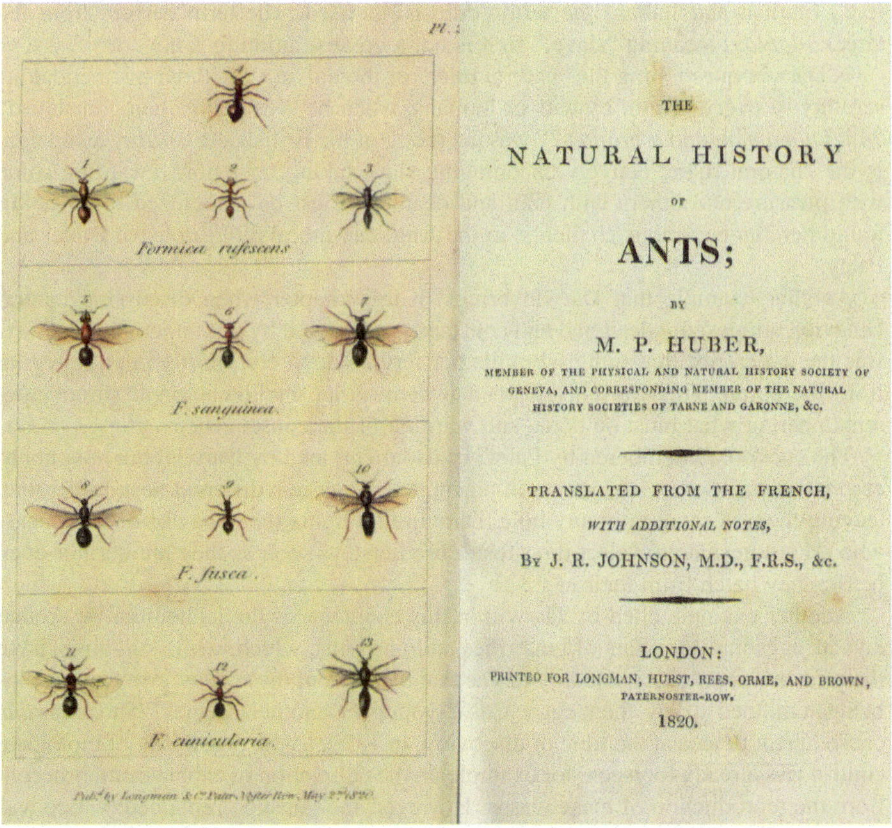

Fig. 18.2 Pierre Huber's book (Recherches sur les Moeurs des Formis Indigenes, Paris, 1810) in English translation, published in 1820. (License: Public Domain Mark. Credit: The natural history of ants / by M.P. Huber. Translated from the French, with additional notes, by J.R.Johson. Wellcome collection https://wellcomecollection.org/works/nddg3m22/items])

participation in it. William Wilberforce (1759–1833), whose son would become Archbishop of Oxford and a strong opponent of Darwin, fought for the passage of the Anti-Slave Trade Act (1807). In addition, as already mentioned, only in 1833 would slavery be abolished in the British colonies, with the declared support of several members of the Darwin family. He recorded his repugnance to the practice during his visit to Brazil at the time.

It is interesting to read in Chap. VII how Darwin talks about the enslavement of ants, avoiding dealing with "slavery." In fact, until the present day, these terms commonly appear in the form of euphemisms, with technical terms such as "esclavagismo" in Portuguese, which has a different spelling than the literal term for slavery (escravismo). Darwin used the most neutral term possible to designate "ants that make slaves," avoiding pejorative terms. In several languages (e.g., English and Italian) the term "dulosis" is used. The term derives from the Greek (*doúlos*) meaning "slave," so it is not a great solution to use.

Is Darwin questioning the "natural ruler" of the universe, without his meticulous scrutiny of every action, by ants or humans, when he sees beings being enslaved? "*Am I not a man and a brother?*" was the motto of the British anti-slavery campaign. Is the "natural ruler" actively encouraging slave-taking behavior, rewarding some with pleasure and others with pain and death? It must be recognized that Darwin found here an enormous challenge to the Anglican moral philosophy of Butler and Paley.

Another example that Darwin brings in this chapter when discussing natural behavior, supposedly designed and constantly stimulated by this supernatural entity, was the queen bee that ruthlessly kills her fertile daughters, shortly after emerging from the comb. This can only happen if there is an intelligent agent stimulating, remembering what must be done, and warning of the consequences.

The cuckoo, remembered by Paley, is also mentioned by Darwin, but now not to emphasize the mother's kindness in placing each egg in a different nest, rather than leaving them at random in any hole. Darwin highlights the behavior of the chicks, who do not hesitate to "eject their foster brothers" as soon as they are born or even before they hatch from their eggs.

Another example cited by Darwin in this chapter was the ichneumon wasp that lays its eggs in caterpillars of butterflies and/or moths, which, while still alive, have their innards slowly devoured by larvae that hatch from these eggs. Are these wasps being reminded to lay their eggs in live young of another species? This example caused great debate at the time of discovery, in 1830, as the possibility of biological control was already foreseen, for example, in the cultivation of cabbage and broccoli, from the reproduction of these wasps. However, the question remained: where was the extreme benevolence of the deity that Butler and Paley spoke of? Was the Deity rewarding the virtues of the cabbage farmer, or the breeding habits of the little wasp, and punishing the moths with pain and death for their small, helpless young?

18.5 Vigilant Natural Government or a Great General Law?

How do the "enslavement instinct," matricide, and infanticide relate to the benevolence of the deity? Enslaving other people would be a behavior planned and continually stimulated and rewarded with pleasure for the strong, and pain and death for the weak, reflecting by analogy the relationship that the "master exercises over his servants, or the civil magistrate over his subjects" mentioned by Butler?

Darwin is not here arguing with zoologists or entomologists, but with the Anglican elite of his day, asking whether such examples were "virtuous" designed and continually stimulated by an intelligent agent. This discussion is not entomological, but theological, in the field of natural theology and moral philosophy.

Darwin is not arguing with the experts but presenting facts that need to be interpreted. Anglicans taught these behaviors in universities as effects of a final cause intentionally designed by an "Intelligent Designer." At the closing of Chap. VII, Darwin states:

> It is far more satisfactory to look at such instincts as the young cuckoo ejecting its foster-brothers, - ants making slaves, - the larvae of ichneumonidae feeding within the live bodies of caterpillars, - not as specially endowed or created instincts, but as small consequences of one general law, leading to the advancement of all organic beings, namely, multiply, vary, let the strongest live and the weakest die (Darwin, 1859/2018, p. 253).

The author is talking about a general law, where morally "censorable" behaviors would not have been "designed" or "planned," but would be a consequence of this law that leads to the optimization of the efficiency of individuals. The fact is that, clearly, by closing the chapter in this way and inserting Butler's epigraph, Darwin seeks an alliance with renowned progressive Anglican ministers, such as the Reverend Charles Kingsley. He had received a copy of *On the Origin of Species* a month before the book's publication, and upon receiving the copy, he immediately responded to Darwin, thanking him for the gift and adding a valuable note. He talks about the harmony between Darwin's theory and religion. Charles Kingsley highlighted the new way of conceiving nature, not as a myriad of independent phenomena that require constant interventions by an intelligent agent, but as a set of mechanisms with laws that govern life, as in the Newtonian tradition.

Kingsley's letter has a biblical reference (an allusion to Paul's Epistle to Romans 3:4), which sounds like an encouragement to Darwin to face (predictable) opposition from other scientists and theologians. He goes on to say that understanding the change of living beings as a general law adds nobility to the Deity, by exposing an ingenious strategy, creating primitive forms capable of self-development, instead of remaining all the time acting to fill gaps in its creation. He ends by saying the framework Darwin drew may be a "loftier thought" (Kingsley, 1859). Darwin responds promptly, thanking him for his reflection and asking for permission to reproduce the phrase in the next edition, which in fact occurred. Eventually "Christian evolution" had gained wide acceptance in Britain, and Darwin ended his life when his theories were considered not only "both permissible and respectable" among educated Christians, but also fully accepted among the Anglican clergy and divines (Moore, 1979).

18.6 The Great General Law, the Nobility of the Divinity, and the Classroom

The new Darwinian interpretation allowed the conclusion that the "intelligent agent" could not be cruel enough to constantly create and encourage disgusting behavior in nature. How grandiose is the behavior of making slaves? Why marvel at the parasitism and infanticide of the cuckoo, at the atrocious way in which certain wasps make the little shoots of butterflies and moths suffer? What is beautiful about a mother murdering her fertile daughters as soon as she sees them? There is no grandeur or nobility in deliberately planning such actions.

Thus, Darwin somehow anticipated his defense of the need for new theoretical frameworks for the Anglican elite, of which he did not expect any reception, even more so after the first critical reviews of his book at the end of 1859. This reflection may have important consequences for the teaching and learning processes of biological evolution today, discussing, from a historical point of view, how moral and religious aspects were made compatible with strictly scientific issues.

If, on the one hand, Darwin undermined the belief in a "natural government" requiring the constant surveillance and intervention of a Deity, on the other hand, he bestowed nobility on the Creator, as Eversley parish vicar Charles Kingsley recalled. The actual creation would have been an original set of simple and well-defined instructions, capable of reaching some automated outputs, as adjustments in living beings. Thus, a "natural law" would be far more ingenious, by providing small changes, than a laborious process of continuous surveillance. The cuckoo's instinct, by parasitizing the nests of other species, may have originated from the behavior of other birds and not as the result of special and intelligent design. Darwin even mentions the rheas, which make collective nests, hatched by a male, and attributes the instinct to the long interval between the laying of each egg by the females. If each female had her own nest, only the last egg would have a chance of being viable. Collective nests would increase the final progeny of all females. Simple and well-defined rules could have profound effects in the long run by turning the habit of oviposition independent of nest building.

Butler defended the idea of a moral philosophy mirroring the divine will, from which benevolence and compassion would derive, criticizing the idea of a Hobbesian nature, controlled by force and violence. Darwin points to fighting in nature as a law that would explain morally repulsive behaviors like ant slavery, wasp infanticide, and bee matricide.

It is well known that this chapter on instinct was originally conceived as a supplement to the chapter on hybridism, with the title "Mental Powers and Instincts of Animals," a topic that Darwin found promising but still understudied. He researched and later published on the subject. The fact is that, in this chapter, Darwin is dialoguing with Anglican moral principles and concludes that the Deity could not be so cruel in planning and constantly instigating parasitism, infanticide, slavery, and matricide. Nevertheless, these behaviors should be the result of a great general law, which has its development algorithm, but which does not allow us to predict its future outcome.

Thus, if Charles Darwin's original discussion did not neglect the moral dimension of knowledge, we must ask ourselves whether it should be fully sublimated in classroom discussions. The teachers discuss with their students the relevance of genetically manipulating embryos, interfering with human reproduction, carrying out experiments on animals, and even including meat in human food. Then, why should we think that the theory of evolution is a topic above good and evil, that does not raise moral doubts in today's students?

Darwin concluded his book with that paragraph with a slight reference to the book of Genesis and speaks precisely of the greatness ("there is grandeur in this view of life") of the countless extremely beautiful and wonderful productions that we can only admire. However, he attributes this result, not to an intelligent agent, but as the long-term result of a rule as simple as the law of gravity. Perhaps Darwin should be remembered in today's classrooms not only for natural selection and its evolutionary consequences but also for his thoughts and epigraphs on moral philosophy.

References

BBC News. (2006). Church apologizes for slave trade. http://news.bbc.co.uk/2/hi/uk_news/4694896.stm

Browne, J. (1989). Botany for gentlemen: Erasmus Darwin and "The loves of the plants". *Isis, 80*, 593–621.

Butler, J. (1736) *The analogy of religion, natural and revealed, to the constitution and course of nature*.

Catsikis, P. J. (2009). *A brilliant burst of botanical imagination: Proserpina and the nineteenth-century evolution of myth*. Ph.D. thesis. University of Glasgow.

Darwin, C. (1859/2018). *A Origem das Espécies. Tradução de Daniel Moreira Miranda. Prefácio, revisão técnica e notas de Nelio Bizzo*. Edipro.

Ferreira, R. (2012). *Cross-cultural exchange in the Atlantic world: Angola and Brazil during the era of the slave trade*. Cambridge University Press.

Gomes, L. (2019). *Escravidão*. Globo Livros.

Huber, P. (1820). *The natural history of ants*. Longman, Rees, Orme, and Brown.

Kingsley, C. (1859). *Darwin correspondence project. Letter DCP-LETT-2534*. Available at: https://www.darwinproject.ac.uk/letter/?docId=letters/DCP-LETT-2534.xml&query=Kingsley

Linné, C. von, & Wahlbom, J. G. (1746). Sponsalia plantarum. : Laurentii Salvii. Available at: https://linnean-online.org/120091/#?s=0&cv=0&z=0.0918%2C0.586%2C0.6586%2C0.6835

von Linné, C. (1749). *Amoenitates Academicæ* (p. 1). Antehac Seorsim Editæ.

Moore, J. (1979). *The post-Darwinian controversies*. Cambridge University Press.

Paley, W. (1802/1829). *Natural theology or evidences of the existence and attributes of the deity*. Lincoln & Edmands.

Seybolt, R. F. (1930). Hunting Indians in Massachusetts: A scouting Journal of 1758. *The New England Quarterly, 3*(3), 527–531. https://doi.org/10.2307/359401

Nelio Bizzo is a senior full professor and scientific coordinator of the Center for Research in Education, Diffusion and Epistemology of Evolution (EDEVO-Darwin / USP) at the Faculty of Education of the University of São Paulo. He is also an adjunct professor at the Federal University of São Paulo, Diadema campus, where he works at undergraduate and graduate levels, and is a collaborating professor at the graduate level at UFABC.

Lucas Marino Vivot is a doctoral student in education and history of science at the Federal University of ABC (UFABC). Holds a degree in Biology from the Federal University of São Paulo (UNIFESP) and a Master's degree in Ecology and Evolution from UNIFESP. Has experience in teaching Science and Biology for Elementary and High School level, as well as for undergraduates at Integrated Faculties of Guarulhos (FiG) and UniBTA University Center.

Chapter 19
Origin's Chapter VIII: Darwin for and Against Hybridism

P. Lorenzano

Abstract The chapter about hybridism written by Darwin falls in the cluster of *The Origin of Species* which deals with the main difficulties of the theory of descent with modification through natural selection. Darwin's goal in analyzing the phenomenon of hybridism was to debate the validity of the existence of a fundamental distinction between species and varieties; thus, this discussion addresses an integral part of the so-called species problem, which includes a set of questions about the definition of the concept of species, that is, of what a species is. In this chapter, we analyze the historical background of this debate, from Linnaeus to Kölreuter and Gärtner, and discuss the structure and arguments present in *Origin*'s Chapter VIII, in which Darwin tackles the problem at hand.

19.1 The Place of Chapter VIII within the Context of the *Origin*

Throughout the *Origin of Species,* Darwin presents what he calls "one long argument" (Darwin, 1859, p. 459). This argument constitutes a narrative without a linear structure, but rather a network made up of successive steps (chapters/partial arguments), each of which summarizes the previous step, and introduces the subsequent ones, where the parts sustain the whole. On the other hand, it is from the complete argument that each part (chapter/partial argument) derives its support and meaning.

The original version of the chapter has been revised. A correction to this chapter can be found at https://doi.org/10.1007/978-3-031-40165-7_26

P. Lorenzano (✉)
Center of Studies in Philosophy and History of Science (CEFHIC), National University of Quilmes (UNQ); National Council for Scientific and Technical Research (CONICET), Buenos Aires, Argentina

M. Elice Brzezinski Prestes (ed.), *Understanding Evolution in Darwin's "Origin"*, History, Philosophy and Theory of the Life Sciences 34, https://doi.org/10.1007/978-3-031-40165-7_19

To substantiate his theory ("the theory of descent with modification through natural selection," Darwin, 1859, pp. 343, 459), Darwin uses different resources in the book, in addition to the more traditional deductive logical arguments and procedures, such as metaphors, analogies, and rhetorical arguments – in the Aristotelian sense of non-demonstrative, but persuasive ones, whose underlying logic is non-monotonic (Regner, 2013).

Besides the abovementioned particular part-whole movement and linking his appeal to explanatory power as a whole, Darwin also makes use of the comparison of his point of view with those of his opponents, the treatment of difficulties/objections, the interplay of the actual and the possible, and the weight of reasons as rhetorical arguments or strategies (Regner, 2007).

The objectives of these strategies are, on the one hand, to show that the difficulties and objections do not have the supposed force or even are not such, but disappear with a deeper analysis of the issue; on the other hand, to defend his proposal both as the best possible explanatory alternative and even as the only possible (rational) explanation.

In Chapters I to V, Darwin establishes the foundations of his theory. In particular, in Chapter IV, based on the previous chapters, he presents what many analyses of his work consider the "short argument" or "core" of the argument in favor of natural selection put forward in *Origin.*

Several authors – beginning with Darwin's contemporary evolutionary theorist, Alfred Russel Wallace (1891), and following by Huxley (1942), Flew (1959), Ghiselin (1969), Lewontin (1978), Ruse (1979), Ospovat (1981) and Mayr (1982, 1991, 2001), among others – reconstruct the "short argument" as a hypothetic-deductive argument with the principle of natural selection as the conclusion, even going so far as to argue that Darwin's theory itself possesses such a deductive structure.

Other authors, however, question the deductive validity of the "short argument" (Kitcher, 1993; Ginnobili, 2006; Regner, 2010), as well as that what is inferred is not the principle of natural selection in all its generality but rather one of its components, namely, differential reproduction (Ginnobili, 2006, 2016).

In any case, the inferences made in the "short argument" would constitute an argument, along with the analogy with artificial selection, in favor of natural selection (Darwin, 1868, p. 9; Gayon, 1998, p. 23) and not a demonstration of that principle, let alone would such an argument represent the structure of Darwin's theory (Ginnobili, 2016, 2018).

Darwin concludes these first chapters by discussing the "laws of variation" (Chapter V).

In Chapter VI, Darwin raises four possible difficulties and objections to his theory. These are the following: "Firstly, why, if species have descended from other species by insensibly fine gradations, do we not everywhere see innumerable transitional forms? Why is not all nature in confusion instead of the species being, as we see them, well defined?" (Darwin, 1859, p. 171); "Secondly, is it possible that an animal having, for instance, the structure and habits of a bat, could have been formed by the modification of some animal with wholly different habit?" (Darwin, 1859, p. 171); "Thirdly, can instincts be acquired and modified through natural selection?"

(Darwin, 1859, p. 172); and, "Fourthly, how can we account for species, when crossed, being sterile and producing sterile offspring, whereas, when varieties are crossed, their fertility is unimpaired?" (Darwin, 1859, p. 172).

Considering that none of them is fatal to his theory, in the same Chapter VI Darwin deals with the first two difficulties and objections, leaving the last two for separate treatment in the next two chapters – Chapter VII, on instinct, and Chapter VIII, on hybridism.

Chapter VIII is thus part of the set of Chapters VI, VII, VIII – in which Darwin gives answers to the possible difficulties and objections that could be raised to his theory (Olby, 2009; Reznick, 2009; Smocovitis, 2009).

More importantly, Chapter VIII constitutes a continuation of Darwin's argument for the transmutation of species and against Natural Theology and the idea that species are products of acts of special creation. In particular, Darwin's main goal in analyzing phenomena of hybridism was to debate the still widely held conception that these phenomena confirmed the existence of a fundamental distinction between species and varieties.

Thus, the discussion of hybridism addresses an integral part of the so-called species problem. The species problem includes a set of questions about the concept of species. One of these questions involves issues around the "definition" of the concept of species, that is, of what a species is (Reznick, 2009; Beatty, 1985). However, it is worth noting that in this context, the term "definition" is used in a rather loose way of giving operational criteria, methods, or ways to distinguish species from varieties, that is, to determine the extension of the concept of species, meaning how it can be determined what are the objects that fall under the concept of species, and how a species differs and is related to varieties, and not in the strict logical sense of providing necessary and sufficient conditions for the application of the concept of species (Ginnobili, 2013).

Darwin himself was aware of this situation, recognizing that even if there was no "definition" of the species concept, nor an essential difference between species and varieties, there could be agreement about the extension of the species concept:

> [. . .] it is no wonder that there should be difficulty in defining the difference between a species & a variety; there being no essential, only an arbitrary difference. In the following pages I mean by species, those collections of individuals, which have commonly been so designated by naturalists. (Stauffer, 1975, p. 98)

As a part of his general argument against the view that "species were immutable productions, and had been separately created" and in favour of the conception that "species undergo modification, and that the existing forms of life are the descendants by true generation of pre-existing forms" (Darwin, 1866, p. xiii), Darwin explores in Chapter VIII the role that sterility plays in the so-called species problem (Olby, 2009).

An important aspect to consider in this discussion is that, according to him, sterility is "a variation which is not inherited," and therefore is a property "unimportant for us" (Darwin, 1859, p. 12), on the one hand. On the other hand, "sterility of hybrids could not possibly be of any advantage to them, and therefore could not have been acquired by the continued preservation of successive profitable degrees of sterility" (Darwin, 1859, p. 245). That means that sterility does not need to be explained by natural selection.

19.2 The Place of Chapter VIII in the Context of Hybridism

It is customary to distinguish two schools which, towards the end of the eighteenth and during the first half of the nineteenth century, used the method of breeding (Roberts, 1929; Mayr, 1982; Bowler, 1989). These schools or traditions, which had very different interests and aims, were that of *animal and plant breeders* (Mayr, 1982), also called *horticulturalists* (Bowler, 1989), and that of *species hybridizers*, or simply *hybridists*.[1]

Of breeders, we might say that they were practical men that wanted to know how new and economically valuable varieties could be created and fixed in offspring. They aimed to improve the productivity of the plants they cultivated or the animals bred – for example, their resistance to cold, the color of their flowers, or the quality of the wool or meat they obtained – and to produce new varieties by *crossing existing ones* that differed in a few characteristics. Among them, we might mention Thomas Andrew Knight (1799, 1824), Alexander Seton (1824), and John Goss (1824).[2]

Hybridists, in turn, possessed an academic background and, starting from the problem of the sexuality of plants, engaged the question raised in the eighteenth century of whether new species could be created by *crossing – hybridizing – preexisting ones*. Important hybridists were Carl Linnaeus (also known after his ennoblement in 1761 as Carl von Linné) (1744, 1755, 1760), Joseph Gottlieb Kölreuter (1761, 1763, 1764, 1766), Augustin Sageret (1826), Arend Joachim Friedrich Wiegmann (1828), William Herbert (1837), Marc Antoine Puvis (1837), Henri Lecoq (1845), and Carl Friederich von Gärtner (1849).[3]

[1] As is known, the first edition of the *Origin* does not contain footnotes or bibliographical references. However, the original long manuscript work on *Natural Selection* (edited by Stauffer in 1975), of which *Origin* would be the new abstract of his views, does. We will indicate in corresponding footnotes when Darwin refers to primary literature mentioned in this chapter.

[2] Darwin (1859) mentions Knight and in Stauffer (1975) he refers to his work. Seton and Goss are not mentioned in either Darwin (1859) or in Stauffer (1975). Darwin (1863) refers to Goss and his work. Darwin does not refer to Seton either in Darwin (1859) or Stauffer (1975). However, we find a reference to Goss' and Seton's results in a work referred by Darwin (Stauffer, 1975), Gärtner (1849, p. 85), but without saying that the publication (simply signed by "G," and with changed indications of the years of experiments) is actually a German translation of Goss' and Seton's papers (G, 1837).

[3] In Darwin (1859), he mentions Linnaeus, Kölreuter, Gärtner, Herbert, and Sageret (even though erroneously named "Sagaret"; *erratum* corrected after the fourth edition of 1866). In Stauffer (1975), Darwin refers to the aforementioned works of Kölreuter, Sageret, Wiegmann, Herbert, Puvis, Lecoq, and Gärtner; he also refers to other works by Linnaeus.

The so-called doctrine of special creation claims that all existing species are an immediate creation of God. In his early writings, Carl Linnaeus accepted this doctrine and the resulting constancy of species (e.g., Linnaeus, 1736, 1737). However, he later proposed what Callender (1988) called the "*modified* or *new* doctrine of special creation": certain hybrids that appear in nature but that can also be produced artificially are fertile and reach the status of new species, that is, there is generation, artificial and natural, of new constant species originated in the crossing of already existing species (Linnaeus, 1744, 1755, and more clearly, elaborately and decisively, 1760; also in the work of his disciples – Linnaeus & Haartman, 1751; Linnaeus & Daldberg, 1755; Linnaeus & Gråberg, 1762).[4]

Darwin considered that, for many naturalists, Linnaeus included, Linnaeus' Natural System "reveals the plan of the Creator" (Darwin, 1859, p. 413), aligning Linnaeus with Darwin's *bête noire* Natural Theology. However, on the one hand, as Bremekamp (1953) argues, Linnaeus was not especially interested in creation. What he intended to express with the aphorism "Species to numeramus quot diversae formae in principio sunt creatae" ("We count so many species as in the beginning different forms were created") (Linnaeus, 1751, § 157, p. 99) is mainly that the species are constant, a view which he modified, as already noted (see also Linnaeus 1764, 1767, 1792). And on the other hand, in order to avoid prejudging the position held on creation by members of the hybridist tradition – noting, moreover, that "those two conscientious and admirable observers, Kölreuter and Gärtner, who almost devoted their lives to this subject" (Darwin, 1859, p. 246), do not develop the subject –,[5] we would like to make the following terminological clarification.

The term "hybridism" is used in the literature to refer both to this tradition that studies the union of distinct species and of its offspring and deals with the question of whether new species can be created by hybridization of preexisting ones, and to the position of those who answer affirmatively.

The first use we may label "hybridism in the broad sense," and the second "hybridism in the narrow sense," that is, "hybridism" as a "theory about the origin of new species by hybridization" (a "theory of speciation through hybridization").

Among those who gave a positive answer to this question, besides the abovementioned Linnaeus and his disciples Haartman, Daldberg, and Gråberg, are Herbert (1837), Puvis (1837), and Lecoq (1845). Whereas among those who accepted the challenge made by Gmelin (1749) of settling the issue experimentally and gave a negative answer to that question are Kölreuter (1761, 1763, 1764, 1766) and Gärtner (1849).

As we shall see, in Chapter VIII, "On Hybridism," and along *Origin*, Darwin refers to authors against and for the theory of speciation through hybridization as well as of both traditions or schools – that of animal and plant breeders, or

[4] Darwin does not mention Linnaeus' followers or their works.

[5] Nor does another hybridist who knew Darwin's work, even if Darwin did not know his own, and who favored hybridism as a theory of speciation through hybridization, Johann (Gregor) Mendel. For the (asymmetrical) relations between Darwin and Mendel, see Lorenzano (2011).

horticulturalists, and that of species hybridizers, or hybridists – and to other naturalists and botanists (Secord, 1985).

It is worth noting here that, on the one hand, some breeders were explicit on the role played by hybrids in the generation of new species and, on the other hand, hybridists also reported what happened in their crossing experiments with individual traits – thus blurring the original distinction (or, at least, some of the terms in which it is formulated).

However, because of the importance that Darwin assigns to Kölreuter and Gärtner, and in order to better place the context of Darwin's discussion of hybridism, we will briefly mention their views and results.

In his main work, which appeared in four parts between 1761 and 1766, Kölreuter reported the successful hybridization of a great number of plant species. Hybridization experiments on plants were considered important in that context because if the progeny showed parental traits, or if hybrid plants could be produced and analogies could be noticed between them and animal hybrids, that would provide powerful support for the theory of the sexuality of plants. Kölreuter had no doubts that hybrids could be produced artificially but was sure that nature had its own means both to prevent their natural creation (Kölreuter, 1761, § 16; 1766, § 20), and to avoid the propagation as new species of those hybrids that had been obtained through experiments (Kölreuter, 1763, § 1, 1766, § 20).

Kölreuter not only obtained hybrid plants but also succeeded even in creating a second generation of true hybrids from self-pollination of these tobacco hybrids, finding a *slight degree of fertility* that was even higher in hybrids of other species.

Despite this, he claimed, together with the *closest affinity of the pollen* – of the species itself, and of the pollen of one of the parental species in the hybrid – the *reversion* of hybrids to one or the other of the original species instead of reproducing as hybrids, their *intermediate* character, the *identity of hybrids* obtained by reciprocal crosses, the *transformation of one species into another* by successive reciprocal cross-pollinations, and the *sterility* of hybrids.[6]

Thus, Kölreuter intended, in addition to supporting the theory of plant sexuality through the production of hybrids, to *oppose* Linnaeus' "hybridism in the narrow sense," but to be in complete *agreement* with the belief in the *constancy of species* and their *sharp differentiation from varieties.*

After having investigated fertilization by means of natural and artificial pollination with pollen from the same plant (Gärtner, 1844), Gärtner intended to show that in some cases fertilization takes place by means of the use of strange pollen (from another species), that is, that hybrid fertilization occurs. In the scholarly book *Versuche und Beobachtungen über die Bastarderzeugung im Pflanzenreich*, Gärtner (1849) summarizes the knowledge of his time on hybridization and describes countless experiments performed by himself and by others and the results obtained.

[6]Furthermore, Kölreuter used the last three results against of the variants "espermatist" and "ovist" of the "preformationism" and of Linnaeus' "two-layer theory," and in support to the belief in the necessity of both seeds – the maternal and the paternal one – for the fecundation.

As Kölreuter previously, Gärtner believed in the sexuality of plants as well. However, contrary to him, he considered, on the one hand, that hybrids are not only artificially produced, but "that also in the free nature hybrids of plants can be originated, and that they appear even more frequently than in the animal kingdom" (1849, p. 4). On the other, Gärtner also considered that it cannot be appreciated as an "essential difference" between the hybrids artificially produced and the hybrids found in nature (1849, p. 12).

In addition, following Kölreuter again, Gärtner considered that exists an essential difference between the species and the varieties, as well as between the hybrids of varieties and the hybrids of species (or "true hybrids," as he also calls them) too.

But contrary to Kölreuter, he did not believe this difference could be established using the distinct fertility among them (while the first ones would be fertile, the same as its descendant, the last ones would be sterile). According to Gärtner, hybrids of varieties differ from hybrids of pure species by being extraordinarily more receptive to the effects of the species' original parents than true hybrids, by having a much greater tendency to return to the original form through the generation in the propagation, by having an incomparably higher variability of forms in the following generations than in hybrids of species, and by habitually possessing male organs undamaged and complete and from there an imperturbable, and even not little frequent, high and increased fertility.

Gärtner's most important results and conclusions were *the contribution of two factors* (one maternal–female and the other paternal–male) in hybrid fertilization; the *impurity of characters* in hybrids; the *inadequacy of a mathematical* (*or chemical*) *treatment* or analysis of fecundation and hybridization; the *spontaneous production of hybrids* in nature; their *non-reproduction like new species*, but their later extinction or return to one of the original species; the *sharp difference between species and varieties* and *between hybrids of species and hybrids of varieties*; the *classification of hybrids of species in mediated or assorted, mixed and decided*; and the *identity of reciprocal crosses*.

Thus, Gärtner, in addition to supporting the theory of plant sexuality through the production of hybrids, also *opposed* Linnaeus' "*hybridism in the narrow sense*" – a conception "that contradicts the nature of pure species" (Gärtner, 1849, p. 15) – and collected evidence in *favour of stability, constancy, and invariability of species* and of their *sharp differentiation from varieties*.

19.3 Darwin's Argumentation in Chapter VIII, "on Hybridism"

By the eighteenth and nineteenth centuries, hybridization had drawn the attention of practical horticulturalists, plant and animal breeders, and hybridists. Darwin was aware of the complex issues raised by hybridism and reserved that discussion for Chapter VIII, devoted exclusively to the subject. In that chapter (Darwin, 1859, pp. 245–278), we can observe several of the argumentative strategies used by Darwin, which "persuade without demonstrating."

As already noted, Chapter VIII is part of both the "long argument" in general – with its part-whole movement and appeal to the explanatory power of his theory as a whole – and the group of chapters in which Darwin deals with the difficulties/ objections presented to his theory.

In order to fulfill the specific objectives of, on the one hand, showing that the difficulties and objections do not have the assumed strength or even are not such, but disappear with a deeper analysis of the question (e.g., by transforming the apparently "unfavorable" evidence into "favorable" – a strategy used mainly in Chapter IX), and, on the other hand, to defend his proposal as the best possible explanatory alternative and even as the only possible (rational) explanation, Darwin carries out a comparison of his point of view with those of his opponents.

Recall that, on the one hand, although many, but not all, of "Darwin's contemporaries thought that each species was the product of an act of special creation and that all species were absolutely reproductively isolated from one another" (Reznick, 2009, p. 190) – including Natural Theologists and some hybridists in the broad sense – this was not so for those who held a theory of speciation through hybridization, that is, hybridism in the narrow sense.

On the other hand, either the assumed sterility of hybrids (Kölreuter) or the gradual decreasing fertility of hybrid crosses (Gärtner) in contrast to the fertility of variety crosses (whose progeny is called "mongrels" by Darwin) were considered the Creator's (for Natural Theologists) (Olby, 2009) or Nature's (for Kölreuter and Gärtner) way of preserving the fixity of species.

Darwin "needed to explain why species, which, according to his theory, had descended from each other, usually could not interbreed" (Smocovitis, 2009, p. 230), given that, on the one hand, sterility or difference in fertility was not a property to be explained by natural selection, since it is not "a specially acquired or endowed quality, but is incidental on other acquired differences" (Darwin, 1859, p. 245), as already mentioned in Sect. 19.1 of this chapter; and on the other hand, the standard explanation of why species did not interbreed, namely, "that barriers to hybridization existed" (Smocovitis, 2009, p. 230) to "prevent confusion of all organic forms," (Darwin, 1859, p. 245), "was [...] not a viable explanation" (Smocovitis, 2009, p. 231) for him (Darwin, 1859, p. 245).

After establishing the fundamental distinction between the sterility of species crosses and hybrid crosses (Darwin, 1859, pp. 245–246), Darwin presented a brief overview "of the varied and complex patterns observed" (Smocovitis, 2009, p. 230) in the phenomena first of sterility and fertility of species crosses and then of hybrid crosses (Darwin, 1859, pp. 246–253; Reznick, 2009).

Darwin relied extensively on the insights and experimental studies of horticulturalists, plant and animal breeders, and hybridists (Smocovitis, 2009). In particular, he looked for relevant knowledge and examples in the works of "the two most experienced observers who have ever lived, namely, Kölreuter and Gärtner," and "should have arrived at diametrically opposite conclusions in regard to [the sterility and fertility of] the very same species" (Darwin, 1859, p. 248).

Darwin compared his point of view with that of them, and used his interpretation of their experimental results, together with the results obtained by "the third most experienced hybridizer [...] Hon. and Rev. W[illiam]. Herbert" (Darwin, 1859, pp. 249–250), to oppose their views (Reznick, 2009).

Herbert "is as emphatic in his conclusion that some hybrids are perfectly fertile – as fertile as the pure parent-species – as are Kölreuter and Gärtner that some degree of sterility between distinct species is a universal law of nature" (Darwin, 1859, p. 250). He showed that some species that Gärtner found to be absolutely sterile when crossed were actually perfectly interfertile. For Darwin, the difference in their results could lie in Herbert's superiority in horticultural skills and the advantage of hothouses (Darwin, 1859, p. 250).

In support of his view, Darwin also relied on the reports and works of the hybridist Augustin Sageret, of animal breeders such as Thomas Campbell Eyton, Edward Blyth, Frederick Wollaston Hutton, and Edward Hewitt, of other plant breeders such as Charles Noble, Gustave Adolphe Thuret, Charles Girou de Buzareingues, and André Thouin, as well as of other naturalists and botanists, such as Peter Simon Pallas, Johann Georg Gmelin, and Christian Konrad Sprengel.[7]

Darwin's conclusions of his overview are, besides that sterility is affected by "close interbreeding" (1859, p. 248) and is "capable of being removed by domestication" (1859, p. 254), "that some degree of sterility, both in first crosses and in hybrids, is an extremely general result; but that it cannot, under our present state of knowledge, be considered as absolutely universal" (1859, p. 254; this last citation also appears in Reznick, 2009, p. 196).

These conclusions, which would show that "different species can sometimes interbreed and [...] produce" fertile hybrids, would go against those who defend the doctrine of special creation and would constitute an argument "that species cannot be defined as discrete units separated by absolute reproductive barriers" (Reznick, 2009, p. 196) and there were no clear lines between varieties, incipient species, and species.

In the next section (Darwin, 1859, pp. 254–263), Darwin discusses the "laws governing the sterility of first crosses and hybrids" based on a summary of the rules and conclusions "chiefly drawn up from Gärtner's admirable work on the hybridisation of plants" (1859, p. 254).

However, contrary to Gärtner's proposal, Darwin makes the following points (see Reznick, 2009, pp. 196–198). First, sterility and fertility are gradual rather than being black or white (Darwin, 1859, p. 255). Second, the capacity of parent species to cross and produce viable hybrids is a separate property from the fertility of those hybrids (Darwin, 1859, p. 256). Third, the capacity of parent species to cross and produce viable, fertile hybrids is "innately variable," that is, it varies between individuals rather than being an absolute property of the species (Darwin, 1859, p. 256). Fourth, there is not a strict correspondence between "systematic affinities," that is, "the resemblance between species in structure and in constitution" present in allied species, and the capacity of crossing (Darwin, 1859, pp. 256–257). Fifth, there is not a correlation

[7] Darwin (1859) mentions all these persons. He refers to their works in Stauffer (1975).

between any character of a plant species and its capability to hybridize with another species (Darwin, 1859, p. 257). Sixth, reciprocal crosses between the same two species do not always yield the same result (Darwin, 1859, p. 258).

Darwin indicates that there are more Gärtner's "rules," but the ones he has already discussed are sufficient to prove his own view (Reznick, 2009, p. 198).

Before he ends this section, Darwin draws an analogy between hybridization and grafting, that is, pointing out "that there is a rude degree of parallelism in the results of grafting and of crossing distinct species" (Darwin, 1859, pp. 262–263).

Finalizing this section, he reiterates his most important conclusions, namely, that "these complex and singular rules" are incidental to or dependent on some other "unknown differences" between species and that species have been not "endowed with sterility simply to prevent their becoming confounded in nature" (Darwin, 1859, p. 260).

In discussing the "causes of the sterility of first crosses and of hybrids" (Darwin, 1859, pp. 263–265), Darwin first mentions some reasons for the sometimes sterility of crosses between members of different species (Darwin, 1859, pp. 263–264). But, at the same time, he acknowledges that the factors governing whether viable hybrids would be fertile and capable of producing offspring were unknown (Darwin, 1859, pp. 264–265). Then, Darwin intended to draw a "parallelism between the effects of changed conditions of life and crossing" (Darwin, 1859, p. 266).

The main conclusions of Darwin's discussion of the "fertility of varieties when crossed and of their mongrel offspring" (Darwin, 1859, pp. 267–272) are that it can be proved that such supposed fertility is not a "universal occurrence" – which is supported by examples from domesticated plants and by observations made by the "hostile witnesses" Gärtner and Kölreuter (Darwin, 1859, p. 270) – and not "form a fundamental distinction between varieties and species" (Darwin, 1859, pp. 271–272).

Contrary to those held by Gärtner (1849), Darwin argues that a comparison between "hybrids and mongrels independently of their fertility" (Darwin, 1859, pp. 272–276) yields the result that the existing differences, if they are consistent, are only of degree rather than absolute rules that discriminate one from the other (Darwin, 1859, pp. 272–275). That constitutes a result that "harmonises perfectly with the view that there is no essential distinction between species and varieties" (Darwin, 1859, p. 276).

Lastly, Darwin presents a summary of the chapter (Darwin, 1859, pp. 276–278), which concludes with the following words:

> Finally, then, the facts briefly given in this chapter do not seem to me opposed to, but even rather to support the view, that there is no fundamental distinction between species and varieties. (Darwin, 1859, p. 278)

There, Darwin also emphasizes that "did not consider [. . .] hybridization itself to be the source of the resulting variation" (Olby, 2009, p. 38), either in general or in the vigor and fertility of the offspring of hybrids. Rather, the changes in conditions of life play in it a fundamental role.

This aspect is significant because of its connection with Darwin's position on the limits of "hybridism in the narrow sense."

Thus, as argued by Callender (1988), though Darwin accepted the existence of hybrid forms of plants completely fertile and relatively stable, he thought that absent other sources of variation, hybridization itself "could not account for the evolution of species" (Callender, 1988, p. 49). The main reason – which Darwin develops in other parts of the *Origin* as well as in other texts – is that hybridization "presupposes already existing differences", thus giving rise, once again, to the question of the origins of such differences (Callender, 1988, p. 49). By trying to explain, then, evolutionary change based on crossings without variation, Callender refers to Darwin's words: "we thus only push the difficulty further back in time, for what made the parents or their progenitors different?" (Darwin, 1868, v. II, p. 252; apud Callender, 1988, p. 49). In the case of the evolution of pigeons, for example, Darwin wrote, in a passage also cited by Callender and expanded here:

> Great as the differences are between the breeds of pigeons, I am fully convinced that the common opinion of naturalists is correct, namely, that all have descended from the rock-pigeon (*Columba livia*), including under this term several geographical races or sub-species, which differ from each other in the most trifling respects. As several of the reasons which have led me to this belief are in some degree applicable in other cases, I will here briefly give them. If the several breeds are not varieties, and have not proceeded from the rock-pigeon, they must have descended from at least seven or eight aboriginal stocks; for it is impossible to make the present domestic breeds by the crossing of any lesser number: how, for instance, could a pouter be produced by crossing two breeds unless one of the parent-stocks possessed the characteristic enormous crop? (Darwin, 1859, p. 23)

That is, "*the fundamental objection to the doctrine of 'evolution' solely by means of hybridization [. . .]*: one is inevitably confronted either with an infinite regress or [with] some version of the doctrine of Special Creation" (Callender, 1988, p. 49; emphasis added). And either possibility is, according to Darwin, untenable.

To summarize the conclusions of this chapter and explain its subtitle, while Darwin acknowledges the importance of "hybridism in the broad sense", i.e, hybridism as the tradition that studies hybrids ("[t]he offspring of the union of two distinct species", Darwin, 1872, p. 435), he clearly rejects "hybridism in the narrow sense", i.e. hybridism as a theory of the origin of new species by hybridization, and defends his own theory, as he reaffirms elsewhere ("the whole subject of hybridism is one of the greatest obstacles to the general acceptance and progress of the great principle of evolution" Darwin, 1876, p. 27).

References

Beatty, J. (1985). Speaking of species: Darwin's strategy. In D. Kohn (Ed.), *The Darwinian heritage* (pp. 265–282). Princeton University Press.

Bowler, J. (1989). *The Mendelian revolution*. The Athlone Press.

Bremekamp, C. E. B. (1953). Linné's views on the hierarchy of the taxonomic groups. *Acta Botanica Neerlandica, 2*(2), 242–253.

Callender, L. A. (1988). Gregor Mendel: An opponent of descent with modification. *History of Science, 26*, 41–75.
Darwin, C. (1859). *On the origin of species by means of natural selection, or the preservation of Favoured races in the struggle for life* (1st ed.). John Murray.
Darwin, C. R. (1863). Vindication of Gärtner—effect of crossing peas. *Journal of Horticulture and Cottage Gardener* (3 February), 93.
Darwin, C. (1866). *On the origin of species by means of natural selection, or the preservation of favoured races in the struggle for life* (4th ed.). John Murray.
Darwin, C. (1868). *Variation in animals and plants under domestication* (Vol. 2 vols). John Murray.
Darwin, C. (1872). *On the origin of species by means of natural selection, or the preservation of favoured races in the struggle for life* (6th ed.). John Murray.
Darwin, C. (1876). *The effects of cross and self fertilisation in the vegetable kingdom*. John Murray.
Flew, A. G. N. (1959). The structure of Darwinism. *New Biology, 28*, 25–44.
G. (1837). Über die durch kreuzende Befruchtung bewirkte Veränderung in der Farbe der Erbsen. *Allgemeine deutsche Garten-Zeitung, 15*, 213–214.
Gärtner, C. F. (1844). *Beiträge zur Kenntnis der Befruchtung der vollkommenen Gewächse*. Schweizerbart.
Gärtner, C. F. (1849). *Versuche und Beobachtungen über die Bastarderzeugung im Pflanzenreich*. K.F. Hering & Comp.
Gayon, J. (1998). *Darwinism's struggle for survival: Heredity and the hypothesis of natural selection*. Cambridge University Press.
Ghiselin, M. T. (1969). *The triumph of the Darwinian method*. University of California Press.
Ginnobili, S. (2006). La selección natural como conjunto de hechos e inferencias. In *Epistemología e Historia de la Ciencia, Selección de Trabajos de las XIV Jornadas* (Vol. 12, pp. 266–275). Universidad Nacional de Córdoba.
Ginnobili, S. (2013). On the origin of «that thing you call "species"». In P. Lorenzano, L. A.-C. P. Martins, & A. C. Regner (Eds.), *History and philosophy of life sciences in the South Cone* (pp. 173–184). College Publications.
Ginnobili, S. (2016). Missing concepts in natural selection theory reconstructions. *History and Philosophy of the Life Sciences, 38*(8). https://doi.org/10.1007/s40656-016-0109-y
Ginnobili, S. (2018). *La teoría de la selección natural: Una exploración metacientífica*. Universidad Nacional de Quilmes.
Gmelin, J. G. (1749). *Sermo Academicus de Novorum Vegetabilium post Creationem divinam exortu*. Erhardt.
Goss, J. (1824). On the variation in the colour of peas, occasioned by cross-impregnation. *Transactions of the Horticultural Society of London, 5*, 234–236.
Herbert, W. (1837). *Amaryllidaceae*. James Ridgway.
Huxley, J. (1942). *Evolution: The modern synthesis*. George Allen and Unwin.
Kitcher, P. (1993). *The advancement of science: Science without legend, objectivity without illusions*. Oxford University Press.
Knight, T. A. (1799). An account of some experiments on the fecundation of vegetables. *Philosophical Transactions of the Royal Society of London, 89*, 195–204.
Knight, T. A. (1824). Some remarks on the supposed influence of the pollen in cross breeding. *Transactions of the Horticultural Society of London, 4*, 278–280.
Kölreuter, J. G. (1761, 1763, 1764, 1766). *Vorläufige Nachricht von einigen das Geschlecht der Pflanzen betreffenden Versuchen und Beobachtungen, nebst Fortsetzungen 1, 2 und 3*. In der Gleditschischen Handlung.
Lecoq, H. (1845). De la Fécondation naturelle et artificielle des Végétaux et de l'Hybridation, considérée dans ses rapports avec l'horticulture, l'agriculture et la sylviculture. . .contentant les moyens pratiques d'opérer l'hybridation et de créer facilement des variétés nouvelles. Audot.
Lewontin, R. C. (1978). Adaptation. *Scientific American, 239*(3), 212–230.
Linnaeus, C. (1736). *Fundamenta botanica*. S. Schouten.

Linnaeus, C. (1737). *Critica botanica*. Ligduni Batavorum apud C. Wishoff.
Linnaeus, C. (1744). Prefatio. In D. Rudberg (Ed.), *Dissertatio botanica de Peloria*, in *Amoenitates academicæ seu dissertationes variæ physicæ, medicæ botanicæ antehac seorsim ediatæ nunc collectæ et auctæ cum tabulis aneis* (Vol. 1, pp. 55–56). Jacobi Palm.
Linnaeus, C. (1760). *Disquisitio de sexu plantarum*. Academia Scientiarum.
Linnaeus, C. (1764). *Genera plantarum* (6th ed.). Laurentii Salvii.
Linnaeus, C. (1767). *Systema Naturae* (12th ed., Vol. 2). Salvius.
Linnaeus, C. (1792). *Praelectiones in ordines naturales plantarum*. Hoffman.
Linnaeus, C., & Daldberg, N. E. (1755). Dissertatio botanica metamorphoses plantarum sistens. In *Amoenitates academicæ seu dissertationes variæ physicæ, medicæ botanicæ antehac seorsim ediatæ nunc collectæ et auctæ cum tabulis aneis* (Vol. 4, pp. 368–286). Jacobi Palm.
Linnaeus, C., & Gråberg, J. M. (1762). *Fundamentum fructificationis*. In *Amoenitates academicæ seu dissertationes variæ physicæ, medicæ botanicæ antehac seorsim ediatæ nunc collectæ et auctæ cum tabulis aneis* (Vol. 6, pp. 279–304). Jacobi Palm.
Linnaeus, C., & Haartman, J. J. (1751). Plantæ hybridæ. In *Amoenitates academicæ seu dissertationes variæ physicæ, medicæ botanicæ antehac seorsim ediatæ nunc collectæ et auctæ cum tabulis aneis* (Vol. 3, pp. 28–62). Jacobi Palm.
Lorenzano, P. (2011). What would have happened if Darwin had known Mendel (or Mendel's work)? *History and Philosophy of the Life Sciences, 33*, 3–48.
Mayr, E. (1982). *The growth of biological thought*. Harvard University Press.
Mayr, E. (1991). *One long argument*. Harvard University Press.
Mayr, E. (2001). *What evolution is*. Basic Books.
Olby, R. (2009). Variation and inheritance. In M. Ruse & R. Richards (Eds.), *The Cambridge companion to the "Origin of species"* (pp. 30–46). Cambridge University Press.
Ospovat, D. (1981). *The development of Darwin's theory: natural history, natural theology, and natural selection, 1838-1859*. Cambridge University Press.
Puvis, M. A. (1837). *De la dégénération et de l'extinction des variétés des Végétaux propagés par les greffes, boutures, tubercules, etc., et de la création des varétés nouvelles par les croisements et les semis*. Huzard.
Regner, A. C. (2007). The rhetoric of science in Darwin's *on the origin of species*. In B. de Sousa Santos (Ed.), *Cognitive justice in a global world* (pp. 225–249). Rowman and Littlefield Publishers, Inc.
Regner, A. C. (2010). The structure of the Darwinian argument in the *origin of species*. In P. Lorenzano, H.-J. Rheinberger, E. Ortiz, & C. Galles (Eds.), *History and philosophy of science* (Vol. 1, pp. 302–328). EOLSS Publishers Ltd.
Regner, A. C. (2013). Reading Darwin through Dascalian eyes. In D. Riesenfield & G. Scarafile (Eds.), *Perspectives on theory of controversies and the ethics of communication: Explorations of Marcelo Dascal's contributions to philosophy* (pp. 29–37). Springer.
Reznick, D. N. (2009). *The Origin then and now: An interpretive guide to the Origin of species*. Princeton University Press.
Roberts, H. F. (1929). *Plant hybridization before Mendel*. Princeton University Press.
Ruse, M. (1979). *The Darwinian revolution: Science red in tooth and claw*. University of Chicago Press.
Sageret, A. (1826). Considérations sur la production des hybrides, des variants et des variétés en général et sur celles de la familie des Curcubitanceés en particulier. *Annales des Sciences Naturelles, Prem. Série, 8*, 294–314.
Secord, J. A. (1985). Darwin and the breeders: A social history. In D. Kohn (Ed.), *The Darwinian heritage* (pp. 519–542). Princeton University Press.
Seton, A. (1824). On the variations in the colours of peas from cross impregnations. *Transactions of the Horticultural Society of London, 5*, 236–237.

Smocovitis, V. B. (2009). Darwin's botany in the Origin of species. In M. Ruse & R. Richards (Eds.), *The Cambridge companion to the "Origin of species"* (pp. 216–236). Cambridge University Press.

Stauffer, R. (Ed.). (1975). *Charles Darwin's natural selection: Being the second part of his big species book written from 1856 to 1858*. Cambridge University Press.

Wallace, A. R. (1891). *Natural selection and tropical nature*. Macmillan.

Wiegmann, A. F. (1828). *Über die Bastarderzeugung im Pflanzenreiche*. Friedrich Vieweg.

P. Lorenzano is a philosopher specialized in history and philosophy of science (HPS), in particular of biology. He is known for his philosophical (synchronic and diachronic) analyses of empirical theories by means of metatheoretical structuralism, and for his historiographical analysis of the origins of genetics. Having a Ph.D. from the Free University of Berlin, he has published the book *Geschichte und Struktur der klassischen Genetik* (Frankfurt am Main: Peter Lang, 1995), and edited *History and Philosophy of Science and Technology* (Lorenzano, Rheinberger, Ortiz & Galles (Eds.), Oxford: EOLSS, 4 Vols., 2010) and *History and Philosophy of the Life Sciences in the South Cone* (Lorenzano, Martins & Regner (eds.), London: College Publications, 2013), among others. He was chair of the Joint Commission of the International Union of History and Philosophy of Science and Technology. He is a member of the Académie International de Philosophie des Sciences as well as of the International Academy of the History of Science. He is full professor, and director of the Center of Studies in Philosophy and History of Science (CEFHIC), at the National University of Quilmes, besides main researcher of the National Scientific and Technical Research Council (CONICET), from Argentina, and editor-in-chief of the journal *Metatheoria*.

Chapter 20
Origin's Chapter IX and X: From Old Objections to Novel Explanations: Darwin on the Fossil Record

Charles H. Pence

Abstract The ninth and tenth chapters of the *Origin* mark a profound, if perhaps difficult to detect, shift in the book's argumentative structure. In the previous few chapters and in the ninth, Darwin has been exploring a variety of objections to natural selection, some more obvious (where are all the fossils of transitional forms?) and some showing careful attention to challenging consequences of evolution (could selection really produce instincts?). Starting in the tenth, however, Darwin turns to showing us what kinds of new and unexpected results evolutionary theory might be able to offer us, again in domains both predictable (extinction) and unexpected (biogeography, embryology). It is notable that it is the fossil record that serves as this pivot point, being both a source of potentially powerful objections to evolutionary theory and home to some of its most compelling new explanations. In this chapter, I present both sets of arguments and consider what role Darwin gave to fossil evidence, in the process attempting to discover why it might have played this unique role in two different parts of Darwin's "long argument" for evolution by natural selection. Geology's special place, I argue, derives at least in part from its particular importance in Darwin's social and intellectual context.

20.1 Introduction

Darwin famously described the *Origin of Species* as "one long argument" (Darwin, 1859, p. 459), and that argument reaches a crucial turning point in the ninth and tenth chapters. As its structure has often been reconstructed, by a variety of commentators (Hodge, 1977, 1989, 1992; Lennox, 2005; Hull, 2009; Lewens, 2009; Waters, 2009; Pence, 2018), a first phase establishes the existence of natural selection by analogy with artificial selection as practiced by agricultural or horticultural breeders; a

C. H. Pence (✉)
Institut supérieur de philosophie, Université catholique de Louvain,
Louvain-la-Neuve, Belgium
e-mail: charles.pence@uclouvain.be

M. Elice Brzezinski Prestes (ed.), *Understanding Evolution in Darwin's "Origin"*,
History, Philosophy and Theory of the Life Sciences 34,
https://doi.org/10.1007/978-3-031-40165-7_20

second phase proposes that natural selection – given the much longer time that it has to act and its ability to work on invisible characters as well as the visible characters on which breeders might focus – could indeed be capable of producing the kinds of adaptations we see around us in the natural world, and (equally importantly) that it can resist the various objections that one might raise against it; and third, and lastly, a kind of responsibility or consilience phase, in which Darwin contends that a whole host of biological cases that might be inexplicable or at least confusing in the absence of evolutionary theory are provided simple and clear explanations in light of evolution by natural selection.

When we see the argument as a whole in this way, Darwin's treatment of the fossil record in Chaps. 9 and 10 sits squarely between the second and third parts of the work. On the one hand, the fossil record poses a significant challenge to natural selection. If there really has been a gradual evolution of every existing living form, then the fossil record should be full of myriad transitional forms marking the history of every evolutionary change: the manifold gradations between dinosaur and bird, or between whales and their terrestrial ancestors. The absence of such fossil evidence is, at least prima facie, a refutation of a gradual theory of evolution. But on the other hand, the fossil record is a perfect example of the kind of power that an evolutionary framework can give us. Explanations of extinction, of the resemblances between living and fossil groups, of the slow and steady rate of change of organisms over time, and more are all to be found in the geological record.

Darwin recognizes the peril that this presents to his nascent theory. "He who rejects these views on the nature of the geological record," he writes, "will rightly reject my whole theory" (Darwin, 1859, p. 342). But in doing so such a critic would renounce geology's explanatory power, the beautifully straightforward way in which "all the other great leading facts in paleontology seem to me simply to follow on the theory of descent with modification through natural selection" (Darwin, 1859, p. 343). In this sense, the geological record takes on double importance in Darwin's work. A reader of Darwin's day would likely have left these two chapters with the feeling either that the objections had been surmounted and the novel explanations were enlightening, or that the theory was unsupported by the fossil record, and thus the proffered explanations were misleading at best. A pivotal moment in the argument, indeed.

20.2 Darwin's Geological Sources and Context

A key target in teaching and learning about the nature of science is the social context in which it develops. Darwin, just as any other scientist (as the other chapters in this volume also make clear), was a creature of his time – and his context was that of mid-nineteenth century life and earth science (though there is healthy debate surrounding the relative importance of the different elements of this context; see Richards & Ruse, 2016). Geology and (the comparatively new discipline of) paleontology were particularly crucial in this regard. Darwin left for his voyage on the

H.M.S. Beagle carrying the first volume of Charles Lyell's *Principles of Geology* and received the next two volumes by post at various stops during the trip (Bowler, 1989, p. 157). He regularly cited it – including throughout these chapters of the *Origin* – as having fundamentally shaped much of his thinking about the history of the earth and of the development of life. It is worthwhile, then, to investigate Lyell's approach and to see what imprints it left upon Darwin's presentation.

Lyell's own work, in turn, is a product of a number of influences. First is what has come to be known as the *uniformitarian* theory in geology, perhaps now best exemplified by the work of the late-eighteenth century Scottish geologist James Hutton (though most commonly, at the time, through the summary of Hutton's work by John Playfair, 1802; Rudwick, 1998, p. 4). According to all such theories, we should seek no great floods or massive upheavals in order to explain the earth's geological past (*contra*, that is, the approach of the opposing position of *catastrophism*, personified by Lyell's own mentor, William Buckland). Rather, the very causes that we see producing geological change around us today – things like volcanic eruptions, erosion, earthquakes, and so on – are sufficient, when extended over a longer history of the earth, to explain the events detailed in the geological record.

This appeal to history sets up what M.J.S. Rudwick has identified as the second major important element of context for Lyell: the idea that the current state of the Earth should be read as the product of the series of contingent events that make up its history. This idea has an intellectual legacy that runs far deeper than I have space to do justice to in a short chapter such as this one.[1] While this could be seen as having some affinity with the uniformitarian model – after all, the uniformitarian posits a physical history for the Earth as well – an emphasis on the uniqueness of these events gives it a decidedly catastrophist flavor. The version of this view that Lyell received from Buckland would have emphasized the extent to which the historical Earth had been a radically *different* place from the Earth of today, manipulated by different kinds of geological causes (like the Biblical Flood, which Buckland hoped to explain and verify geologically) and (*contra* Hutton, who had written in a time before the discovery and widespread appreciation of extinction) different kinds of living species. But Lyell would leave this catastrophist model behind, and, Rudwick persuasively argues,

> [...] what eventually convinced him that the geological deluge was a chimera was the cumulative weight of specific empirical cases, in which the phenomena could be still better explained without recourse to any recent catastrophe. (Rudwick, 1998, p. 5)

Thus, Lyell's *Principles* became the manifesto for a renewed uniformitarian system – a uniformitarianism more radical than Hutton's, on which not only the *kinds* of causes acting in geology remain the same over geologic time, but even their *intensity* has been constant deep into the Earth's past (Cannon, 1976). As the young Darwin

[1] At the very least, landmarks for this view prior to Lyell's work include Buffon's *Histoire naturelle* (1749; for an especially illuminating analysis of the biological context, see Sloan 1987) and the anatomical works of Georges Cuvier (1817; see Rudwick (1997).

sets out on the *Beagle* voyage, he immediately goes about describing the environment around him in a Lyellian vein (Hodge, 1983, pp. 13–16), looking, for instance, for explanations of extinction and the fossil record that are generally consistent with Lyell's paradigm (Darwin, 1835b). While Darwin the biologist might be read as maturing by a series of divergences from his mentor on the question of species creation and extinction (Pence, 2022, pp. 2–6), Darwin the geologist stayed a relatively devout Lyellian for the rest of his career.

Darwin is therefore approaching his study of geology with a quite rich social and intellectual heritage, a heritage that was the object of dispute and the subject of transition during this period. To take one illustrative example, consider one of Darwin's personal mentors, the geologist Adam Sedgwick, whose "geology walks" may well have inspired some of the examples that Darwin would later use in these very chapters (Darwin, 1958, pp. 69–71; Secord, 1991). Sedgwick would remain, throughout his career, an opponent of uniformitarianism, and in fact would, for lack of empirical evidence, become increasingly skeptical of general geological theories as a whole, whether they be uniformitarian or catastrophist (Barrett, 1974, pp. 149–150). He was, on the one hand, instrumental in the development of Darwin's reputation as a professional scientist, having communicated some of Darwin's early results from South America to the Geological Society and thus assured his reputation upon his return from the voyage (Darwin, 1835a), but, on the other, a staunch opponent of his former student's theory of evolution (Sedgwick, 1860).

Just after Darwin's return from the *Beagle* voyage, Lyell, as president of the Geological Society, elected Darwin its secretary. He and Lyell would remain fairly regular correspondents for decades. Darwin's first major professional role within the scientific community, then – as evidenced by these interactions with Sedgwick and Lyell, and his membership in the Geological Society – was as a geologist. These relationships can be extremely illuminating for a broader perspective on Darwin's intellectual milieu (Manier, 1978), especially in the early years of his career as he was initially crafting the theory of evolution. First, we see clearly the extent to which the hard and fast divisions between scientific disciplines with which we are familiar today were simply absent in the nineteenth century. It was not in the slightest unusual for Darwin to be professionally respected both as a geologist and as a naturalist, a member of the Geological Society who is nonetheless not often remembered today for his geological writings. As James Secord notes, "it is easy to forget that the most extraordinary decision he ever made was to devote his life to the study of the natural world by becoming a geologist" (Secord, 1991, p. 133).

More generally, this period of five years, from Darwin's return to London in 1837 until his departure for his home in Down in 1842 and subsequent life as a somewhat reclusive rural man of science, marked the peak of Darwin's daily social engagement as an active, highly connected, prolific (or perhaps better, overworked) member of the British scientific community. It is not easy even to survey the theoretical developments that Darwin undertook during these years, which included much of the early construction of natural selection in his notebooks (Hodge, 2009) – but in any case, a constant throughout the entire period was his extensive personal and practical engagement with geology, inside and outside the environs of the

Geological Society. In an important sense, then, natural selection was born in a thoroughly geological context. After he left London, his social and professional context changed radically – as Rudwick notes, he ceased being involved in geological fieldwork, published in geology only the remaining studies from the *Beagle* voyage, and was much less influential at the Geological Society (Rudwick, 1985, p. 458). But the mark of geology had already been stamped on the theory of natural selection.

20.3 Taming the Fossil-Record Objection

Nonetheless, evolution constituted a relatively dramatic break with the geologists to whom Darwin had turned for inspiration. Even a staunch uniformitarian like Lyell had made room in his theory for the special, divine creation of living beings, and he remained skeptical of and cautious about, though not at all hostile to, evolution by natural selection (Lyell, 1863, pp. 407–421; Rudwick, 1998, p. 13). In the ninth chapter of the *Origin*, Darwin canvasses a number of the objections to his new theory which might be raised on geological grounds, objections with which he was all too familiar.

First and foremost is simply the vast amount of time that seemed to be required for evolution by natural selection to produce the degree of diversity that we see in the living world. For all that Lyell's theory had tried to radically extend our horizon for a "plausible" age of the Earth, Darwin's incessant emphasis on the *slow* and *gradual* character of the changes in natural selection seemed to mean that even the vast period of time that Lyell required for geological change would not suffice for the development of life in a Darwinian manner. Darwin readily saw that being able to come to grips, even intuitively, with the passage of this much time would be a challenge for many of his readers, just as it was for Darwin himself. "The consideration of these facts," he noted, "impresses my mind almost in the same manner as does the vain endeavour to grapple with the idea of eternity" (Darwin, 1859, p. 285).

In one of the few numerical calculations present in the entirety of the *Origin*, Darwin derives the amount of time that it must have taken for the denudation of the Weald, a major geologic feature in southwest England with which he was personally familiar, and estimates this to be around three hundred million years. While today's assessment would cut this value roughly in half, this figure would immediately become the target of intense criticism. William Thomson (later to become Lord Kelvin) would calculate, following thermodynamic models, and postulating an Earth that had begun its life as a single molten mass of rock, cooled in the vacuum of space, that the Earth could be no more than around four hundred million years old (Thomson, 1862) – apparently insufficient for an extremely gradual theory of evolution by natural selection to take place.[2] Darwin was thus quite right to be

[2] He would later revise the estimate down by around a factor of ten, making the problem that much worse for Darwin (Kelvin, 1895).

worried that his assumptions about the age of the Earth would prove hard to swallow, and he could only go so far toward assuaging these worries in the *Origin*.

Setting aside the question of the overall age of the Earth, we can turn to the content of the fossil record itself. First, Darwin launches a lengthy argument to the effect that the fossil record must necessarily be of a lower quality than we might have predicted by superficially surveying our current collections. From a nature of science perspective, it is interesting to note that in support of his larger, theoretical goals, Darwin needs to actually *dispute* the quality of a certain kind of evidence – here, the completeness of the fossil record. There can be very good reasons, at least occasionally, for scientists to question the accuracy, precision, or completeness of established scientific evidence.

In essence, Darwin is contending that there are a host of conditions that need to be met in order for a fossil to wind up excavated, acquired, and cataloged in the collections of paleontologists. Especially in the nineteenth century, only a small area of the world had been carefully surveyed for fossils – largely Europe and a few portions of North America. For the sake of argument, let us assume that we have thoroughly explored the fossils present in these regions that have so far been the focus of paleontological efforts (a fairly implausible assertion). It would still be the case, Darwin claims, that fossils would have been deposited in these areas only for a small fraction of the history of the Earth, as the physical conditions for fossil production are relatively rare. Darwin points out that effectively only in places where large layers of sediment are being deposited (particularly, then, in areas undergoing seafloor subsidence) should we expect significant fossilization to take place. There will thus be massive temporal gaps between apparently adjacent geological strata – a fact which, he notes, is confirmed by comparative geological studies across Europe, which find certain layers to be "missing" in some areas.

Now, let us assume both that the fossils in a given area have been well explored, and that some of the conditions for fossilization were present in that area. Even this might not be enough to provide detailed evidence of transitional forms. For that, we would need the process of fossil accumulation to be in action, constantly, for a period longer than the time that it takes for the species in question to diverge. Again, given how committed Darwin is to the slow and gradual action of evolution by natural selection, he thinks that this will be unlikely – it would imply that a very precise set of conditions remained operative, in the same place, for much longer than we have any right to expect. And even in the presence of optimal conditions in an optimally explored area, we might need to look across different strata to find evidence of transitional forms – in which case it would become progressively harder for us to recognize that we were actually looking at the particular common ancestor or transitional form for which we were searching.

In spite of all of these difficulties, Darwin notes that we have nonetheless been surprisingly successful at discovering more and more diverse fossilized forms. Darwin outlines a number of pronouncements from the paleontological literature insisting that a particular form has not been or could never be found, and observes that often, within a relatively short time after those assertions are published, the fossil in question is unearthed after all. For instance, Darwin notes that a number of claims to the discovery of the oldest known form of life had been successively revised, finding simpler and simpler forms in lower and lower strata. This success has continued into the modern day; particularly striking is the example of the fossil *Tiktaalik*, a proto-tetrapod species that was discovered by, essentially, a targeted search of geological regions that would be likely to yield rock formations of the relevant age (Daeschler et al., 2006).

Finally, Darwin also picks up, in Chap. 9, a more general potential difficulty with evolutionary theory that is all too familiar in today's classrooms. If we ask whether or not transitional forms exist in the fossil record, we have to remember just what it is that we might mean by a transitional form in the first place. "I have found it difficult," Darwin writes, "when looking at any two species, to avoid picturing to myself, forms *directly* intermediate between them" (Darwin, 1859, p. 280). Of course, this is a natural tendency: because we do not know what the common ancestor of two organisms might have looked like (nor, at a quick glance at the descendants, can we tell how long ago it might have lived), it is all too easy to forget that common ancestors will almost certainly not resemble some kind of "average" between two existing organisms. It is thus the case that even if we were in possession of the very fossil specimen that was the common ancestor of two extant groups, we might be incapable of recognizing its importance.

Darwin summarizes his response to all of these potential objections by developing a metaphor for the evidential quality (or lack thereof) of the geological record that is worth quoting at length:

> For my part...I look at the natural geological record, as a history of the world imperfectly kept, and written in a changing dialect; of this history we possess the last volume alone, relating only to two or three countries. Of this volume, only here and there a short chapter has been preserved; and of each page, only here and there a few lines. Each word of the slowly-changing language, in which the history is supposed to be written, being more or less different in the interrupted succession of chapters, may represent the apparently abruptly changed forms of life, entombed in our consecutive, but widely separated, formations. On this view, the difficulties above discussed are greatly diminished, or even disappear. (Darwin, 1859, pp. 310–311)

20.4 Geology in Evolution's Service

With that, then, we move from the ninth to the tenth chapter – from geology as a source of objections and problems to geology as a way in which we might confirm evolution's action, using evolution as a source for new, unexpected, and powerful explanations of the history of life. There is, to be sure, something of a paradox here, underscored by the long quote above. Darwin has just spent a chapter explaining to us the manifold ways in which the geological record might fail to give us a precise picture of the history of life on Earth. To turn around and use this same evidence to ground compelling explanations that will offer support for evolutionary theory will thus require sophisticated and careful argument.

Two particularly striking features of the fossil record, Darwin argues, can be readily explained by evolution (and can only be explained with difficulty from a non-evolutionary perspective). First is the structure of resemblances between fossils. Fossil groups tend to be intermediate in character between living groups, with fossils recapitulating the kind of "group-within-group" structure familiar from Linnaean taxonomy. This makes perfect sense, of course, if those intermediate groups are actually ancestral to living species. Darwin will proceed to construct a similar argument that evolutionary theory offers an explanation for taxonomic classification itself in Chap. 13 of the *Origin*. Darwin writes that these group resemblances offer us, "in short, such evidence of the slow and scarcely sensible mutation of specific forms, as we have a just right to expect to find" (Darwin, 1859, p. 336).

Only slightly less important is evolution's explanation of extinction. For Lyell, as Darwin will discuss in more detail in Chap. 11, extinction was explained by the redistribution of climatic conditions across the surface of the globe. One could still be a consistent uniformitarian in saying that, while the nature and intensity of the causes of geological change remain the same over time, their distribution across the Earth could differ, thus causing the extinction of species no longer capable of living in their former habitats (Lyell, 1832, pp. 129–130). Darwin agreed, but thought that this could only tell part of the story – for he believed that he had seen cases on the *Beagle* voyage of extinctions without any corresponding change in climatic conditions (Hodge, 1983, pp. 21–22).

As Darwin notes, extinction would also follow as a straightforward consequence of natural selection. If the organic world really is as finely balanced as Darwin's invocations of the struggle for life seem to argue, then the slightest change to the relationships between organisms or environmental conditions will likely cause some groups to be favored and others disadvantaged, "and the consequent extinction of less-favored forms almost inevitably follows" (Darwin, 1859, p. 320). Darwin's understanding of species as the product of common descent with modification also explains the fact that extinction is permanent. Once a species has disappeared, even if another similar species were to arise, it would not be the same species – once the "link of generation has been broken" (Darwin, 1859, p. 344), a group cannot be recovered.

A number of other, smaller features of the geological record are also explicable by evolution. We find in this chapter Darwin's mature view on the concept of *progress*. A question that had haunted him throughout his writings (see, e.g., Ruse, 1996, pp. 145–154), Darwin struggled with the tension between the apparent progress toward "higher" organization over the history of life (with human beings at the pinnacle) and the implication of his theory that there was no way to derive a global tendency toward a "goal" or "direction" in evolution. Natural selection offers us, at best, a kind of local or piecemeal progress, with descendant organisms having had some kind of advantage that enabled them to spread – but with no guarantee that those advantages would "accumulate" in any particular way. This would still produce something that resembles progress, but it will be a progress of a strange sort. As Darwin puts it,

> I do not doubt that this process of improvement has affected in a marked and sensible manner the organisation of the more recent and victorious forms of life, in comparison with the ancient and beaten forms; but I can see no way of testing this sort of progress. (Darwin, 1859, p. 337)

While we might thus expect that today's organisms would outcompete those of the past – they have, after all, accumulated different advantages that let them survive while others perished – there would be no way, in general, to say in advance what form that success might take, and thus no guiding, overall notion of improvement or progress.

Lastly, with the exception of a handful of "living fossil" species, the fossil record gives us extensive evidence of the slow but constant modification of organisms over the course of time. More precisely, it shows us that while almost all organisms are constantly changing, those rates of change are themselves variable, with some groups known to vary faster and others more slowly. This is, Darwin notes, entirely explicable on an evolutionary basis, but difficult to explain if we think that this variability is some kind of intrinsic feature of organisms or species. A group that is faced with more diverse interactions (whether with other organisms or a more complex environment) will have a more diverse collection of selective pressures to which to respond, and by extension more opportunities for specialization, division of labor, and the other processes that Darwin thinks are crucial for driving speciation.

20.5 Geology and the Argument of the *Origin*

It is interesting to note that the fossil record is one of the only facets of the *Origin*'s "long argument" that plays both the roles of a generator of objections and a generator of positive evidence. In that sense, the transition that I noted above from the second, objection-refuting portion to the third, consilience-building part of the book, which takes place between Chaps. 9 and 10, could easily pass without notice. But its significance should not be understated. Darwin often lamented that he was too hemmed in by the constant pressure to respond to objections. As he wrote to his friend and colleague John Stevens Henslow, in response to Sedgwick's scathing review of the *Origin*, if it was permissible "(& a great step) to invent the undulatory theory of Light – that is, hypothetical undulations in a hypothetical substance the ether," he saw no reason why he could not be permitted to

> invent hypothesis of natural selection...& try whether this hypothesis of natural selection does not explain (as I think it does) a large number of facts in geographical distribution – geological succession – classification – morphology, embryology &c. &c. (Darwin 1860)

Darwin's two chapters on geology thus also encapsulate the ambivalence that he felt, torn between the apparent requirement to respond to all his objectors and his desire to focus on the exciting, fruitful consequences of an evolutionary worldview.[3]

[3] The idea that Darwin's responses to objections cloud the force of his argument has often motivated philosophers of biology to read the *first* edition of the *Origin*, prior to the integration of many such responses. This opinion goes at least as far back as Darwin's son Leonard, who wrote to R. A. Fisher that he wondered whether "it would be worth republishing the first edition of the *Origin of Species*...*because* it was written before my father had been subject to *any criticism whatever*" (L. Darwin to Fisher, [late-September 1926?], Bennett, 1983, p. 81, original emphasis).

The importance of these two chapters of the *Origin* comes as no surprise. The fossil record has been and remains a crucial source of evidence, dispute, and debate with regard to evolutionary theory. It is likely the first thing that a new student of evolutionary biology would think of when asked to identify the base of evidence that supports an evolutionary explanation of the history of life. As Darwin makes apparent, however, the relevance of this evidence to the phenomena, or the relationship between evolutionary theory and the fossil record, is much more complex than it might at first appear. Many of the obvious, most simplistic inferences that we might draw – like the idea that we will confirm evolutionary theory in every lineage, just by tracing out the history of transitional forms between two groups of interest – will not only be falsified (there are many groups for which this will be impossible), but would even serve as objections to the coherence of evolutionary theory itself (in the absence of such sequences of transitional forms, why not reject the evolutionary explanation outright?). The ways in which the fossil evidence supports evolution must thus be selectively and cautiously argued. Darwin clearly saw this double-edged role for geology in evolution and embraced it, making it a central part of the argument of the *Origin*.

References

Barrett, P. H. (1974). The Sedgwick-Darwin geologic tour of North Wales. *Proceedings of the American Philosophical Society, 118*(2), 146–164.

Bennett, J. H. (Ed.). (1983). *Natural selection, heredity, and eugenics: Including selected correspondence of R.A. Fisher with Leonard Darwin and others*. Clarendon Press.

Bowler, P. J. (1989). *Evolution: The history of an idea*. University of California Press.

Cannon, W. F. (1976). Charles Lyell, radical Actualism, and theory. *British Journal for the History of Science, 9*(2), 104–120.

Cuvier, G. (1817). *Le règne animal distribué d'après son organisation, pour servir de base à l'histoire naturelle des animaux et d'introduction à l'anatomie comparée, tome I*. Deterville.

Daeschler, E. B., Shubin, N. H., & Jenkins, F. A. (2006). A Devonian tetrapod-like fish and the evolution of the tetrapod body plan. *Nature, 440*(7085), 757–763.

Darwin, C. R. (1835a). *Extracts from letters addressed to Professor Henslow*. Privately printed. Kees Rookmaker (Ed.) http://darwin-online.org.uk/

Darwin, C. R. (1835b). *The position of the bones of Mastodon (?) at Port St Julian is of interest.* CUL-DAR42.97–99. Kees Rookmaker (Ed.) http://darwin-online.org.uk/

Darwin, C. R. (1859). *On the origin of species*. 1st ed. John Murray.

Darwin, C. R. (1958). *The autobiography of Charles Darwin, 1809–1882, with original omissions restored.* Nora Barlow (Ed.). Collins.

Darwin, C. R. (1860). Letter 2971 – Darwin, C. R. to Henslow, J. S., 8 May [1860]. Darwin Correspondence Project. https://www.darwinproject.ac.uk/entry-2791

Buffon, G.-L. L. (1749). *L'Histoire Naturelle, générale et particulière, avec la description du Cabinet du Roi*. Imprimerie Royale.

Hodge, M. J. S. (1977). The structure and strategy of Darwin's 'long argument'. *British Journal for the History of Science, 10*(3), 237–246. https://doi.org/10.1017/S0007087400015685

Hodge, M. J. S. (1983). Darwin and the Laws of the animate part of the terrestrial system (1835–1837): On the Lyellian origins of his Zoonomical explanatory program. *Studies in History of Biology, 6*, 1–106.

Hodge, M. J. S. (1989). Darwin's theory and Darwin's argument. In M. Ruse (Ed.), *What the philosophy of biology is: Essays for David Hull* (pp. 136–182). Kluwer Academic Publishers.

Hodge, M. J. S. (1992). Darwin's argument in the origin. *Philosophy of Science, 59*(3), 461–464. https://doi.org/10.1086/289682
Hodge, M. J. S. (2009). The notebook programmes and projects of Darwin's London years. In M. J. S. Hodge & G. Radick (Eds.), *The Cambridge companion to Darwin* (2nd ed., pp. 44–72). Cambridge University Press.
Hull, D. L. (2009). Darwin's science and Victorian philosophy of science. In M. J. S. Hodge & G. Radick (Eds.), *The Cambridge companion to Darwin* (2nd ed., pp. 173–196). Cambridge University Press.
Kelvin, W. T. (1895). The age of the earth. *Nature, 51*(1323), 438–440. https://doi.org/10.1038/051438c0
Lennox, J. G. (2005). Darwin's methodological evolution. *Journal of the History of Biology, 38*, 85–99. https://doi.org/10.1007/s10739-004-6511-4
Lewens, T. (2009). The *Origin* and philosophy. In M. Ruse & R. J. Richards (Eds.), *The Cambridge companion to the "Origin of Species"* (pp. 314–332). Cambridge University Press.
Lyell, C. (1832). *Principles of geology* (Vol. II). John Murray.
Lyell, C. (1863). *The geological evidences of the antiquity of man, with remarks on theories of the origin of species by variation*. John Murray.
Manier, E. (1978). *The young Darwin and his cultural circle*. D. Riedel Publishing Company.
Pence, C. H. (2018). Sir John F. W. Herschel and Charles Darwin: Nineteenth-century science and its methodology. *HOPOS, 8*(1), 108–140. https://doi.org/10.1086/695719
Pence, C. H. (2022). *The rise of chance in evolutionary theory: A pompous parade of arithmetic*. Academic Press.
Playfair, J. (1802). *Illustrations of the Huttonian theory of the earth*. Cadell and Davies.
Richards, R. J., & Ruse, M. (2016). *Debating Darwin*. University of Chicago Press.
Rudwick, M. J. S. (1985). *The great Devonian controversy: The shaping of scientific knowledge among gentlemanly specialists*. University of Chicago Press.
Rudwick, M. J. S. (1997). *Georges Cuvier, fossil bones, and geological catastrophes: New translations & interpretations of the primary texts*. University of Chicago Press.
Rudwick, M. J. S. (1998). Lyell and the principles of geology. *Geological Society, London, Special Publications, 143*(1), 1–15. https://doi.org/10.1144/GSL.SP.1998.143.01.02
Ruse, M. (1996). *Monad to man: The concept of progress in evolutionary biology*. Harvard University Press.
Secord, J. A. (1991). The discovery of a vocation: Darwin's early geology. *British Journal for the History of Science, 24*(2), 133–157. https://doi.org/10.1017/S0007087400027059
Sedgwick, A. (1860). Objections to Mr. Darwin's theory of the origin of species. *The Spectator, 33*(March), 285–286.
Sloan, P. R. (1987). From logical universals to historical individuals: Buffon's idea of biological species. In *Histoire du concept d'espèce dans les sciences de la vie* (pp. 101–140). Editions de la Fondation Singer-Polignac.
Thomson, W. (1862). On the secular cooling of the earth. *Transactions of the Royal Society of Edinburgh, 23*(1), 157–169. https://doi.org/10.1017/S0080456800018512
Waters, C. K. (2009). The arguments in *The Origin of Species*. In M. J. S. Hodge & G. Radick (Eds.), *The Cambridge companion to Darwin* (2nd ed., pp. 120–143). Cambridge University Press.

Charles H. Pence is Chargé de cours at the Université catholique de Louvain, in Louvain-la-Neuve, Belgium, where he directs the Center for Philosophy of Science and Societies (CEFISES). He also serves as a coeditor of the journal *Philosophy, Theory, and Practice in Biology* (*PTPBio*). His work centers on the philosophy and history of biology, with a focus on the introduction and contemporary use of chance and statistics in evolutionary theory.

Chapter 21
Origin's Chapter XI and XII: "Seed! Seed! Seed!": Geographical Distribution in on the Origin of Species

Tina Gianquitto

Abstract The two chapters on "Geographical Distribution" in *On the Origin of Species* reinforce a key premise of Charles Darwin's evolutionary theorizing: how the distribution of species around the globe supports his hypothesis of descent from a common ancestor. The chapters also address other main evolutionary premises: the role of the individual in shaping both the future of a species as well as future species; the importance of the migration/dispersal of individuals through both "accidental" and "occasional" processes; and the "subsequent modification and multiplication of new forms" arising from that migration. Seed dispersal offers a compelling explanation for how related populations appear in geographically distinct areas of the globe. But in the context of the 1859 publication of *Origin*, it also served as a crucial counterpoint to the challenges presented by the popular antievolutionary theory of separate centers of creation. Proponents of this theory, such as Louis Agassiz, argued that species originated simultaneously in widely separated areas of the globe via multiple creative acts. Separate creation rationalized racist polygenist theories undergirding slavery in the United States and in global politics of enslavement and racial difference. In Darwin's work, seeds instead illuminate a world of relations linked by physiology, geography, and time.

21.1 Introduction

Charles Darwin's two chapters on "Geographical Distribution" (11 and 12) in *On the Origin of Species* are filled with the stories of seeds. Seeds fly with the wind, traverse the land, and sail the oceans. To Darwin, the seed was not just an object, a promise of a plant, but it was an *individual,* and, like other individuals, the seed was both *of* a moment and timeless. Darwin follows seeds to explore the role that geographical isolation and migration play in the development of species in particular

T. Gianquitto (✉)
Colorado School of Mines, Golden, CO, USA
e-mail: tinagian@mines.edu

M. Elice Brzezinski Prestes (ed.), *Understanding Evolution in Darwin's "Origin"*, History, Philosophy and Theory of the Life Sciences 34,
https://doi.org/10.1007/978-3-031-40165-7_21

environments. Together, these chapters reinforce a key premise of Darwin's evolutionary theorizing; namely, how the distribution of species around the globe supports his hypothesis of descent from a common ancestor. Further, the "Geographical Distribution" chapters address several other main evolutionary premises: the role of the individual in shaping both the future of a species as well as future species; the importance of the migration/dispersal of individuals through both "accidental" and "occasional" processes; and the "subsequent modification and multiplication of new forms" arising from that migration. A major focus of the chapters is seed dispersal, and this discussion examines how Darwin's early experiments with seed salting and seed dispersal played a role in the development of his thinking on the geographical distribution in *Origin*. These investigations into the ability of a seed to remain viable in salt water long enough to cross oceans occupied Darwin from his early *Notebooks*, especially *Questions and Experiments* (1839–1844), through to his more formal exploration of the subject in seed-salting experiments that began in 1854. The portrait of the seed-as-individual that emerges in these studies of seeds resounds in Darwin's evolutionary imagination.

Seed dispersal offers one compelling explanation of how related populations appear in geographically distinct areas of the globe. But in the context of the 1859 publication of *On the Origin of Species*, the humble seed also serves as a crucial counterpoint to a persistent challenge to evolutionary theory: the theory of separate centers of creation. Proponents of this anti-evolutionary counter theory, such as the popular naturalist Louis Agassiz, argued that species originated simultaneously in widely separated areas of the globe via multiple creative acts. Separate creation theory also rationalized racist polygenist theories that undergirded both slavery in the United States and in global politics of enslavement and racial difference. While these matters might seem far removed from scientific inquiry, they were very much on Darwin's mind as he wrote these chapters. He specifically targeted Agassiz's ideas about racial difference in his first published accounts of his seed-salting experiments in the *Gardener's Chronicle*, stating that his experiments "might naturally appear childish to many", but instead

> [...] have a direct bearing on a very interesting problem, which has lately, especially in America, attracted much attention, namely, whether the same organic being has been created at one point or several on the face of our globe. (Darwin, 1855, p. 356)

In Darwin's chapters on geographical distribution, these organic beings illuminate a world of relations linked by physiology, geography, and time.

21.2 Migration, Modification, and Multiplication

Geographical distribution was so important to Darwin's evolutionary theorizing that he devoted 2 of the 14 chapters of the book to the topic, or about 13%, more than any other single topic (Bowler, 2009, p. 153). These chapters guide the reader across oceans and mountain ranges, to islands and glaciers, presenting concrete evidence of Darwin's premise, that:

> If [. . .] in the long course of time the individuals of the same species, and likewise of allied species, have proceeded from some one source; then I think all the grand leading facts of geographical distribution are explicable on the theory of migration [. . .] together with subsequent modification and the multiplication of new forms. (Darwin, 1859, p. 408)

The great actors – migration, modification, and multiplication – link organisms across the globe into great families all "descended from the same parents" (Darwin, 1859, p. 407).

To begin his argument, Darwin states his first "great fact" about geographical distribution: contrary to popular belief, climate and other physical conditions do not determine species relationships. Instead, geography, especially barriers to species migration, is the "best predictor of biogeographical relationships" (Costa, 2009, p. 347) Darwin argues, for instance, that while the "Old World" and the "New" share "the most diversified conditions" – from deserts to marshes and plains to mountains – their "living productions" are vastly different (Darwin, 1859, p. 346). At the same time, areas with "considerably different climate" will possess species "incomparably more closely related to each other," than areas around the globe with "nearly the same climate" (Darwin, 1859, p. 347). Similar climes, in short, need not support similar biota, while regions varying in climactic and physical conditions may – and do.

These observations led Darwin to the second fact of geographical distribution: "barriers of any kind or obstacles to free migration are related in a close and important manner to the differences between the productions of various regions" (Darwin, 1859, p. 347). The greater the barrier, the greater the differences, as is the case with separated land masses (Australia, Africa, and South America) or separated bodies of water, such as the eastern and western shores of South and Central America, divided "only by the narrow, but impassable, isthmus of Panama" (Darwin, 1859, p. 348). The opposite is also true: areas open to free migration, without any insurmountable barriers, exhibit similarities in inhabitants (Helgeson, 2017, p. 340). Here, Darwin calls attention to the role of ocean islands as "halting places" for "emigrants" in supporting or preventing migration (Darwin, 1859, pp. 357, 348). In the next chapter, Darwin returns to oceanic islands, considering especially the forms he encountered in the Galapagos Islands, as he explores the migratory conditions under which these islands became "tenanted" by both distinct and allied species (Darwin, 1859, p. 400).

Darwin reserves his most evocative language for his third "great fact": species in confined areas are both different and related. His strategy here is worth exploring in detail as it leads to the main argument of *Origin* – descent with modification from a common ancestor. To get there, readers are invited to walk along with the naturalist – perhaps a young Darwin during his voyage on the *Beagle*: "Nevertheless the naturalist in traveling, for instance, from north to south never fails to be struck by the manner in which successive groups of beings, specifically distinct, yet clearly related, replace each other." Sensory perception is enlisted in the argument to make the imaginative scene more real for the reader: the naturalist "hears from closely allied, yet distinct kinds of birds, notes nearly similar" and "he sees their nests similarly constructed, but not quite alike, with eggs coloured in nearly the same manner." By the end of the paragraph, the reader has become a co-investigator with the naturalist, who has now moved from "he" to "we": "We ascend the lofty peaks of the Cordillera, and we find an alpine species [. . .]"; "we see [. . .]"; "we look [. . .]"; "we find" (Darwin, 1859, p. 349).

By bringing the reader along with him, Darwin's conclusions will be difficult to escape:

> We see in these facts some deep organic bond, prevailing throughout space and time, over the same areas of land and water, and independent of their physical conditions[...]. This bond, on my theory, is simply inheritance. (Darwin, 1859, p. 350)

As Hegelson explains: "Darwin's theory thus provides [. . .] an explanation; geographical proximity (accessibility) and taxonomic similarity go together because both are consequences of recent common ancestry" (Helgeson, 2017, p. 344). As Darwin writes:

> On these views, it is obvious that the several species of the same genus, though inhabiting the most distant quarters of the world, must originally have proceeded from the same source, as they have descended from the same progenitor. (Darwin, 1859, p. 351)

Where no physical or ecological barriers to migration exist, "almost any amount of migration is possible" during "the vast geographical and climate changes which have supervened since ancient times" (Darwin, 1859, p. 351). Migration is the reason similar species inhabit "the most distant quarters of the world," but time is the essential ingredient, necessary for the "slow process of modification through Natural Selection"to operate and produce "groups of modified descendants (Darwin, 1859, p. 350).

After laying out these "facts" of distribution – geography rather than climate, the presence or absence of barriers, and the coexistence of sameness and difference – Darwin takes on two other debates within the naturalist community concerning the appearance of identical or closely related species occupying distinct geographical areas of the globe (disjunct distribution) (Kinch, 1980, p. 92). As will be discussed below, Darwin's primary concern was to address "separate centers" advocates, who argued that the same species appeared simultaneously in more than one location and who viewed species as immutable and inviolable, created by divine plan and fiat to inhabit distinct geographical locales. Since this theory allowed the creation of similar or identical species in geographically distant locations around the globe, separate creationists had no oceans or continents to bridge and did not have to account for the migration and gradual adaptation of species to new environments. Instead, according to Darwin, it required something far more problematic – "the agency of a miracle" (Darwin, 1859, p. 352).

Proponents of single origin theories, on the other hand, held that species migrate around the globe "from a single birthplace" (Darwin, 1859, p. 354) and included Darwin's friends Charles Lyell and Joseph Hooker. These naturalists, therefore, had to explain this problem of dispersal for their perspective to triumph: How *did* individual organisms navigate vast distances and imposing geographical barriers, such as mountains and oceans, to populate new locales? To Darwin's eye, many single-origin theorists, while perhaps stopping short of requiring miracles, nevertheless also resorted to fanciful solutions to answer the question of distribution. Edward Forbes, Charles Lyell, and even Hooker had to conjure geological features such as land bridges and continental extensions (once extant, now destroyed) to facilitate the dispersal and migration of species. Forbes, for instance, "insisted that all the islands in the Atlantic must *recently* have been connected with Europe or Africa, and Europe likewise with America" (Darwin, 1859, p. 357, emphasis added).

Darwin complained that such arguments "hypothetically bridged over every ocean, and have united almost every island to some mainland." Darwin's problem was not with speculating on the geological processes that may or may not have united islands to continents in the distant past. Indeed, he agreed with Lyell and others that subsidence and elevation of continents was "a dominant geological process governing earth history" (Costa, 2009, p. 357). Nor did he categorically oppose the concept of land bridges; he accepted, for example, the likelihood of a land connection between North America and Asia. What he opposed was the "ad hoc invocation" of former land bridges and continental extensions "without fully considering other means that might account for observed plant and animal distributions". The problem is time: "we are not authorized in admitting such enormous geographical changes *within the period of existing species*." One day, Darwin argues, "we shall be able to speculate with security on the former extensions of land" – a challenge answered in part by twentieth-century continental drift theory – but,

> I do not believe it will ever be proved that *within the recent period* continents which are now quite separate, have been continuously, or almost continuously, united with each other, and with the many existing ocean islands. (Darwin, 1859, pp. 357–358, emphasis added)

Darwin, therefore, argued that if land bridges and continental extensions do not answer the problem, the "varied means" of migration available to countless organisms do, and in these two chapters on geographical distribution, Darwin shows how organisms from seeds to fish to mammals move around the world.

21.3 Seed Time

Darwin's challenge throughout *Origin*, as seen in these chapters on distribution, is representational: how to picture to readers the vast yet incremental movement of countless individual forms across time and space that are effectively invisible except in the imagination (Beer, 2000, p. 9). As Devin Griffiths argues, Darwin's science is

> [...] narrative and comparative, [...] producing [...] histories for the reader by carefully tracing their analogy to other events (in the domestic, contemporary world) that could be described from life. (Griffiths, 2016, p. 11)

This is vibrantly illustrated in Darwin's discussion of the migration of organisms during the glacial period where Darwin asks his readers to conceive of a world on the move in response to changing environmental conditions. Suffering and invasion are the dominant motifs, conditions with which readers might sympathize:

> Hence, it seems to me possible, bearing in mind that the tropical productions were in a suffering state [due to the cooling climate and no escape routes] and could not have presented a firm front against intruders [. . .]. (Darwin, 1859, p. 377)

Darwin also understands that he faces a particular descriptive challenge when delineating long-ago migrations for his readers: terrestrial nonhuman mammals, despite their obvious mobility, "have not been able to migrate", whereas plants, despite their obvious rootedness, "have migrated across the vast and broken interspace" (Darwin, 1859, p. 353). "I shall here confine myself to plants," Darwin explains, and the pages that follow are awash in seeds as Darwin illustrates for readers why similar species can be found in widely dispersed places around the globe. Seeds and plants also gave Darwin an advantage in his argument: widespread plant literacy enabled Darwin's audience to comprehend the imaginative leaps he made throughout.

Seeds in these chapters are on the move. They float overseas on logs and hitch rides in the digestive tracts of birds and fish (Darwin, 1859, p. 362). They get transported in mud stuck to the feet of wading birds and in the wool and fur of passing animals. They hitch rides on drift timber, in the crops of birds, on icebergs, and on the wind (Darwin, 1859, pp. 360, 361, 363). They move in response to environmental pressures of warming or cooling (Darwin, 1859, p. 377). They colonize when the conditions are favorable and are displaced when they are not. They are both natives and immigrants, facing the trials of migration. As Darwin wrote in an early publication on seeds: "But when the seed is sown in its new home then, as I believe, comes the ordeal; will the old occupants in the great struggle for life allow the new and solitary immigrant room and sustenance?" (Darwin, 1855) Seeds are, in short, the minute engines of global species dispersal.

Darwin wanted to comprehend plants crossing both space and time: "Ask Entomologists," he writes in *Notebook B*, "whether they know of any case of introduced plant, which any insects hav[e] become *attached to*" (Darwin, 1837–1838, p. 218, emphasis by the author). What does it mean to become "attached to" a plant? As Carla Hustak and Natasha Myers argue, in the context of Darwin's plant studies to be "attached to" something means to be invested and involved in it, to be inextricably bound up in its success or failure (Hustak & Myers, 2012, p. 77). For a plant to survive in a new environment it needs to thrive, and to thrive it needs to reproduce. To do this, plants generally need to convince a pollinator or two to become "attached to" it, and in this, time and an element of chance are required, especially in evolutionary unions. Thus, sea-carried seeds, thrown up on a shore, might find just enough soil to grow. If successful, they change the relationships of the organisms they encounter over time, bringing them into "new relations with each other" (Darwin, 1859, p. 351) – perhaps the pollinator attaching itself to an introduced plant. As Elizabeth Grosz notes, evolution is, for Darwin, "the emergence in time of biological innovation and surprise," the possibility of "becoming, and becoming-other" (Grosz, 2004, pp. 19–20).

Darwin saw the potential of seeds early in his evolutionary theorizing, and his *Notebooks* and correspondence are brimming with thoughts and observations about seed dispersal, many of which ultimately made their way into these chapters (Black, 2009, p. 193). In one query from his *Questions & Experiments* (1839–1844), "Soak all kinds of seeds for a week in Salt. artificial water.–," Darwin challenges himself to determine how long a soaked seed remains viable (Darwin, 1839–1844). The longer seeds survived in salt water, defined by germinating after marinating, the more likely it is that similar species would be found at great distances from each other. Darwin proposed an experiment to test a hypothesis he had previously formulated: "see if there are any species same as T. Del Fuego & C. Of Good Hope show *possibility* of transport." At first, he is unsure: "If some cannot be explained, more philosophical to state we do not know how transported. –" (Darwin, 1837–1838, p. 94). By Notebook D (~1838), Darwin is more confident: "*When* I show that islands would have no plants were it not for seeds being floated about [. . .]" (Darwin, 1838, emphasis added). By the time he wrote his *Essay* in 1844, as Janet Browne observes, the

> [. . .] crux of Darwin's system was the proposition that species could spread [. . .] as far as barriers, the means of transportal, and the preoccupation of the land by other species would permit. (Browne, 1983, pp. 196–197)

When Darwin shared his ideas on migration with Hooker, the botanist was skeptical, writing to Darwin in early 1844:

> It is I think high time to throw overboard laying much stress on the subject of the *migration of seeds*, except in the case of lands we know to have been recently formed, or, from devastating causes, to be recently clothed with vegetation. (DCP-LETT-737; letter from J. D. Hooker to Darwin, February 23–March 6, 1844)

Hooker challenged Darwin again in 1847, writing that

> [. . .] the more I see the less I am inclined to take migration as a sufficient agent in effecting the strange similarity between the Alpine Floras of V.D.L. [Van Diemen's Land] N.Z. & that of Fuegia. (DCP-LETT-1097; letter from J. D. Hooker to Darwin, June 16, 1847).

In his *Introductory Essay to the Flora of New Zealand* (1853), Hooker largely dismissed Darwin's argument, writing:

> I cannot think that those who, arguing for unlimited powers of migration in plants, think existing means ample for ubiquitous dispersion, sufficiently appreciate the difficulties in the way of necessary transport. (Hooker, 1853–1855, v. 2, p. xx)

Indeed, Hooker only gradually converted to Darwin's theory of migration, preferring instead to explain species dispersal through the agency of land bridges and continental extensions. Not the least problem, Hooker argued, is that prolonged exposure to salt water would certainly kill seeds.

Therein was Darwin's challenge. He needed experimentation to prove his theorizing, (Costa, 2015, p. 135) and so on 7 April 1855, Darwin told Hooker: "I have begun my seed-salting experiments." (DCP-LETT-1661; letter from Darwin to J. D. Hooker, April 07, 1855). A week later, he announced these seed-salting experiments in a short notice in the *Gardener's Chronicle and Agricultural Gazette* (April 14, 1855):

> I have begun making some few experiments on the effects of immersion in sea-water on the germinating powers of seeds, in the hope of being able to throw a very little light on the distribution of plants, more especially in regard to the same species being found in many cases in far outlying islands and on the mainland. (Darwin, 1855)

Darwin turned his house and gardens into a sort of miniature globe, where seeds were exposed to as many varied conditions as Darwin could manage. He confided again in Hooker later that April: "If you knew some of the experiments (if they may be so called) which I am trying, you would have a good right to sneer for they are so *absurd* even in *my* opinion that I dare not tell you" (DCP-LETT-1671; letter from Darwin to J. D. Hooker, April 24, 1855).

In one experiment, Darwin placed seeds of different types in separate bottles containing two to three ounces of artificially produced seawater with salt provided by a chemist. He put some of the bottles outside in the shade, other bottles were placed in the cellar of Down House where they were subjected to more constant temperatures. Since Darwin was investigating ocean-going seeds, he added snow to his saltwater tanks, to test the resiliency of seeds in colder temperatures. The smell of putrid water was apparently so remarkable that Darwin was astonished that any seeds survived to germinate (Darwin, 1855). He removed soaked seeds, planted them in glass tumblers, and kept them on the mantle of the chimney in his study so that he could "see the seeds all the time, before & after germination" (DCP-LETT-1667; letter from Darwin to J. D. Hooker, April 13, 1855).

Throughout his experiments, Darwin confided to his friends that he was alternately astonished and delighted and made "rather low" by the progress of his seed experiments (DCP-LETT-1678; letter from Darwin to William Darwin Fox, May 07, 1855). He regaled Hooker with accounts of his trials:

> Everything has been going wrong with me lately; the fish at the Zoolog. Soc. ate up lots of soaked seeds, & in imagination they had in my mind been swallowed, fish & all, by a heron, had been carried 100 miles, been voided on the banks of some other lake & germinated splendidly, – when lo & behold, the fish ejected vehemently, & with disgust equal to my own, *all* the seeds from their mouths. (DCP-LETT-1681; letter from Darwin to J. D. Hooker, May 15, 1855)

As an expert on plant distribution, Hooker supplied Darwin with valuable examples and statistics demonstrating range and migration patterns mainly from the Southern Hemisphere, New Zealand, India, and the polar regions, from his expeditions to the regions in 1839–1843. Hooker also introduced Darwin to Asa Gray, the Harvard botanist who was an expert on plant distributions in North America and who provided Darwin with valuable statistics that fed into the picture of the moving world that Darwin was developing. The contact was valuable to both men: Gray was also at work on questions of species distribution, in his case trying to identify the cause of similarities between the floras of eastern Asia, especially Japan, and North America (Hung, 2013). He would also become Darwin's fiercest advocate of evolutionary theory in the United States – and his most valuable ally in the upcoming battle with his Harvard colleague, Louis Agassiz.

As he recounts the results of these experiments and the conclusions that he draws about the history of species migration in the chapters on the geographical distribution in *Origin* (see Darwin, 1859, pp. 358–60), Darwin notably uses conditional language repeatedly; indeed, the conditional is the operative mode though much of *Origin.* The "inextricable web of affinities" (Darwin, 1859, p. 434) binding the natural world are prone to the contingencies "if," "when," "might," and "could be" of any given process. Darwin reminds Hooker that he is not trying to establish details on specific species nor is he trying to prove that *all* plants germinate after immersion in salt water, writing,

> I shd. think a few seeds of the leading orders, or a few seeds of each of the classes mentioned by you with albumen of different kinds wd suffice to show the *possibility* of considerable sea-transportal. (DCP-LETT-1669; letter from Darwin to J. D. Hooker, April 19, 1855)

Since he only needed to prove that it was plausible that saturated seeds could grow into a plant, all Darwin had to do was to identify *occasional* means of dispersal: if even only *one* might survive migration – a seed here or there escaping on an ocean current or in the crop of a bird, landing, and thriving – it will help prove his point.

Accident and occasion collide in Darwin's thinking – one alone is not enough. This is most evident in the efforts Darwin made to bracket possibility and probability in his discussion of seed dispersal mechanisms in the chapters on geographical distribution. He writes in the first edition (1859), for instance: "I must now say a few words on what are called accidental means, but which more properly *might be* called occasional means of distribution" (Darwin, 1859, p. 358). By the sixth edition of *Origin* (1876), however, Darwin has changed the passage to "[...] but which more properly *should be* called occasional means of distribution" (Darwin, 1876, p. 324). Similarly, Darwin revisited his cautionary explanation of how seeds that might otherwise sink after submersion in salt water instead float on currents to distant destinations; in the first edition, he writes:

> [...] it occurred to me that floods *might* wash down plants or branches, and that these *might* be dried on banks, and then by a fresh rise in the stream be washed into the sea. (Darwin, 1859, p. 359)

By 1876, these accidents morphed into occasional occurrences: "floods *would often* wash into the sea dried plants or branches with seed capsules or fruit attached to them" (Darwin, 1876, p. 325).

Even more striking are the examples that Darwin adds in the 1876 edition to support his claim. Consider that in both editions, of 1859 (p. 363) and 1876 (p. 328), Darwin writes:

> Although the beaks and feet of birds are generally quite clean, I can show that earth sometimes adheres to them: in one instance I removed [...] earth from one foot of a partridge, and in the earth was a pebble quite as large as the seed of a vetch.

But a pebble is not a seed, and in 1859, Darwin was forced to speculate:

> Thus seeds might occasionally be transported to great distances; for many facts could be given showing that soil almost everywhere is charged with seeds. (Darwin, 1859, p. 363)

By 1876, Darwin had accumulated the examples he needed to argue probability instead of possibility. After noting the found pebble, he adds:

> Here is a better case: The leg of a woodcock was sent to me by a friend, with a little cake of dried earth attached to the shank [...] and this contained a seed of the toad-rush (*Juncus bufonius*) which germinated and flowered. (Darwin, 1876, p. 328)

And another: Naturalist George Swaysland provided Darwin with anecdotes of "little cakes of earth" attached to migratory birds (Darwin, 1876, p. 328). And yet another:

> Professor Newton [Cambridge zoologist] sent me the leg of a red-legged partridge...with a ball of hard earth attached to it. The earth had been kept for three years, but when broken, watered, and placed under a bell glass, no less than 82 plants sprung from it. (Darwin, 1876, p. 328)

As these quotations show, even in a landscape dominated by contingency, Darwin is nevertheless keen to purge it of *accident*. The shift in terminology from "accidental" to "occasional" is key: if a "means of distribution" is accidental, then Darwin cannot build a theory on it. Darwin's language is reminiscent of Lyell. In a section of *Principles* dedicated to the transporting effects of tides and currents, Lyell argues that the principle currents of the globe – the same ones that Darwin uses in his hypothesis on seed dispersal – "[...] depend on no temporary or accidental circumstances, but on the laws which preside over the motions of the heavenly bodies" (Lyell, 1837, v. 2, p. 392). Darwin agrees: "the currents of the sea are not accidental, nor is the direction of prevalent gales of wind" (Darwin, 1859, p. 364). And yet, a flooding stream that tosses a seed-swollen branch on the shore, where it is taken up by a wave and thrust into an ocean current, is purely accidental. But if Darwin can demonstrate that accidental events are also occasional, he can show actual processes at work in the world, as Lyell recommends. So accidental means of transport (birds, floods, etc.) plus geological time (the "long lapse" of ages) equals biogeographical distribution:

> [...] the several above means of transport, and that several other means, which without doubt remain to be discovered, have been in action year after year, for centuries and tens of thousands of years, it would I think be a marvelous fact if many plants had not thus become widely transported. (Darwin, 1859, pp. 363–4)

21.4 The Context: Darwin, Gray, Agassiz, and Seed Dispersal

What is at stake in Darwin's discussion of migrating seeds, whirlwind-borne fish, and the "peculiar" birds of the Galapagos (Darwin, 1859, pp. 384, 390)? The key theoretical issues explored in these chapters are fairly clear: Darwin is interested in understanding the principles that govern distribution, including, for instance, disjunct distribution, as well as the appearance of endemic species on isolated islands who nevertheless demonstrate "close taxonomic affinity to species of the nearest mainland" (Costa, 2015, p. 135). Darwin is thus making a potent argument for descent with modification from a common ancestor. What is harder to discern is the magnitude of the context. As noted above, Darwin understood that the greatest threat to his theory generally came from proponents of separate "spheres of creation," and even more specifically from one of its more vocal proponents, the Swiss-American naturalist Louis Agassiz. Indeed, although Agassiz is only mentioned once in the chapters on geographical distribution, in a brief note on the distribution during glacial periods (Darwin, 1859, p. 365), his presence looms over the entire discussion of the dispersal of species through migration (Irmscher, 2013, p. 139).

Why Agassiz? In the first place, Louis Agassiz was already well-known for his detailed studies of fossil fish and living marine animals, as well as for his articulation and popularization of ice age theory when he arrived in America in 1846. He took both scientific and public audiences by storm when he delivered the prestigious Lowell Institute Lectures in Boston, and he stayed for good after he was asked to head Harvard's newly formed Lawrence Scientific School (Ellis, 2018, p. 65). Agassiz, insatiably hungry for accolades by many accounts, cultivated his celebrity with American audiences by always having something to say, whether it be about fossil fish or American racial politics. He involved the public in his research through both public lectures and appearances, and also through direct requests to the broad public for specimens that he could use to chronicle America's diverse – and unique – flora and fauna. In this, he tapped into an expanding nation's keen interest in participating in scientific knowledge-making, especially through natural history collecting. Like Darwin, Agassiz amassed vast collections of specimens from amateur collectors around the country and the world for the Museum of Comparative Zoology, which he founded at Harvard. Agassiz was inescapable, and his opinion mattered.

Perhaps more to the point, Agassiz's vocal and adamant support of separate creation theory put him directly in Darwin's crosshairs. Unlike Darwin, Agassiz believed in the fixity of species, in a static world where organisms sprang from – and stayed in – place. He was the polygenist foe of Darwin's monogenism, believing that species were created separately by a divine Creator. Agassiz understood the history of the earth as one of stasis, where movement came only in the form of catastrophic action (such as an ice age) leading to the extinction of one form and its replacement with another. Further, Agassiz's experiences in America puzzlingly transformed the naturalist from a once relatively indifferent monogenist to a vocal supporter and promoter of racist, pro-slavery pseudo-scientists, such as Josiah Nott, George Gliddon, and Samuel Morton (Irmscher, 2013, p. 219). Irmscher, in a textured and detailed discussion of Agassiz's imbrication with the complex and politicized racial landscape of antebellum and post-Civil War America, argues that Agassiz felt racial difference on a deeply personal level. The insistence on delineating racial difference that blossomed in Agassiz in America became a way to assert "the continuity of his identity as a [white] European" in a country threatened by racial mixing and the specter of "complete social equality between black and white" (Irmscher, 2013, p. 249).

By 1850, Agassiz was making the case for polygenism in several of his U.S. lectures and publications, including in the Lowell lectures, and at a meeting of the American Association for the Advancement of Science held, notably, at Charleston, South Carolina on March 15, 1850. At the AAAS meeting, Agassiz clearly stated his views: While he had "never denied" that the "races of men [...] constituted one brotherhood" in humanity, when "viewed zoologically, the several races of man were well marked and distinct." Since "the races of men," like the rest of the natural world, occupy multiple and unique "zoological provinces," they, likewise, "did not originate from a common centre, nor from a single pair" (American Association for the Advancement of Science, 1850, p. 107). Shortly after the meeting, Agassiz published his most definitive statement of his racial views, the article "Geographical Distribution of Animals" in which he denied the biblically sanctioned unity of "all living beings upon earth" emanating from "one common centre of origin" (Agassiz, 1850, p. 181). Migration, intercrossing, and modification, key principles of Darwinian evolution, were anathema both at the individual level and at the level of natural law:

> It is inconsistent with the structure, habits, and natural instincts of most animals, even to suppose that they could have migrated over any great distances. It is in complete contradiction with the laws of nature, and all we know of the changes our globe has undergone, to imagine the animals have actually adapted themselves to their various circumstances during their migration. (Agassiz, 1850, p. 184)

Species in Agassiz's schema were fixed, static. Individual organisms were akin to Platonic Forms, the simulacra of a divine thought, which is itself forever fixed and unchangeable. For Agassiz, "species" exist "as ideas, which represent the actual relations individuals bear to the world" (Wilkins, 2009, p. 112). They are immune to the variations affecting individuals; they are immutable, unchangeable over time. As he writes in *Methods of Study in Natural History* (1863), species are "based upon a positive, permanent, specific principle, maintained generation after generation with all its essential characteristics" (Agassiz, 1863, p. 136).

21.5 Conclusion: Back to the Seed

Adrian Desmond and James Moore have persuasively argued that Darwin's hatred of slavery and its scientifically suspect reliance on plurality or polygenetic theory in part motivated his study into the common origin of all species, from seeds to humans (Desmond & Moore, 2009, p. 309). Given Agassiz's preeminence and the increasing popularity of his views, Darwin understood that he would have to take on the naturalist directly. Darwin rightly feared that Agassiz's views would be taken up by "the slave-holding Southerns" (DCP-LETT-1352; letter from Darwin to W. D. Fox, September 4, 1850), and indeed, after his AAAS talk, Josiah Nott had written excitedly to Samuel Morton, a fellow polygenist: "With Agassiz in the war the battle is ours" (Lurie, 1954, p. 237). And so, we are brought back to the humble seed, which Darwin enlisted as a tiny warrior to battle against Agassiz's separate creationism.

Darwin had Agassiz directly in his sights when he wrote to the *Gardener's Chronicle* in 1855 about experiments that bore directly on an "interesting problem" captivating America, "namely, whether the same organic being has been created at one point or several on the face of our globe." Darwin sought Gray's help here, sending him the short article soon after it was published. If Hooker had been initially skeptical about the role of migration in species dispersal, Gray immediately understood both the text and the subtext of Darwin's experiments: "Why has nobody thought of trying the experiment before! [. . .] I shall have it nearly all reprinted [. . .] as a *nut* for Agassiz to crack" (DCP-LETT-1707; letter from Asa Gray to Darwin, June 30, 1855). Unlike in Darwinian evolution, where individuals have the power to exert tremendous change to a species *over time*, neither time nor individuals in Agassiz's scheme can affect the shape of a species. Species, stationary and immutable, outlive their individual constituents, which are conversely transient and immaterial (Wilkins, 2009, p. 113). Darwin, instead, draws our attention to the individual, how it acts with and reacts to others, and how it possesses the power to change the whole. In the Darwinian frame, "species" are provisional, "useful for naming groups of interacting individuals" at specific points in time (Menand, 2001, p. 123). In his chapters on geographical distribution, Darwin gave power and import to these individual migrants, thrown into the world by chance and struggling to find a foothold in new worlds, thereby founding new life, and new species in the process.

References

Agassiz, L. (1850). The geographical distribution of animals. *Christian Examiner and Religious Miscellany, 48*(2), 181–204.

Agassiz, L. (1863). *Methods of study in natural history*. Ticknor and Fields.

American Association for the Advancement of Science. (1850). *Proceedings of the American Association for the Advancement of Science* (Vol. 3). The Permanent Secretary.

Beer, G. (2000). *Darwin's plots: Evolutionary narrative in Darwin, George Eliot and nineteenth-century fiction*. Cambridge University Press.

Black, M. (2009). Darwin and seeds. *Seed Science Research, 19*(4), 193–199.

Bowler, P. (2009). Geographical distribution in the origin of species. In M. Ruse & R. Richards (Eds.), *The Cambridge companion to the origin of species* (pp. 153–172). Cambridge University Press.

Browne, J. (1983). *The secular ark: Studies in the history of biogeography*. Yale University Press.

Costa, J. T. (2009). *The annotated origin: A facsimile of the first edition of on the origin of species*. Belknap Press.

Costa, J. T. (2015). Sailing the backyard *Beagle*. In C. Boulter et al. (Eds.), *Darwin-inspired learning* (pp. 131–146). Sense Publishers.

Darwin, C. (1837–1838). *Notebook B*: [Transmutation of species]. CUL-DAR121. Transcribed by Kees Rookmaaker. In J. van Wyhe (Ed.), *The Complete Work of Charles Darwin Online*. http://darwin-online.org.uk/.

Darwin, C. (1838). *Notebook D*: [Transmutation of species (7-10.1838)]. CUL-DAR123. Transcribed by Kees Rookmaaker. In J. van Wyhe (Ed.), *The Complete Work of Charles Darwin Online*. http://darwin-online.org.uk/.
Darwin, C. (1839–1844). *Questions & experiments notebook*. CUL-DAR206.1. Transcribed by Kees Rookmaaker. In J. van Wyhe (Ed.), *The Complete Work of Charles Darwin Online*. http://darwin-online.org.uk/.
Darwin, C. (1855). Does sea-water kill seeds? *Gardeners' Chronicle and Agricultural Gazette, 21*, 356–357. In Darwin Correspondence Project, "Letter no. 1684," https://www.darwinproject.ac.uk/letter/?docId=letters/DCP-LETT-1684.xml.
Darwin, C. (1859). *On the origin of species by means of natural selection, or the preservation of favoured races in the struggle for life* (1st ed.). John Murray.
Darwin, C. (1876). *The origin of species by means of natural selection, or the preservation of favoured races in the struggle for life* (3rd ed.). John Murray.
Desmond, A., & Moore, J. (2009). *Darwin's sacred cause: Race, slavery, and the quest for human origins*. University of Chicago Press.
Ellis, C. (2018). *Antebellum Posthuman: Race and materiality in the mid-nineteenth century*. Fordham University Press.
Griffiths, D. (2016). *The age of analogy: Science and literature between the Darwins*. Johns Hopkins University Press.
Grosz, E. (2004). *The Nick of time: Politics, evolution, and the untimely*. Duke University Press.
Helgeson, C. (2017). Pattern as observation: Darwin's "great facts" of geographical distribution. *European Journal for Philosophy of Science, 7*(2), 337–351.
Hooker, J. D. (1853–1855). *Flora Novæ-Zelandiæ*. 2 vols. Lovell Reeve.
Hung, K.-C. (2013). *Finding patterns in nature: Asa Gray's plant geography and collecting networks (1830s–1860s)* (PhD thesis). Harvard University.
Hustak, C., & Myers, N. (2012). Involutionary momentum: Affective ecologies and the sciences of plant/insect encounters. *Differences, 23*(3), 74–118.
Irmscher, C. (2013). *Louis Agassiz: Creator of American science*. Houghton, Mifflin, Harcourt.
Kinch, M. P. (1980). Geographical distribution and the origin of life: The development of early nineteenth-century British explanations. *Journal of the History of Biology, 13*(1), 91–119.
Lurie, E. (1954). Louis Agassiz and the races of man. *Isis, 45*(3), 227–242.
Lyell, C. (1837). *Principles of geology* (Vol. 3 vols). John Murray.
Menand, L. (2001). *The metaphysical Club: A story of ideas in America*. Farrar, Strauss, and Giroux.
Wilkins, J. S. (2009). *Species: A history of the idea*. University of California Press.

Tina Gianquitto is an associate professor of environmental humanities at the Colorado School of Mines. She has published a book on women, nature, and science, *"Good Observers of Nature": American Women and the Scientific Study of the Natural World, 1820–1885* (2007), a coedited collection, *America's Darwin: Darwinian Evolution and U.S. Literary Culture* (with Lydia Fisher; 2014), as well as articles on Darwinian botany, evolutionary science and women's social reform, and women's participation in botanical correspondence networks. She has been awarded fellowships from the National Endowment for the Humanities, the American Council of Learned Societies, and the U.S. Fulbright Scholar Program. Current projects include a digital site (www.herbaria3.org) on plant stories in the Anthropocene and a book that examines the influence Darwin's plant studies had on galvanizing responses to evolutionary theory in the late nineteenth-century United States.

Chapter 22
Origin's Chapter XIII. The Meaning of Classification, Morphology, Embryology, and Rudimentary Organs to the Theory of Descent with Modifications

Aldo Mellender de Araújo

Abstract "From the first dawn of life, all organic beings are found to resemble each other in descending degrees, so that they can be classed in groups under groups." This is the first sentence of the 13th chapter of *The Origin of Species* (1859). Darwin's objective was to stress that a natural system of classification, both for plants and animals, must be based on embryology, as well as on adult morphology. To support his ideas, he first discusses the different ways organisms have been classified by their predecessors. For instance, classifying organisms by their affinities is easy in most groups according to him, since we can list a number of characters in common that they must show. The geographic distribution was also used, especially for birds; other naturalists extended this practice to insects and plants. For Darwin this is an illogical method, as also is the comparative value of the higher levels of taxonomy, such as families, orders, and so on; all this can be shown as arbitrary. As he called his theory "descent with modification," a natural system of classification of species should be based on the affinities of what they inherited from a common ancestor; that is, a classification should be genealogical.

Darwin begins this chapter by discussing the problem of classification of organisms with a very simple idea, which, in fact, is derived from his own theory. As the extant living beings are derived from ancestors both immediate and remote, so that their relatedness is in descending degree, a classification system based on such fact would be ideal, a truly *natural system*. As he does in other chapters of *On the Origin of Species* (1859), he examines how different naturalists have dealt with this problem. Perhaps the oldest form of classifying the organisms was by their similarities on one side and by their differences on the other. Another system is to list a series of

A. M. de Araújo (✉)
Departament of Genetics, Institute of Biosciences, Federal University of Rio Grande do Sul, Porto Alegre, Brazil

M. Elice Brzezinski Prestes (ed.), *Understanding Evolution in Darwin's "Origin"*, History, Philosophy and Theory of the Life Sciences 34,
https://doi.org/10.1007/978-3-031-40165-7_22

characters common to a certain group, say, the mammals or the birds, and so on. These systems suffer from a common flaw, clearly identified by him:

> [...] many naturalists think that something more is meant by the Natural System; they believe that it reveals the plan of the Creator; but unless it be specified whether order in time or space, or what else is meant by the plan of the Creator, it seems to me that nothing is thus added to our knowledge.[...] I believe that something more is included; and that propinquity of descent, – the only known cause of the similarity of organic beings, – is the bond, hidden as it is by the various degrees of modification, which is partially revealed to us by our classifications. (Darwin, 1859/1964, pp. 413–414)

One common mistake made by old naturalists was to compare those parts of an organism that are of specific importance in one environment, as being important candidates for classification. “Nothing can be more false,” he says; let us think for instance of the body of a shark and of a dolphin (although they differ in the orientation of the tail fin): their bodies are adapted to swim fast, they are *analogical characters* driven by natural selection to solve the same problem, fast locomotion (Darwin goes further at this point at page 427 of this chapter). An opposite attitude is to consider variables as rudimentary organs: “organs in this condition are often of high value in classification” (Darwin, 1859/1964, p. 416). That is the case of rudimentary teeth in the upper jaws of young ruminants, which, together with certain rudimentary bones in the leg,[1] bring Ruminants and Pachyderms into close genealogical (phylogenetic, as we say today) affinity (today they are grouped in the same order the Artiodactyla, which have originated about 80 million years ago – Vaughan et al., 2015).

Another point that Darwin stressed is the value of using several characters, instead of one. Here, is a sentence he wrote that was totally true then as it is now:

> [...] it has been found that a classification founded on any single character, however important that may be, has always failed. (Darwin, 1859/1964, p. 417)

We can trace a parallel of that position of Darwin with our present knowledge of genes. When genes started to be used to build phylogenies, by the end of the twentieth century, it was soon shown that a phylogeny based on a single gene, or a few ones, was worthless to support a classification of species. This was called the problem of gene genealogies versus species genealogies. For example, let us have an ancestral species that in a certain gene has six alleles derived from one single allele by mutations; suppose the corresponding genealogy of these alleles is as follows in Fig. 22.1.

Now suppose this ancestral species splits into two species, and that one of these species loses genes 3, 5, and 6, by natural selection or genetic drift, while the other loses genes 1, 2, and 4. Let us call the first species, A (with genes 1, 2, and 4) and the other, B (with genes 3, 5, and 6). Suppose now that species A gives rise to species A_1, which has only gene 2 (loss of genes 1 and 4, by selection or drift). Finally, imagine a new speciation event in species B, which originates species B_1 (with only

[1] This bone is known as the astragalus, a bone of the foot, which is especially diagnostic in the order Artiodactyla (Vaughan et al., 2015).

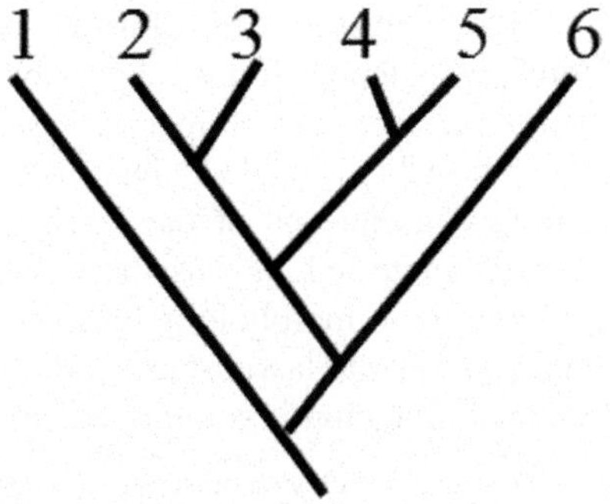

Fig. 22.1 Genealogy of alleles in a fictional ancestral species

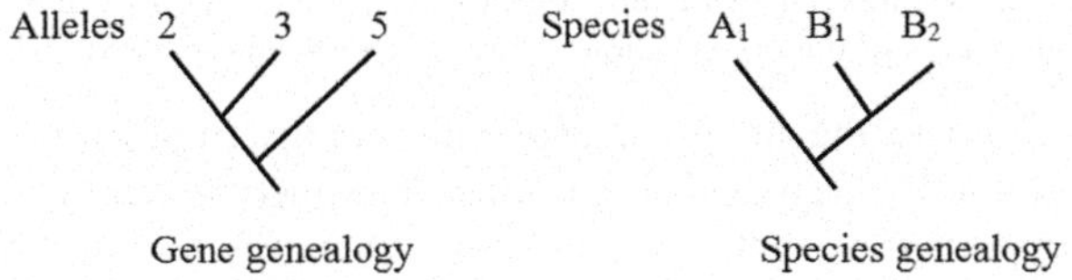

Fig. 22.2 Left: final gene tree resulting of selection or drift from the fictional ancestral species (represented in figure 23.1). Right: final species tree resulting from the same processes. (Both figures adapted from Herron & Freeman, 2014, p. 774)

gene 3) and B_2 (with only gene 5). Figure 22.2 represents the final gene tree (at left) and the species tree (at right): they are different.

In his search for good characters to use in classification, Darwin emphasizes that characters of the embryo should be so informative as are those of the adult individual. He explores this point later in the chapter. Some naturalists have proposed the use of geographic distribution as an important source of information to use in classification, a point which Darwin disagrees. A natural system of classification should be founded on the "theory of descent with modification"; moreover, before sending the reader to the single figure of the *Origin*, he declares:

> [. . .] I must explain my meaning more fully. I believe that the arrangement of the groups within each class, in due subordination and relation to the other groups, must be strictly genealogical in order to be natural; but that the amount of difference in the several branches or groups, though allied in the same degree in blood to their common progenitor, may differ greatly, being due to the different degrees of modification which they have undergone; and this is expressed by the forms being ranked under different genera, families, sections, or orders. (Darwin, 1859/1964, p. 420)

It is remarkable how Darwin explored his theory to multiple consequences. One very interesting point about this is explained when he imagines a genealogy of all human languages, present at the time as well as extinct:

> If we possessed a perfect pedigree of mankind, a genealogical arrangement of man would afford the best classification of the various languages now spoken throughout the world; and if all extinct languages, and all intermediate and slowly changing dialects, had to be included, such an arrangement, I think, be the only possible one. (Darwin, 1859/1964, p. 422)

This expresses a fundamental property of ideal genealogies (phylogenies, we might say), that is, if it were possible to arrange all forms of life, present and extinct, in the precise order of appearance, this universal genealogy would be unique. This could possibly be called "the principle of uniqueness of phylogenies." This is a very strong consequence of the "theory of descent." Moreover, this theory helps any person interested in classification to correctly include those species with great differences in morphology between males and females,[2] or, as is the case of many arthropods, with larval stages. All the power and robustness of the theory of descent with modification is summarized in this sentence:

> Therefore we choose those characters which, as far as we can judge, are the least likely to have been modified in relation to the conditions of life to which each species has been recently exposed. Rudimentary structures on this view are as good as or even sometimes better than, other parts of the organisation. (Darwin, 1859/1964, p. 425)

Before going on to the importance of morphology in separating groups from groups, Darwin briefly discusses the question of extinction and of natural selection. Extinction is important by widening the interval between two or more groups; it is the case of the Crustaceans, "[. . .] for here the most wonderfully diverse forms are still tied together by a long, but broken, chain of affinities" (Darwin, 1859/1964, pp. 431–432).

One could add to this example the phylogeny of our own species, today represented by only *Homo sapiens*; however, when we also take extinct species, we have a more interesting scenario: we found at least eight more species from the genus *Homo* connecting to *Australopithecus* and other genera (see, for instance, Fig. 20.22 in Herron & Freeman, 2014, p. 787).

As for natural selection, it is worth letting Darwin speak for himself:

> Finally, we have seen that natural selection, which results from the struggle for existence, and which almost inevitably induces extinction and divergence of character in the many descendants from one dominant parent-species, explains that great and universal feature in the affinities of all organic beings, namely, their subordination in group under group. (Darwin, 1859/1964, p. 433)

Maybe here he contradicts himself since natural selection is responsible for the development of analogical characters between two species distantly related. It is possible that such contradiction results from his description of the process of new species formation: for him, a new variety, better adapted to the environment, substitutes the old one and further becomes a new species. In brief, he describes the process as if two varieties compete for resources in the same environment, what is presently known as *sympatric speciation*. Our present knowledge of speciation is, of course, far richer than in Darwin's time; we now know, for instance, that there are

[2]There is one very interesting example of an extreme difference between males and females of a species. This is the case of an animal living in northeastern Atlantic Ocean *Bonellia viridis,* known as the "green spoonworm" (Annelida; Polychaeta): females have a body 15 cm long, while the male has a body size of 1–3 mm long, living inside the reproductive organs of the female (Gilbert & Epel, 2015, p. 8).

other processes of speciation, where natural selection is necessarily not involved (a good text on this topic is Coyne & Orr, 2004).

The next subject treated in this chapter is morphology, the most interesting department of natural history, says Darwin, indeed "its very soul" (Darwin, 1859/1964, p. 434). He compares the structure of the hand of a human being, formed for grasping, with that of a mole, for digging, the wing of a bat, for flying together with many other similar structures of other vertebrates. They can be of different sizes, with different parts growing at different rates; however, they are all homologies.[3] This is well explained by his theory of descent with modification. Darwin's disappointment at other explanations is clearly stated here:

> Nothing can be more hopeless than to attempt to explain this similarity of patterns in members of the same class, by utility or by the doctrine of final causes.[4] (Darwin, 1859/1964, p. 435)

The power of his doctrine can still be viewed when comparing homologous structures in members of different orders of insects, as for instance, the mouth parts. They can be greatly different, as the long proboscis of a sphinx moth[5] with that of a domestic fly or the complex organization of a beetle's mouth parts. To stress the importance of natural selection, a series of "whys" are made, of which the flower one is worth mentioning:

> [...] why should the sepals, petals, stamens, and pistils in any individual flower, though fitted for such widely different purposes, be all constructed on the same pattern? (Darwin, 1859/1964, p. 437)

It is impressive his ability to be clear and straight in his argumentation:

> Naturalists frequently speak of the skull as formed of metamorphosed vertebrae: the jaws of crabs as metamorphosed legs; the stamens and pistils of flowers as metamorphosed leaves; but it would in these cases probably be more correct, as Professor Huxley[6] has remarked, to speak of both skull and vertebrae, both jaws and legs, etc., as having been metamorphosed, not one from the other, but from some common element. (Darwin, 1859/1964, p. 438)

To emphasize the importance of embryology in separating groups by relationship, Darwin tells a story involving Agassiz, according to which,

---

[3]The clear distinction between homology and analogy was due to the British anatomist Richard Owen, Darwin's contemporary (Hall, 1999).

[4]The doctrine of final causes here referred to is also known as teleology. For a long time, it was attributed to the Greek philosopher Aristotle, which, in fact, used the expression "final cause" with a very distinct meaning. A good discussion of this philosophical subject is in chapter 3 of Mayr (2004).

[5]The name sphinx moth was given due to the posture of their caterpillars when at rest, which resemble the sphinx of Giza. This group of moths belong to the lepidopteran family Sphingidae, with over 1200 species worldwide.

[6]Thomas Henry Huxley (1825–1895), known as "Darwin's bulldog," was a close friend and supporter. He was the grandfather of Julian Huxley, an important biologist of the twentieth century, and Aldous Huxley, well-known science fiction writer.

[. . .] having forgotten to ticket the embryo of some vertebrate animal, he cannot now tell whether it be that of a mammal, a bird, or a reptile. (Darwin, 1859/1964, p. 439)

It is true indeed, that adult animals can differ greatly in some traits, while the respective embryos are very similar at the same age of development.

The embryo in the course of development generally rises in organization: I use this expression though I am aware that it is hardly possible to define clearly what is meant by the organization being higher or lower. But no one probably will dispute that the butterfly is higher than the caterpillar. In some cases, however, the mature animal is generally considered as lower in the scale than the larva, as with certain parasitic crustacean. (Darwin, 1859/1964, p. 441)

How could a naturalist explain these facts (the one on Agassiz and the quotation above)? He is very confident in answering: "I believe that all these facts can be explained, as follows, on the view of descent with modification" (Darwin, 1859/1964, p. 443). In the following paragraph, he discusses a dictum of his time, where breeders explained the origin of certain monstrosities in domestic animals originating early in embryonic life; he is cautious about this and argues that:

[...] we have little evidence on this head – indeed the evidence rather points the other way; [...] We see this plainly in our own children; we cannot always tell whether the child will be tall or short, or what its precise features will be. The question is not, at what period of life any variation has been caused, but at what period of life it is fully displayed. (Darwin, 1859/1964, p. 443)

He clearly distinguishes the differences between effects whose origins are in "the male and female sexual elements," or, in present terms, whose origins are in the genes of male or female, or both, from the differences which started to appear during embryonic development. He has called these "two principles," which, he says, "[...] if their truth be admitted, will, I believe, explain all the above specified leading facts of embryology" (Darwin, 1859/1964, p. 444). It is almost impossible to ask oneself how it would be if Darwin had read Mendel's paper published in 1866. The fact is that in the sixth and last edition of the *Origin of Species* (1876), which Darwin had revised himself, this is still obscure, not a single line being dedicated to Mendel, even though the section on embryology is much expanded. As for embryology, the name of Karl Ernst von Baer (1792–1876) does not appear in the first edition; however, his name appears three times in the sixth edition, one of them in chapter 14, which corresponds to the 13th of the first edition (in the sixth edition there is one added chapter, "VII – Miscellaneous objections to the theory of natural selection"). Karl von Baer was one of the most important embryologists of his time, not only because he was the first to describe the ovum of a mammal, but also by his four laws of development[7] (the name "law" was commonly used when certain regularities appear

[7] First law: general characters of the group to which an embryo belongs appear in development before the specialized characters. Second law: less general structures form in development after more general (this law is practically a continuation of the first). Third law: during development embryos progressively diverge from embryos of other groups. Fourth law: embryos of higher animals resemble embryos and not adults of other animals (Hall, 1999, p. 70).

in the natural world). He was interested in comparative embryology, particularly whether "'higher' animals repeated in their development adult stages of 'lower' animals" (Hall, 1999, p. 70).

The last pages of the chapter are dedicated to the "rudimentary, atrophied, or aborted organs" (Darwin, 1859/1964, p. 450). These characters for him were very common in nature; for example, males of mammals show rudimentary mammae; in some snakes it is possible to find rudiments of the hind limbs; in our own species, the vermiform appendix, also called cecal appendix; this is a remnant of an ancient functional organ (although now it is known as a reservoir of beneficial bacteria for our digestive system). Our coccyx is now part of a tail our ancestors possessed. Curiously, there are also molecular vestigial traits: on human chromosome 6 there is a DNA sequence which is very similar to that present in chimpanzees, gorillas, and orangutans, our closest relatives (we all belong to the family Hominidae). In the great apes this sequence expresses an enzyme, while in humans it is not expressed, because there are 92 base-pair nucleotides absent (this is called a "deletion" – see Herron & Freeman, 2014, pp. 42–43).

Back to Darwin, it is interesting how he comments on a rudimentary pistil in plants where the flowers have separate sexual structures:

> In plants with separate sexes, the male flowers often have a rudiment of a pistil; and Kölreuter[8] found that by crossing such male plants with a hermaphrodite species, the rudiment of the pistil in the hybrid offspring was much increased in size; and this shows that the rudiment and the perfect pistil are essentially alike in nature. (Darwin, 1859/1964, p. 451)

In other cases, as in the snapdragon (genus *Antirrhinum*), the fifth stamen is totally absent, or in Darwin's words, it is "utterly aborted." There are those cases where the rudimentary structure is easily seen in the embryo, not in the adult, as teeth in the upper jaw of ruminants and whales. Darwin states this as a universal rule: "a rudimentary part or organ is of greater size relatively to the adjoining parts of the embryo than in the adults; so that the organ at this early age is less rudimentary" (Darwin, 1859/1964, p. 453). He concludes this section with this comment:

> I have now given the leading facts with respect to rudimentary organs. In reflecting on them every one must be struck with astonishment: for the same reasoning power which tells us plainly that most parts and organs are exquisitely adapted for certain purposes, tells us with equal plainness that these rudimentary or atrophied organs, are imperfect and useless. (Darwin, 1859, p. 453)

Darwin was aware that his theory of descent with modification could easily explain the presence (or absence) of such structures. He completes by saying:

> I believe that disuse has been the main agency; that it has led in successive generations to the gradual reduction of various organs, until they have become rudimentary – as in the case of

---

[8] Joseph Gottlieb Kölreuter (1733–1806) was a German botanist who studied mainly the process of fertilization and hybridization in plants. He was the first to discover the phenomenon of self-incompatibility, in plants which are hermaphrodites but not self-pollinated; they need the pollen from another plant.

> the animals inhabiting dark caverns, and of the wings of birds inhabiting oceanic islands, which have seldom been forced to take flight, and have ultimately lost the power of flying. Again, an organ useful under certain conditions, might become injurious under others, as with the wings of beetles living on small and exposed islands; and in this cases natural selection would continue slowly to reduce the organ, until it was rendered harmless and rudimentary. (Darwin, 1859/1964, p. 454)

Here, as in other pages of the *Origin*, Darwin is using a principle accepted without restrictions by his contemporary naturalists: the principle (or law, as Lamarck as called it in his 1809 book) of use and disuse and of the inheritance of acquired characters (later in a mature work, from 1815 on, he modified both laws slightly and added two others – see Martins, 2007, p. 195). Darwin used both, the law of use/disuse and the inheritance of acquired characters in his theory of heredity, called "pangenesis," in a long chapter of *Variation of Animals and Plants under Domestication* (Darwin, 1868/1896). This subject has been largely discussed in the literature on the history of science; one of the most illuminating is the one of Martins (2015), which gives many details on August Weismann's experiments to test the inheritance of mutilations in mice.

Almost at the end of the chapter, Darwin makes use of an interesting metaphor:

> Rudimentary organs may be compared with the letters in a word, still retained in the spelling, but become useless in pronunciation, but which serve as a clue in seeking for its derivation. (Darwin, 1859/1964, p. 454)

The Origin of Species was a fundamental book to understand how organisms change over time. The process that Darwin imagined could explain these changes was natural selection. The irony is that the book he was preparing was another one, a big book on Natural Selection. The circumstances forced him to publish in a hurry a condensed version.

In the last years of the nineteenth century, the process of natural selection was criticized, particularly because there was no clear demonstration of it in nature or laboratory. The fact of evolution was accepted, not the process. It is known, however, that in 1898 an American biologist called Hermon Bumpus received in his laboratory a sample of house sparrows (*Passer domesticus*) weakened by a storm, which he thought could be used to test the hypothesis of natural selection using morphological data, namely, eight size variables, six of them skeletal. He published a paper on these data in 1899, where he concluded that the differential survival of the birds was due to natural selection (Johnston et al., 1972). This paper became a classic on natural selection, even though many reinterpretations have been done (for instance, see Pugesek & Tomer, 1996 and references therein). New analyses of natural selection, in butterflies, appeared in the Appendix of Punnett's book (Punnett, 1915); by this time there was no mathematical model of natural selection, the first being the one by Ronald A. Fisher (1922). Evolution went to the field and laboratory with Theodosius Dobzhansky, in the 1930 decade (Dobzhansky & Queal, 1938). Finally, after many meetings, symposia, persuasion tactics, a new theory of evolution appeared in the 1940s; this was called the Evolutionary Synthesis, Modern Synthesis, or Synthetic Theory of Evolution. New evolutionary processes were proposed, in addition to natural selection: mutation (with recombination), gene flow, and genetic drift (the historical development of this theory is well discussed in Mayr & Provine, 1980/1998). The modern synthesis is still valid, although a recent rival theory has been launched (Pigliucci & Müller, 2010; Laland et al., 2015).

References

Coyne, J. A., & Orr, H. A. (2004). *Speciation*. Sinauer Associates, Publishers.

Darwin, C. R. (1859/1964). *On the origin of species: A facsimile edition of the first edition, with an introduction by Ernst Mayr*. Harvard University Press.

Darwin, C. R. (1868/1896). *Variation of animals and plants under domestication* (2 vols). D. Appleton and Company.

Dobzhansky, T., & Queal, M. L. (1938). Genetics of natural populations. I. Chromosome variation in populations of *Drosophila pseudoobscura* inhabiting isolated mountain ranges. *Genetics, 23*, 239–251.

Fisher, R. A. (1922). On the dominance ratio. *Proceedings of the Royal Society of Edinburgh, 42*, 321–341.

Gilbert, S. F., & Epel, D. (2015). *Ecological developmental biology – The environmental regulation of development, health, and evolution* (2nd ed.). Sinauer Associates, Publishers.

Hall, B. K. (1999). *Evolutionary developmental biology* (2nd ed.). Kluwer Academic Publishers.

Herron, J. C., & Freeman, S. (2014). *Evolutionary analysis* (4th ed.). Pearson Education.

Johnston, R. F., Niles, D. M., & Rohwer, S. A. (1972). Hermon Bumpus and natural selection in the house sparrow *Passer domesticus*. *Evolution, 26*(1), 20–31.

Laland, K. N., Uller, T., Feldman, M. W., Sterelny, K., Müller, G. B., Moczek, A., Jablonka, E., & Odling-Smee, J. (2015). The extended evolutionary synthesis: Its structure, assumptions and predictions. *Proceedings of the Royal Society B, 282*, 20151019. https://doi.org/10.1098/rspb.2015.1019

Lamarck, J. B. (1809/1963). *Zoological philosophy*. Hafner Publishing.

Martins, L. A. P. (2007). *A Teoria da Progressão dos Animais, de Lamarck* (Coleção "Scientiarum Historia et Theoria" Vol. 1). Booklink Publicações Ltda.

Martins, L. A. P. (2015). Weismann e a transmissão de caracteres adquiridos: os cientistas podem mudar de ideia. In J. Ahumada, A. Nicolás Venturelli, & S. S. Chibeni (Eds.), *Filosofia e Historia de la Ciencia en el Cono Sur: Selección de Trabajos del IX Encuentro y las XXV Jornadas de Epistemología e Historia de la Ciencia* (pp. 557–564). Editorial de la Universidad Nacional de Córdoba.

Mayr, E. (2004). *What makes biology unique? Considerations on the autonomy of a scientific discipline*. Cambridge University Press.

Mayr, E., & Provine, W. B. (Eds.). (1980/1998). *The evolutionary synthesis: Perspective on the unification of biology (with a new preface by Ernst Mayr)*. Harvard University Press.

Pigliucci, M., & Müller, G. B. (Eds.). (2010). *Evolution: The extended synthesis*. The MIT Press.

Pugesek, B. H., & Tomer, A. (1996). The Bumpus house sparrow data: A reanalysis using structural equation models. *Evolutionary Ecology, 10*, 387–404.

Punnett, R. C. (1915). *Mimicry in butterflies*. Cambridge University Press.

Vaughan, T. A., Ryan, J. M., & Czaplewski, N. J. (2015). *Mammalogy* (6th ed.). Jones & Bartlett Publishers.

Aldo Mellender de Araújo Graduated in Natural History at the Universidade Federal do Rio Grande do Sul, Porto Alegre, Brasil. Doctorate in Genetics at the same university. Postdoctoral in Ecological Genetics at the University of Liverpool, U.K. and at Universidade Estadual de Campinas, Campinas, São Paulo, Brasil. Main research area: the evolutionary biology of *Heliconius* butterflies. Since the 1990s, also worked in the area of history and philosophy of evolutionary ideas. In 1996 and 1998 was a visiting scholar at Cornell University, NY, with the historian of evolutionary biology and population genetics William B. Provine (1942–2015) for short periods

Chapter 23
Origins' Chapter XIV: The Good Old Habit of Summarizing

Gerda Maisa Jensen, Bruno F. Lima, and Marcelo Monetti Pavani

Abstract The concluding chapter of *On the Origin of Species* presents a comprehensive overview of the "one long argument" expounded by Darwin throughout the book. This chapter highlights critical elements of his exposition and the underlying objectives. It follows a detailed examination of the insights offered by the text concerning Darwin's historical context and the humble way in which he submits his theory for scrutiny, not only by experts but also by his general readership. Building upon the Nature of Science (NOS) pedagogical construct, particular emphasis will be placed on aspects that embody the values inherent in scientific inquiry.

23.1 Introduction

Having traversed 13 chapters and covered over 450 pages (in the first edition of 1859), the reader reaches the 14th and concluding chapter of *On the Origin of Species*, titled "Recapitulation and Conclusion." In a manner befitting an exceptional instructor, throughout the previous chapters, Darwin has taken the reader by the hand and guided him or her through his theory, employing analogies, metaphors, and a plethora of illustrative examples. Through these elucidations, he has shed light on his ideas concerning species transmutation and its underlying causes. Consequently, arriving at this point, an active reader can raise pertinent questions even if the chapter's title is self-evident: its content would summarize the key components of the argument and provide a conclusion. However, is it truly that straightforward? Is not there anything else to learn? Does not he introduce any additional contents of the theory or change his communication strategy?

G. M. Jensen (✉) · B. F. Lima · M. M. Pavani (✉)
Group of History of Biology and Teaching, Institute of Biosciences, University of São Paulo, São Paulo, Brazil
e-mail: gemajen@gmail.com; brunofancio@usp.br; marcelopavani@alumni.usp.br

M. Elice Brzezinski Prestes (ed.), *Understanding Evolution in Darwin's "Origin"*, History, Philosophy and Theory of the Life Sciences 34,
https://doi.org/10.1007/978-3-031-40165-7_23

The history of *The Origin* is well-known, and detailed accounts of its publication have been extensively covered elsewhere (e.g., Browne, 2006; Ruse, 2009). Briefly, after drafting a sketch in 1842 and a more extended essay in 1844, it was not until 1856 that Darwin commenced the writing of a substantial treatise to present his theory of species changes by common descent and natural selection to the world. Two years later, before finishing the work, he received a letter from Alfred Russel Wallace (1823–1913), which outlined a concept of natural selection strikingly like his own (see Chap. 9 of this volume by Viviane Carmo and Lilian Martins for further details). Naturally, there was no more time for a lengthy treatise. Darwin needed to condense his ideas into a concise text. *As* C. Kenneth Waters stated (2009, p. 121), the *Origin* itself can be viewed as an abstract. Hence, Chapter XIV can be considered an abstract within an abstract.

To grasp Darwin's line of argumentation in *The Origin*, Kenneth Waters outlines a three-part overarching structure of the book. According to him (Waters, 2009, p. 125), Chapters I–IV present observations from natural history along with an analogical argument derived from artificial selection. Chapters VI–IX address a diverse range of challenges confronting Darwin's perspective, while the third component consists of Chapters V and X–XIII, elucidating how Darwin's viewpoint can account for various sets of facts. In Waters' structure, Chapter XIV does not form an integral part of Darwin's argumentative framework.

On the other hand, Michael Ruse posits that *The Origin* is a rhetorical work aimed at persuading readers about "something unseen –life history leading up to the present" (2019, p. 10). This notion aligns well with the "one long argument" concept, which has been extensively examined in the literature (e.g., Ruse, 1979; Largent, 2009; Waters, 2009; White et al., 2021; Lennox, 2021).

With these considerations, in this work, we will examine the concluding chapter of Darwin's book. Initially, we will show how he changed the main objectives of the chapter, and its structure, during the process of its writing, from the dialogue sought with his usual interlocutors. The basis of this interpretation will be shown in some of his letters from the period. Then we will discuss the structuring parts of the published chapter. Additionally, we will explore aspects of the Nature of Science (NOS) inside the text and their potential uses for science and biology classroom approaches.

23.2 On the Writing of Chapter XIV: Darwin's Expectation Regarding the Reception of the Book

By examining Darwin's correspondence, we can gain insights into his expectations regarding how readers would receive the book. Some of these letters allowed us to establish a connection with the intricate process of crafting its final chapter. As mentioned, *The Origin* has its roots in a shift of concepts, from a long treatise being written 2 years ago to a quick summary to be published as soon as possible. His correspondence shows us that Darwin himself appeared somewhat disenchanted by this notion. In a letter to his cousin, William Fox, dated July 1858, he wrote:

> After all, I am now beginning to prepare an abstract of my Species Theory. By an odd coincidence, Mr Wallace in the Malay Archipelago sent to me an Essay containing my exact theory; & asking me to show it to Lyell. The latter & Hooker have taken on themselves to publish it in Linnean Journal, together some notes of mine written very many years ago; & both of them have urged me so strongly to publish a fuller abstract, that I have resolved to do it, & shall do nothing till completed: it will be published, probably, in Journal of Linn. Socy. & I shall have separate copies & will send you one. – It is impossible in abstract to do justice to subject. – (DCP 2312)

Three months later, in a letter from October 1858, now to his former Botany professor at Cambridge, Joseph Dalton Hooker (1817–1911), Darwin acknowledged his difficulty in creating a concise abstract but appeared to adopt a more optimistic perspective regarding the summarizing nature of the work:

> I am working most steadily at my Abstract; but it grows to an inordinate length; yet fully to make my view clear, (& never giving briefly more than a fact or two & slurring over difficulties) I cannot make it shorter. It will yet take me three or four months; so slow do I work, though never idle. You cannot imagine what a service you have done me in making me make this abstract; for though I thought I had got all clear, it has clarified my brains much, by making me weigh relative importance of the several elements. – (DCP 2335)

The decision to publish a smaller book did not prove to be detrimental. As highlighted by Hodge (2013), *The Origin* achieved remarkable success in terms of sales from the moment of publication on November 24, 1859. Darwin himself commented on the anticipated reception of the work in a letter to his editor, John Murray, dated March 31, 1859:

> The book *ought* to be popular with a large body of scientific & semi-scientific readers, as it bears on agriculture & history of our domestic productions & on whole field of Zoology, Botany & Geology. – I have done my best, but whether it will succeed I cannot say – I have been quite surprised at finding how much interested strangers & acquaintances have become with the subject. Only some small portions are at all abstruse [...] (DCP 2441)

Due to his significantly deteriorated health, Darwin experienced a delay in the composition of his final chapter, as he confided in a letter to Hooker on March 15th, 1859:

> I shall tomorrow finish my last chapter, (except a Recapitulation) on Affinities, Homologies, Embryology &c & the facts seem to me to come out very strong for mutability of species. – I have been much interested in working out this chapter. I shall now, thank God, begin looking over old first chapters for press. – But my health is now so very poor, that even this will take me long [...]. (DCP 2432)

At the time of another correspondence with John Murray on May 6, 1859, Chapter XIV had not yet been written:

> I will send in middle or latter part of next week, the six first chapters; & now the succeeding chapters have been copied, & I can send exact number of words, *or the M.S. itself* if you like, *of the whole*, with the one exception of the last & very short chapter, not yet written. (DCP 2459)

Eight days later, on May XIVth, 1859, Darwin corresponded with Murray once more, expressing his concerns regarding the length of the chapter:

> My dear Sir
> I highly approve of the specimen sheet. I am sorry to hear the Book will be so big. Moreover please observe that I estimated the last chapter at 3000 words. I have now written half & the number is this much, & I think the chapter will run into 6000 or 7000 words. – I am sorry for my blunder. This may make you put another line in each page. – (DCP 2462)

Between May and September of 1859, Darwin successfully completed Chapter XIV. However, before sending it to the editor, he shared the chapter with Charles Lyell (1797–1875), fully aware that convincing him of the concept of species transmutation would be challenging. Darwin recognized the immense significance of Lyell's "verdict" and as evidenced by his letter to Lyell on September 20, 1859, considered Chapter XIV to be a crucial element deserving of Lyell's thoughtful consideration.

> Although your previously felt doubts on the immutability of species, may have more influence in converting you (if you be converted) than my Book; yet as I regard your verdict as far more important in my own eyes & I believe in eyes of world than of any other dozen men, I am naturally very anxious about it. Therefore, let me beg you to keep your mind open till you receive (in perhaps a fortnights time) my latter chapters which are the most important of all on the favourable side. The last chapter which sums up & balances in a mass all the arguments contra & pro, will, I think, be useful to you. – (DCP 2492)

After a span of 10 days, Darwin eventually forwarded the manuscript to Lyell. In a letter dated September 30, 1859, Darwin underscored the significance of Chapter XIV and its pivotal role, highlighting the importance of Lyell's appreciation and evaluation of its contents:

> I sent off this morning the last sheets, but without index which is not yet in type. I look at you as my Lord High Chancellor in Natural Science, & therefore I request you, after you have finished, just to *re* run over the *heads* in the Recapitulation-part of last Chapter. I shall be deeply anxious to hear what you decide (if you are able to decide) on the balance of the pros & contras given in my volume & of such other pros & contras as may occur to you. (DCP 2496)

In Lyell's response, which arrived 4 days later, on October 4th, 1859, he offers interesting observations regarding the text. Notably, his lengthiest commentary pertains to the discussion surrounding the development of the eye:

> The first page of this most important summary gives the adversary an advantage by putting forth so abruptly & crudely such a startling objection as the formation of "the eye" not by means analogous to human reason or rather to some power immeasurably superior to human reason but to superinduced variations like those of which a cattle breeder avails himself. Pages would be required thus to state an objection & remove it – It would be better as you wish to persuade to say nothing? Leave out several sentences & in a future edition bring it out more fully – Between the throwing down of such a stumbling block in the way of the reader & the passage to the working ants in p 460 there are pages required – & these ants are a bathos to him before he has recovered from the shock of being called upon to believe the eye to have been brought to perfection from a state of blindness or pur-blindness by such variations as we witness – I think a little omission would greatly lessen the objection ableness of these sentences if you have not time to recast & amplify. (DCP 2501)

Indeed, Darwin acknowledged and embraced Lyell's considerations. He made adjustments by retaining the example of ants while omitting the discussion on the formation of the vertebrate eye when referencing complex organs on the first page of Chapter XIV (p. 459) of *The Origin*. The notion of the vertebrate eye's structure, famously exemplified by Paley, resided as a longstanding pillar of Natural Theology (Lustig, 2009, p. 110). William Paley (1743–1805), a clergyman and member of the Anglican Church, authored the book *Natural Theology* in 1802 (Paley, 1802/1819) positing that the intricate design and complexity found in nature could only be attributed to a divine creator. Paley dedicated numerous pages in Chapter III of his work to scrutinizing the structure of the vertebrate eye, suggesting that its complexity was beyond natural explanation and required the intervention of a supernatural designer. While Darwin does discuss in Chapter VI of *The Origin* how natural selection could account for the formation of the eye, he embraced Lyell arguing that including this discussion in the Recapitulation would unnecessarily prolong the chapter and potentially confuse rather than clarify matters for the reader.

The analysis presented here of Darwin's correspondence between July 1858 and October of 1859 offered compelling evidence that (1) before the publication of *The Origin*, Darwin conceived that his audience encompassed both scientific and nonscientific readers, namely, people with knowledge about animals and plants due to their professional occupations, necessitating an approach that catered to both groups within his argumentation; (2) Darwin underwent a change of perspective regarding the significance of summarizing his theory, acknowledging its ability to enhance the clarity of his own argumentation and increase its persuasive impact; and (3) Darwin recognized the rhetorical value of Chapter XIV, particularly when addressing the most resistant audience, personified by Charles Lyell.

23.3 On the Reading of Chapter XIV: Analyzing Its Structure

The unconventional structure of argumentation employed in *The Origin* has sparked debates among scholars (Hodge, 1977; Waters, 2009; Lennox, 2021). Typically, one would anticipate a presentation of the case followed by an immediate presentation of substantial supporting evidence. However, Darwin takes an intriguing approach: after introducing the theory of natural selection in Chapters I–IV, he devotes four chapters (VI–IX) to enumerating various facts that pose challenges to his theory. Only then, in Chapters X–XIII, does Darwin provide evidence in favor of natural selection. Lennox (2021, p. 148) highlights that this strategy of "accentuating the negative" is particularly explicit in Chapter XIV. While one might expect the recapitulation in the chapter to follow the organization of the presentation throughout the book, this is not the case. Darwin initiates his "abstract of the abstract" by recollecting the challenges faced by his theory. In this exploration, we will navigate through the chapter and delve into Darwin's line of thought.

23.3.1 The Main Difficulties of the Theory

What are the challenges inherent in the theory of Natural Selection, and more importantly, how can they be addressed? The first eight pages of the chapter thoroughly address the "many and grave objections" that the theory faces (Darwin, 1859, pp. 459–466). These objections, as indicated by the title of Chapter VI, are systematically and comprehensively discussed throughout successive chapters until Chapter IX.

This is the opportune moment to delve into Darwin's process of condensing the myriad of objections into four distinct categories -complex organs and instincts; issues of fertility and sterility among species and varieties; geographical distribution; and the fourth, which may be named the quarrel of intermediate forms – and learn how Darwin conceived the ways in which all of them could be overcome.

23.3.1.1 Complex Organs and Instincts

Commencing with Darwin's own words:

> Nothing at first can appear more difficult to believe than that the more complex organs and instincts should have been perfected, not by means superior to, though analogous with, human reason, but by the accumulation of innumerable slight variations, each good for the individual possessor. (Darwin, 1859, p. 459)

What unifies the combination of structure and behavior as a single objection? Let us consider, for example, the intricate eyes found in both complex vertebrates and diverse visual structures among invertebrates, as well as the remarkable instincts exhibited, such as the construction of honeycombs by bees. What ties them together is their inherent complexity. Complexity poses a significant challenge, as articulated by William Paley and other theologians who contend that such intricacy can only be attributed to divine creation, surpassing human reasoning. However, is there an alternative explanation within nature to account for complexity? The answer lies in the gradual and protracted process of accumulating countless minor variations. It is not just any answer, but rather a distinct one: not a supernatural rationale, but a natural one.

Why does it present such a challenge to grasp? According to Darwin, the difficulty lies in the fact that accepting natural selection as the driving force behind complexity necessitates embracing three consecutive presuppositions. Firstly, one must acknowledge that gradations in the development of any organ or instinct are beneficial "in their own way," meaning they are not flaws but rather confer advantages upon their possessors. Secondly, it is crucial to recognize that all organs and instincts exhibit some degree of variability, even if it is ever so slight – a fact that has been empirically confirmed. And finally, one must acknowledge the existence of a struggle for survival, which leads to the preservation of each advantageous structural or instinctual variation (Darwin, 1859, p. 459).

Just as Darwin asserted in his time, we can unequivocally affirm that "the truth of these propositions cannot [. . .] be disputed" (Darwin, 1859, p. 459). Advancements in technology and techniques, particularly in the field of molecular biology, have equipped us with the means to comprehend how the evolution of complex structures can occur (Bizzo, 2018, p. 452). Nevertheless, we are confronted with the emergence of creationist movements that advocate for the theory of intelligent design. According to their viewpoint, the intricacy of living organisms can only be elucidated by the deliberate planning and craftsmanship of a superior, divine intellect. These movements, typically associated with religious denominations, are prevalent in the United States and are gaining momentum in countries like Brazil.[1]

23.3.1.2 Fertility and Sterility Among Species and Varieties

Darwin's second challenge involved investigating the contrasting patterns of sterility and fertility observed in hybrids resulting from interbreeding between different species or varieties. To fully grasp Darwin's perspective, it is essential to delve into the specific meanings he ascribed to these terms within the context of his own era rather than applying present-day definitions. By unravelling Darwin's conceptual framework, we can gain a deeper understanding of his observations and the complexities surrounding hybrid sterility and fertility in the context of evolutionary biology.

To grasp the essence of Darwin's arguments, we must first delve into his usage of the term "hybrid," also referred to as "mongrel." In Darwin's framework, a hybrid denotes the result of interbreeding distinct animal or plant "species." The act of hybridization was understood as the union or mixing of two different species or stocks.[2] Notably, the definitions of "species" and "stocks" (or varieties) were not fixed, prompting Darwin to meticulously outline each case. When species are initially crossed, they exhibit nearly universal sterility, while varieties, when initially crossed, display nearly universal fertility. It is crucial to note that the visible traits acquired through hybridization tend to be nonheritable, primarily due to the sterility commonly associated with hybrids.

The understanding of the term "species" in the nineteenth century aligns with the definition provided by British biologist William Bateson (1861–1926). According to

---

[1] For a more comprehensive exploration of these subjects, readers are encouraged to consult the publications of Branch et al. (2010) and Oliveira and Cook (2019).

[2] In the 1886 edition of Webster's Complete Dictionary of the English Language, the term "hybridizable" is defined as "capable of producing a hybrid by union with another species or stock." Notably, the dictionary entry includes a quote from J.D. Hooker, renowned botanist and friend of Darwin, who states, "Hybridizable genera are rarer than is generally supposed, even in gardens, where they are so often operated upon under circumstances most favorable to the production of hybrids" (Webster 1886, p. 647).

Bateson, plants or animals exhibiting distinct hereditary characteristics, even if differentiated by a single hereditary trait, were considered separate species and designated with distinct Linnaean binomials. The contemporary understanding of the term "species" took shape after the Evolutionary Synthesis (as explored in Chap. 26 by Thierry Hoquet in this volume). This perspective, known as the "biological species concept," asserts that members of the same species can have the ability to reproduce and produce viable offspring. However, as discussed by Roberto Rozenberg in Chap. 14 of this volume, defining the concept of species proves challenging both in the past and in the present.

Darwin, in his exploration, delved deeper into the matter, highlighting that the presence of variability among living organisms forms the true foundation for the ambiguity of terms like "species" and "variety" within a taxonomic context. According to Darwin, what truly exists are individuals perpetually undergoing variation.

Darwin acknowledges that hybrids can exhibit both sterility and fertility, challenging the notion of universal sterility. He states that "the fertility of varieties and of their mongrel when intercrossed cannot be considered as universal" because "the constitutions of their reproductive systems should have been profoundly modified" (Darwin, 1859, p. 461). In the case of sterility, Darwin offers a potential explanation: the reproductive organs of first-cross hybrids, both male and female, are initially in optimal condition. However, when subjected to new conditions, "[...] their constitutions can hardly fail to have been disturbed from being compounded of two distinct organisations" (Darwin, 1859, p. 461). In other words, breeding different varieties can disrupt the hybrid's reproductive mechanisms, resulting in infertility.

In contrast, hybrids, particularly in domestic settings, can exhibit fertility. Darwin highlights that most crossed varieties, aimed at acquiring desirable traits, are produced under conditions of domestication, which appears to reduce or eliminate sterility. Even in Chapter I, while exploring horticulture, Darwin remarked:

> Sterility has been said to be the bane of horticulture; but on this view we owe variability to the same cause which produces sterility; and variability is the source of all the choicest productions of the garden. (Darwin, 1859, p. 9)

Hence, Darwin suggests that hybridization can have beneficial effects:

> The vigour and fertility of all organic beings are increased by slight changes in their conditions of life, and that the offspring of slightly modified forms or varieties acquire from being crossed increased vigour and fertility. (Darwin, 1859, p. 461)

For instance, hybrid vigour can result in the offspring of a cross growing longer or taller than either of its parents.

The term "domestication" pertains to the process of bringing wild species under human management. The scientific practice of hybridizing crop plants, as understood today, can be traced back to the pioneering work of Joseph Gottlieb Kölreuter (1733–1806), a German botanist. In 1761, Kölreuter published a seminal work titled "Preliminary report of some experiments concerning the sex of plants" (*Vorläufige Nachricht von einigen das Geschlecht der Pflanzen betreffenden Versuchen*). When discussing hybrid plants, Darwin often referred to Kölreuter's research. In fact,

Darwin's personal library contains a digitized copy of Kölreuter's book, complete with handwritten annotations made by Darwin himself along the margins (Fig. 23.1).

It is crucial to note that in Chapter I, Darwin employed the concept of artificial selection to establish a conceptual connection with the selection process that takes

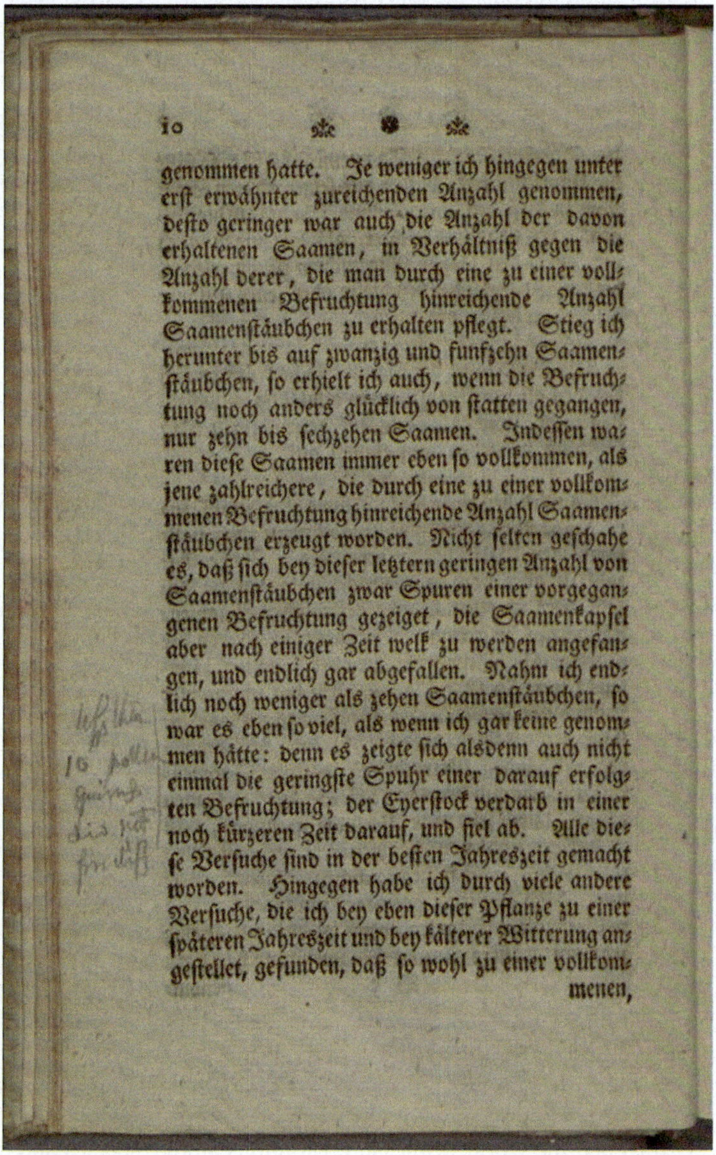

10

genommen hatte. Je weniger ich hingegen unter erst erwähnter zureichenden Anzahl genommen, desto geringer war auch die Anzahl der davon erhaltenen Saamen, in Verhältniß gegen die Anzahl derer, die man durch eine zu einer vollkommenen Befruchtung hinreichende Anzahl Saamenstäubchen zu erhalten pflegt. Stieg ich herunter bis auf zwanzig und funfzehn Saamenstäubchen, so erhielt ich auch, wenn die Befruchtung noch anders glücklich von statten gegangen, nur zehn bis sechzehen Saamen. Indessen waren diese Saamen immer eben so vollkommen, als jene zahlreichere, die durch eine zu einer vollkommenen Befruchtung hinreichende Anzahl Saamenstäubchen erzeugt worden. Nicht selten geschahe es, daß sich bey dieser letztern geringen Anzahl von Saamenstäubchen zwar Spuren einer vorgegangenen Befruchtung gezeiget, die Saamenkapsel aber nach einiger Zeit welk zu werden angefangen, und endlich gar abgefallen. Nahm ich endlich noch weniger als zehen Saamenstäubchen, so war es eben so viel, als wenn ich gar keine genommen hätte: denn es zeigte sich alsdenn auch nicht einmal die geringste Spuhr einer darauf erfolgten Befruchtung; der Eyerstock verdarb in einer noch kürzeren Zeit darauf, und fiel ab. Alle diese Versuche sind in der besten Jahreszeit gemacht worden. Hingegen habe ich durch viele andere Versuche, die ich bey eben dieser Pflanze zu einer späteren Jahreszeit und bey kälterer Witterung angestellet, gefunden, daß so wohl zu einer vollkommenen,

Fig. 23.1 Digitalized copy of Kölreuter's book on hybridization, 1761, with side notes by Darwin. (Charles Darwin's Library)

place in nature. In this regard, it is worth highlighting that artificial selection could only come into play following the domestication of plants and animals, which took place thousands of years ago. This ancient practice laid the foundation for the application of selective breeding techniques in order to shape and modify various traits in living organisms.

Darwin drew upon not only the knowledge passed down by earlier scholars, such as the contributions of Kölreuter, but also conducted his own experiments with the assistance of individuals within his close circle. This circle included his children, employees, and workers residing on his property. The outcomes of these research endeavors led him to draw certain conclusions and prompted further investigations, which he subsequently addressed in subsequent publications. Noteworthy among these works, published after *The Origin*, are his studies on orchid fertilization (Darwin 1862) and the exploration of broad-ranging themes concerning variation in domesticated organisms (Darwin 1868).

23.3.1.3 Geographical Distribution and Intermediate Forms

Darwin identified two additional challenges in his quest to explain the intricacies of evolutionary theory. Firstly, he grappled with the task of elucidating the present-day geographical distribution of species. Secondly, he sought to account for the apparent absence of living intermediate forms or their documentation in the fossil record. These issues posed significant hurdles in his endeavor to provide a comprehensive understanding of the natural world:

> As on the theory of natural selection an interminable number of intermediate forms must have existed, linking together all the species in each group by gradations as fine as our present varieties, it may be asked, Why do we not see these linking forms all around us? Why are not all organic beings blended together in an inextricable chaos? (Darwin, 1859, p. 462)

In addressing the first matter, Darwin openly acknowledges the complexity involved in unravelling this puzzle. Given the premise that all extant species trace their lineage back to shared ancestors and considering their distribution across disparate and geographically isolated regions, it becomes evident that these species must have traversed from one location to another.

Is there any factor that could facilitate the movement of living organisms across the planet? And why do we not find intermediate forms in different regions, what Darwin referred to as "intermediate regions"? In his exploration of these questions, Darwin puts forth compelling arguments. He proposes that during the Earth's past migration might have been aided by climate and geographical factors, such as the existence of land bridges during glacial periods. While he concedes that the problem of geographical distribution remains unresolved, he contends that his observations can help alleviate its complexity to some extent. It is worth noting that Darwin extensively travelled the world, witnessing variations among species and the differential success of certain varieties in specific locations. These observations informed his reflections on the issue of geographical distribution and played a crucial role in the development of his concept of natural selection. Notably, the study of the Galapagos finches quickly came to mind as an illustrative case in point.

Regarding the existence of intermediate forms, Darwin begins his exploration of this topic with a straightforward statement:

> We have no right to expect (excepting in rare cases) to discover *directly* connecting links between them (existing forms), but only between each and some extinct and supplanted form. (Darwin, 1859, p. 462 – emphasis by the author)

Darwin presents a series of concise arguments addressing the challenge of directly uncovering transitional links between extant and extinct organisms. He acknowledges that at any given time, only a limited number of species are undergoing significant changes, making it unlikely to find intermediate forms for all species. Some species remain relatively unchanged for extended periods, and evolutionary changes occur gradually over time. Additionally, the initial varieties of an intermediate region may be replaced by other varieties from neighboring regions. Furthermore, varieties that are more abundant will experience more rapid modification and improvement compared to those with smaller populations. Despite these considerations, if transitional links do exist but are now extinct, the question arises as to why they are not evident in the fossil record.

As we approach the conclusion of the review of the theory's challenges, Darwin regarded the absence of geological records as the most significant hurdle. To address this concern, he dedicated two pages to responding to potential objections and summarizing key considerations related to this difficulty. His arguments encompassed several factors, such as the limited extent of geological exploration worldwide and the constraints on fossilization for all living organisms. Furthermore, Darwin pointed out that when a new fossil is unearthed, it would inevitably be classified as a distinct species rather than recognized as an intermediary connecting two other groups.

However, one of the primary challenges in grasping the concept of natural selection lies in the cognitive difficulty of comprehending the immense age of planet Earth. This is not a predicament unique to Darwin's era. Since the seventeenth century, Natural History museums across Europe have amassed an extensive collection of mineral specimens and fossils, which have been diligently studied to determine their composition and age. These investigations have revealed that these remnants do not always resemble the plants and animals currently in existence. Consequently, a spirited debate unfolded between two factions of scholars: the uniformitarians, including James Hutton (1726–1797), John Playfair (1748–1819), Constant Prévost (1787–1856), and Charles Lyell (1797–1875), and the catastrophists, such as Georges Cuvier (1769–1832), Louis Agassiz (1807–1873), William Buckland (1784–1856), Adam Sedgwick (1785–1873), Alcide d'Orbigny (1808–1857), and Élie de Beaumont (1798–1874). The former argued for an ancient Earth with geological processes unfolding over vast periods of time, while the latter advocated the notion that significant catastrophes, such as floods and volcanic eruptions, swiftly shaped the Earth's crust.

Darwin held the belief that life had evolved through a series of distinct stages, each unfolding gradually over an extensive span of time. Consequently, the age of the Earth itself became a subject of contention and debate. On one side were proponents of stratigraphy, which involved the study of rock layers, while on the other side were those who favored catastrophic events as major drivers of Earth's history. Today, we have conclusive evidence that the Earth is over 4.5 billion years old (despite the attempts of some deniers to refute this fact). However, during Darwin's era, the age of the Earth remained an open question, subject to ongoing discussion and examination.

23.3.2 The Advantages of the Theory

In the subsequent section of the chapter (Darwin, 1859, pp. 467–480), Darwin dedicates 13 pages to summarizing the overall merits and specific evidence that support his theory. This recapitulation serves as a theoretical consolidation of his ideas, building upon the foundational concepts presented in the initial five chapters. Drawing upon the time-honored practice of revisiting previously discussed themes, Darwin thoroughly examines topics such as variation under domestication and in the wild, the underlying causes of variation, the effects of artificial selection, the phenomenon of hybrid vigor, the causal factors behind the artificial selection, and finally, the pivotal concept of natural selection and its association with the struggle for existence. This comprehensive recapitulation serves to reinforce the key arguments underpinning Darwin's theory.

The readers of this volume will undoubtedly observe that individuals inherit certain traits from their parents. However, even within the confines of the same family (in the colloquial sense), individuals exhibit variations both amongst themselves and in comparison to their parents. Darwin astutely noted that an abundance of variability exists within the realm of domestication, and he further ascertained that this phenomenon extends to the natural world. Subsequently, he embarks upon a three-step elucidation to provide a compelling justification for the existence of natural selection.

In his comprehensive analysis, Darwin embarked upon elucidating the causes of variability within the realm of domestication. Firstly, he astutely observed that while variability may appear spontaneous, it is in fact not. He discerned that various factors such as living conditions play a pivotal role in stimulating variability. Moreover, he recognized that this variability is subject to intricate laws, including the correlation of growth, the influence of use and disuse, and the direct impact of the physical conditions of life. The remarkable diversity witnessed in domestic productions can be attributed to the transmission of these modifications to successive generations over extended periods of time. Indeed, all indications point towards the perpetuation of this variability as long as the environmental conditions remain unchanged.

As posited by the author, it is not the human who actively generates variability but rather "unintentionally exposes organic beings to new conditions of life" (Darwin, 1859, p. 467). In this context, nature becomes the driving force behind the emergence of variability, while the farmer assumes the role of selectively favoring desired variations, thereby giving rise to diverse breeds of animals and plants. Furthermore, Darwin contends that these very principles operate within the natural realm as well.

Moving on to the issue of variability, Darwin proceeds to address the intricate matter of the struggle for existence and the subsequent process of natural selection. In his own words:

> More individuals are born than can possibly survive. A grain in the balance will determine which individual shall live and which shall die, - which variety or species shall increase in number, and which shall decrease, or finally become extinct [...] But the struggle will often be very severe between beings most remote in the scale of nature. The slightest advantage in one being, at any age or during any season, over those with which it comes into competition, or better adaptation in however slight a degree to the surrounding physical conditions, will turn the balance. With animals having separated sexes there will in most cases be a struggle between the males for possession of the females. The most vigorous individuals, or those which have most successfully struggled with their conditions of life, will generally leave most progeny. (Darwin, 1859, pp. 467–468)

Darwin proceeds to summarize the contents of the third chapter, focusing on the concept of the Struggle for Existence and its role in the process of natural selection. According to Regner (1995), he incorporates terms like geometric growth and supports his claims with calculated evidence, thereby enhancing the scientific rigor of his work. Additionally, Darwin employs metaphors derived from military terminology to illustrate his point: "But success will often depend on having *special weapons* or *means of defence*, or on the charms of the males; and the slightest advantage *will lead to victory*" (Darwin, 1859, p. 468– emphasis added).

Moving on to the third aspect, Darwin examines various natural phenomena and patterns, including sexual selection, that can be elucidated by his theory. He also provides a comprehensive review of the geological evidence. By the conclusion of this section within the chapter, readers should have gained a thorough understanding of the supporting evidence for natural selection. The remaining portion of the chapter delves into discussions concerning the skepticism surrounding species transmutation and the far-reaching implications of Darwin's theory.

23.3.3 Causes of the General Belief in the Immutability of Species

On the end of page 480, having presented his conviction in the preceding evidence supporting gradual species transformations over time, Darwin poses a crucial question: "Why [. . .] have all the most eminent living naturalists and geologists rejected this view of the mutability of species?" (Darwin, 1859, p. 480).

Darwin proceeded to address his own inquiries in the following manner. Firstly, he compiled a catalogue of assertions that lacked support based on the scientific understanding of his era, such as the presumed sterility of species during crossbreeding or the notion of a relatively brief history of the Earth. He contended that the dismissal of his theory was not rooted in its scientific soundness or unsoundness but rather stemmed from a human reluctance to embrace it.

Secondly, Darwin highlights the peculiarities of scholars, naturalists, and geologists, attributing their reluctance to a simple resistance to "[. . .] admitting any great change of which we do not see intermediate steps" (Darwin, 1859, p. 481) He draws a parallel to the debates provoked by Charles Lyell's investigations into a rock formation, emphasizing the challenge of comprehending the vast timescales involved, such as "millions of years," and perceiving the cumulative impact of countless minor variations across countless generations. As previously mentioned, this challenge persists even in contemporary society, evident in the general public's understanding and discussions about evolution with students at elementary education levels.

Thirdly, Darwin acknowledges that he did not anticipate persuading seasoned naturalists who had firmly entrenched ideas developed over many years. Nevertheless, he highlights, somewhat ironically, that a lack of openness to new concepts is fundamentally unscientific. He notes that "It is so easy to hide our ignorance under such expressions as the 'plan of creation, 'unity of design,' etc., and to think that we give an explanation when we only restate a fact" (Darwin, 1859, p. 482). This subtle critique emphasizes the importance of embracing scientific inquiry and remaining open to the possibility of alternative explanations.

23.3.4 The Conclusion of the "Recapitulation and Conclusion"

In the concluding section of chapter XIV, Darwin articulates his hope that his book would leave an impression on open-minded naturalists and future generations: "I look with confidence to the future, to rising and young naturalists, who will be able to view both sides of the question with impartiality" (Darwin, 1859, p. 482). With this sentiment, Darwin initiates the concluding remarks of *The Origin*, which encompass some of the most renowned and captivating quotes in the annals of scientific history (Ruse, 2019).

In the final six pages of the chapter, and indeed the entire book, Darwin delves into the remaining two subjects alluded to in the subheading of the chapter: *how far the theory of natural selection may be extended* and the *effects of its adoption on the study of Natural History*.

Despite acknowledging the skepticism, resistance, and debates that would surround his theory, Darwin expresses his confidence that the adoption of Natural Selection could "cause a considerable revolution in Natural History" (Darwin, 1859, p. 484). Certain problems, such as the persistent arguments between systematists regarding the distinction between varieties and species, could be resolved. Darwin suggests that precise categorization would become less important as classifications would now reflect the evolutionary process from a common ancestor, providing insight into the genealogy of specific groups of organisms (Darwin, 1859, pp. 484–5). According to Richards (2009), many systematists previously attributed a theological explanation to species classification, following the views of Linnaeus. They believed that a "natural" classification mirrored God's plan, with nature being rational just as God is and scholars being rational as well, reflecting the likeness of God. However, for Darwin, classifications would now reflect the process of modification from a common ancestor, revealing the genealogical relationships within a particular group of organisms.

Furthermore, Darwin postulated that by recognizing that all living beings are interconnected through a chain of descent, potentially tracing back to a single ancestral form, several recurring problems in biology would acquire a clearer significance. These problems include understanding the affinity and relationships between different species, comprehending the existence of communities and species that mutually benefit from living together, deciphering the meaning behind various morphological and adaptive traits, and explaining the presence of rudimentary and vestigial organs. Ultimately, Darwin expressed his hope that this new perspective on the interconnectedness of living organisms would invigorate and enhance the study of natural history, making it even more captivating and intellectually stimulating.

Darwin also anticipated the emergence of novel avenues of scientific inquiry. He envisioned research into the causes and laws governing variation, the interrelationships between growth patterns, the consequences of the use and disuse of organs, and the direct influence of external conditions on organisms. Darwin believed that the study of domesticated plants and animals within the framework of selection would greatly enhance our understanding of biological processes. Additionally, he saw great value in investigating aberrant yet "living" fossils, which could provide insights into ancient life forms. Furthermore, Darwin recognized the potential of embryology in unveiling the prototypes of major groups of organisms, shedding light on their evolutionary origins.

In relation to the field of Geology, Darwin speculated that advancements in this discipline would reveal the Earth's crust as a haphazard and infrequent accumulation. Such a perspective would suggest that species arose and disappeared due to gradual and ongoing processes rather than through miraculous acts of creation or catastrophic events.

Lastly, Darwin foresaw the potential development of new areas of inquiry, including Psychology. He recognized that natural selection would have a significant impact on the improvement and preservation of mental capacities and instincts in animals. In a somewhat enigmatic statement, he expressed his anticipation that "light will be thrown on the origin of man and his history" (Darwin, 1859, p. 488). This marks the sole instance in the book where Darwin explicitly links natural selection to human evolution, a topic that would receive much greater attention in his subsequent work, *The Descent of Man* (1871).

A remarkable work such as *On the Origin of Species* deserves a fitting conclusion, and Darwin certainly delivers in this regard. The final two paragraphs exhibit an outstanding blend of clarity, directness, and even a touch of poetic elegance, showcasing Darwin's prowess as a masterful writer.

In the penultimate paragraph, Darwin provides a justification for viewing species as lineal descendants of a few ancestral beings rather than as independently created entities. He argues that this perspective bestows a sense of nobility upon species, as all living organisms today are the survivors of ancestors who successfully left descendants. Since many species have become extinct, the mere existence of any given species is a cause for celebration. Darwin's profound respect for all living beings is evident in this passage; each organism represents a unique evolutionary process spanning thousands, if not millions, of years, and they deserve their place on this planet alongside us. Darwin boldly posits that considering the immense expanse of the past, we can anticipate an equally vast future wherein species will continue to thrive, evolving toward greater "corporeal and mental endowments" and striving for perfection. Some might argue that, in the present day, Darwin would feel deeply disappointed by the widespread devastation, destruction, and extinction caused by human beings to other species and natural environments. This optimistic view of creatures' future could be related to Darwin's belief that natural selection has a *purpose*, working "solely by and for the good of each being" (Darwin, 1859, p. 489) to lead living beings toward perfection. In other words, it is a clear *teleological sentence*: the author is giving an objective to the process he is describing.

In the final paragraph, Darwin contemplates the intricate complexity of the natural environment, attributing it to the operation of underlying laws that govern the world around us:

> Growth with Reproduction; Inheritance, which is almost implied by reproduction; Variability from the indirect and direct action of the external conditions of life, and from *use and disuse*; a Ratio of Increase so high as to lead to a Struggle for Life, and as a consequence to Natural Selection, entailing Divergence of Character and the Extinction of less-improved forms. (Darwin, 1859, pp. 489–490; emphasis added)

It is interesting, as argued by Waters (2009), that Darwin mentions use and disuse together with Natural Selection, giving support to the claim that, for the author, Natural Selection could be the most important, but not only law responsible to the transmutation of species. One point of interest here, however, is Darwin's concern in making clear that, like gravity and other Newtonian laws which govern the physical world, the biological world is also controlled by laws, revealing a scientific rigor proper of his time.

The final sentences of the paragraph exhibit exquisite prose and serve as a tribute to the marvels of the natural selection process:

> [. . .] from the war of nature, from famine and death, the most exalted object which we are capable of conceiving, namely, the production of the higher animals, directly follows. (Darwin, 1859, p. 490)

Once again, Darwin emphatically conveys the profound significance of Natural Selection, acknowledging its seemingly ruthless nature while underscoring its pivotal role in shaping the vast array of life forms that inhabit our planet.

The book finishes with one of the most famous quotations of the entire work:

> There is grandeur in this view of life, with its several powers, having been originally breathed into a few forms or into one; and that, whilst this planet has gone cycling on

> according to the fixed law of gravity, from so simple a beginning endless forms most beautiful and most wonderful have been, and are being, evolved. (Darwin, 1859, p. 490)

From the second edition (1860) onwards, Darwin slightly changed the end of this sentence, making an explicit reference to God: "There is grandeur in this view of life, with its several powers, having been originally *breathed by the Creator into* a few forms or into one" (Darwin, 1860, p. 490 – emphasis added). Why would he do that? Perhaps due to the pressure made by religious debates, stimulated by his work?

No matter the answer, it is undeniable that the final two paragraphs of the book present Darwin's most eloquent and poetic defense of the significance of natural selection and the transformation of living beings. It is impossible to read these passages without a smile on one's lips and without being filled with the sensation that there is indeed a profound "grandeur in this view of life."

23.4 On the Teaching with *Origin's* Chapter XIV: Aspects of Nature of Science (NOS)

Stuck with this volume's aims of promoting direct engagement with *The Origin*, this section explores facets of Darwin's scientific practice through the pedagogical construct "Nature of Science" (NOS). Literature extensively explored the topic (e.g., Kampourakis & McComas, 2010). In this text, we adopt the definition put forth by McComas (2020):

> The nature of science is a fertile hybrid arena which blends aspects of various social studies of science including the history, sociology, and philosophy of science combined with research from the cognitive sciences such as psychology into a rich description of what science is, how it works, how scientists operate as a social group and how society itself both directs and reacts to scientific endeavors. (McComas, 2020, p. 4)

How can the reading *The Origin* serves as a tool to teach and learn aspects of the Nature of Science (NOS)? Interacting with the primary sources of past science is just one of the various approaches to addressing NOS in the science classroom (McComas, 2020). Considering *Origin's* Chapter XIV an abstract of an abstract, can it be utilized for the same purpose in a teaching environment? The answer of it involves a selection of aspects of science most suitable for the material. McComas (2017, p. 73) prioritizes nine key aspects of NOS for science education, as follows:

1. Science is distinct from technology and engineering.
2. Science is tentative but durable.
3. Science cannot address all questions.
4. Empirical evidence is required.
5. Science shares methods (induction, deduction, etc.).
6. Law/theory distinction.
7. Creativity is vital in science.
8. Subjectivity is a frequent element in science.
9. Social and Cultural elements impact science.

The Origin is especially suitable to address some of these aspects, such as empirical evidence (item number 4 on the list of McComas' list), diverse scientific methods (5), creativity (7), and relatedness with cultural and social factors (9).

Arguably one of the most prominent characteristics of the natural sciences is its reliance on empirical evidence to substantiate explanations and theories. And, of course, any natural fact or phenomenon, consciously or not, turns into evidence under a theoretical framework. It is characteristic of Darwin's argumentative strategy the vast array of facts he marshals in support of his theory. In Chapter XIV, he presents us with evidence drawn from the fields of geology, embryology, and comparative anatomy, as we can read in the passage:

> The framework of bones being the same in the hand of a man, wing of a bat, fin of the porpoise, and leg of the horse, – the same number of vertebræ forming the neck of the giraffe and of the elephant, – and innumerable other such facts, at once explain themselves on the theory of descent with slow and slight successive modifications. The similarity of pattern in the wing and leg of a bat, though used for such different purpose, – in the jaws and legs of a crab, – in the petals, stamens, and pistils of a flower, is likewise intelligible on the view of the gradual modification of parts or organs, which were alike in the early progenitor of each class (Darwin, 1859, p. 479)

However, the purpose of those pieces of evidence is not a matter of simply accumulation of facts. The structure of the *Origin*, as pointed out by Jonathan Hodge in 1977 and discussed by other scholars,[3] is guided by *vera causa*, or the true cause, of the observed phenomenon. In accordance with the contemporary philosopher of Darwin, John Herschel, to establish a *vera causa* is necessary (1) to recognize the existence of the cause, (2) show that the cause is adequate or capable of producing the effects under explanation and, (3), that that cause is responsible for the observed effects (Waters, 2009, p. 124).

Darwin presents the existence and adequacy of natural selection as the cause of species changes in the first four chapters of *The Origin*. In Chapters VI–IX, he argues for adequacy and responsibility of the cause. In Chapters V, X–XIII, arguments exclusively address the responsibility of natural selection in explaining many groups of facts (Waters, 2009, p. 125). Upon examining Chapter XIV, disregarding its peculiar introductory style featuring the "emphasis on the negative" discursive tool, we discover the same argumentative framework structured throughout the chapters of the book, now condensed into a few paragraphs.

First, Darwin establishes artificial selection:

> Under domestication we see much variability. [. . .] Man does not actually produce variability; he only unintentionally exposes organic beings to new conditions of life, and then nature acts on the organisation, and causes variability. But man can and does select the variations given to him by nature, and thus accumulate them in any desired manner. He thus adapts animals and plants for his own benefit or pleasure. (Darwin, 1859, pp. 466–467)

[3] As Waters reminds us, the presence of the ideal of *vera causa* in Darwin's book, pointed out by Jonathan Hodge in 1977 and rediscussed in his publications of 1989 and 1992, as well as in his own chapter at the *Cambridge Companion to Darwin* (2009) and in chapters authored by Philip Sloan, Gregory Radick and David Hull (Waters, 2009, p. 142, note 11).

and designates it as the "great agency in the production of the most distinct and useful domestic breeds" (Darwin, 1859, p. 467). Then, through an analogy with artificial selection, he establishes the transition to the concept of natural selection (the first step in Waters' scheme):

> There is no obvious reason why the principles which have acted so efficiently under domestication should not have acted under nature. In the preservation of favoured individuals and races, during the constantly-recurrent Struggle for Existence, we see the most powerful and ever-acting means of selection. The struggle for existence inevitably follows from the high geometrical ratio of increase which is common to all organic beings. [. . .] More individuals are born than can possibly survive. A grain in the balance will determine which individual shall live and which shall die, – which variety or species shall increase in number, and which shall decrease, or finally become extinct. [. . .] The slightest advantage in one being, at any age or during any season, over those with which it comes into competition, or better adaptation in however slight a degree to the surrounding physical conditions, will turn the balance. (Darwin, 1859, pp. 467–468)

Following this, Darwin asserts the capability of natural selection to generate adaptations, showing the adequacy of the cause (natural selection) to produce the effects (species change) – Waters' step two:

> If then we have under nature variability and a powerful agent always ready to act and select, why should we doubt that variations in any way useful to beings, under their excessively complex relations of life, would be preserved, accumulated, and inherited? Why, if man can by patience select variations most useful to himself, should nature fail in selecting variations useful, under changing conditions of life, to her living products? What limit can be put to this power, acting during long ages and rigidly scrutinising the whole constitution, structure, and habits of each creature, – favouring the good and rejecting the bad? (Darwin, 1859, p. 469)

Lastly, Darwin concludes his argument by showing a list of "special facts and arguments in favor of the theory" (Darwin, 1859, p. 469), concerning classification (p. 470), adaptation (p. 471), homology (p. 473), behavior (p. 474), the fossil record (p. 475), geographical distribution (p. 476), vestigial organs (p. 478), and anatomy (p. 479). Darwin argues that natural selection is responsible for this massive display of facts, fulfilling the last step of Waters' scheme.

According to Michael Ruse (2019, p. 12), the third part of Darwin's argument is an attempt to create a "consilience of inductions" – an interconnected web of interpretations based on facts, all pointing toward Natural Selection as the driving force behind species transmutation. In this way, Darwin made his case as a skilled detective:

> When you are trying to prove the truth of an unseen claim – molecules or murderers – you work through circumstantial evidence, trying to find clues across the spectrum. (Ruse, 2019, p. 12)

This feature of Chapter XIV may be subject to planning class activities that further explore the interconnectedness between observation data and theory, one important aspect of the Nature of Science (NOS).

Another aspect of the Nature of Science (NOS) that can be facilitated by using *The Origin* in the classroom is creativity. Kampourakis and McComas (2010, pp. 641–646) delve into the creative way Darwin draws upon concepts from other disciplines to construct his arguments through analogical reasoning. Chapter XIV is useful to emphasize the inventive and unconventional methods Darwin employs, especially where he mobilizes rhetorical tools to enhance the persuasiveness of his argument.

As previously mentioned, the chapter commences with an emphasis on the collection of facts that could potentially challenge the theory. Lennox (2021) explores possible reasons for this unusual emphasis on the negative. Lennox traces Darwin's value of negative evidence back to the 1837–1838 *Notebooks*, 1842 *Sketch,* and 1844 *Essay.* Nevertheless, he argues that negative evidence received more relevance in 1859 *The Origin.* Contrasting Darwin's Recapitulation and Conclusion section of the two earlier manuscripts with the published book, Lennox (2021, p. 157) observes that in the *Sketch* the recapitulation does not mention any negative evidence (which is only briefly discussed in the text), while the recapitulation in the *Essay* follows the order of presentation in the published text (selection, then difficulties – also briefly – and then positive evidence). It is only in the published version that the Recapitulation and Conclusion chapter presents negative evidence preceding the theme of Natural Selection itself. Lennox concludes that the strategy of initially exposing the weaknesses of an argument before defending it was a means to "ensure a fair hearing for his positive case" (Lennox, 2021, p. 157). Darwin's keen awareness of the objections his theory would face, and his creative approach to disarming his opponents is reminiscent of the tactics employed by a skilled lawyer. As philosopher and historian of biology, Anna Carolina Regner highlighted Darwin's intentions:

> [. . .] Starting with an examination of objections gives Darwin strategic supremacy over his opponents. Anticipating the identification of objections, he shows that he is aware of them, recognizes their importance, takes them into account, and that, despite this, they do not offer insurmountable obstacles to the defense of his theory. (Regner, 1995, p. 79)

Further evidence of Darwin's rhetorical process and creative spirit can be found in his use of dialogue marks throughout the text. Gillian Beer (2000) highlights the unique relationship that Darwin establishes between himself as the speaker and the audience when he employs the rhetorical device of referring to both parties collectively as "we." Rather than saying, "You will understand, from my arguments. . .," Darwin positions himself within the discourse, merging his perspective with that of the audience. This inclusion approach fosters a sense of shared understanding and encourages a deeper engagement with his ideas.

David J. Depew also offers valuable insights into the rhetorical impact of what he identifies as "*The Origin*'s 'we.'" According to Depew, Darwin consistently employs this rhetorical device throughout the book to recapitulate previously discussed points (Depew, 2009, p. 242). Depew suggests that Darwin's use of this discursive tool indicates his intention to present *The Origin* as a "rhetorical performance on a public stage" (p. 243). At that "stage," Darwin's theories would face scrutiny from diverse audiences, including both the scientific community and the general public.

According to Depew, Darwin's use of the inclusive pronoun "we" addresses mixed audiences to engage the general, nonspecialized public and encourage them to participate in evaluating scientific matters. At the same time, the specialized readers,

> Are asked to judge the issue as if they were overhearing an argument addressed to an open-minded but nonexpert third party, thereby prying these experts away from claims they might otherwise dismiss out of hand by virtue of their professional identities. (Depew, 2009, p. 242)

This rhetorical device would prove particularly valuable in a chapter named "Recapitulation." In fact, it is in chapter XIV that we find its extensive use, appearing in 112 passages. As an illustrative example, the following excerpt shows Darwin's adeptness at merging himself with the reader:

> Such is the sum of the several chief objections and difficulties which may justly be urged against my theory; and I have now briefly recapitulated the answers and explanations which can be given to them. I have felt these difficulties far too heavily during many years to doubt their weight. But it deserves especial notice that the more important objections relate to questions on which we are confessedly ignorant; nor do we know how ignorant we are. We do not know all the possible transitional gradations between the simplest and the most perfect organs; it cannot be pretended that we know all the varied means of Distribution during the long lapse of years, or that we know how imperfect the Geological Record is. Grave as these several difficulties are, in my judgment they do not overthrow the theory of descent with modification. (Darwin, 1859, pp. 465–466)

This strategic approach indicates Darwin's keen ability to recognize the heterogeneity of his audience and navigate their specific concerns, showcasing his rhetorical expertise in creating an inclusive and communal discursive environment that bridges the gap between his argument and its evaluators.

These distinct features of *The Origin*, which are condensed and made more explicit in Chapter XIV, along with the historical background of the chapter's writing process in connection to Lyell, reach the third aspect of NOS we pointed out: the interconnectedness between social, cultural, and scientific endeavors.

The *Vestiges of the Natural History of Creation*, a book anonymously published in 1844 but whose authorship was attributed to the editor Robert Chambers, met enormous repercussions at the time while facing resistance from the scientific community. Strong reactions came from influential figures like geologist Adam Sedgwick and natural philosopher William Whewell, who were both influences on Darwin himself (Secord, 2000, p. 223; Hueda & Martins, 2014). Darwin seriously took those reactions into consideration when crafting his line of arguments, aiming to be both persuasive and tactful. He needed to ensure that his theory would resonate with the intellectual community to which he belonged while being mindful of not jeopardizing the positions of authority held by many of his peers. This example serves to illustrate to students that science does not exist in isolation from its social context, and the formulation of scientific ideas often takes into account the anticipated reception and the social dynamics within the scientific community.

23.5 Final Thoughts: *One* Long Argument?

At the conclusion of the book, the underlying theme of the "one long argument" highlighted by Darwin in the opening sentence of Chapter XIV should be evident: species undergo transmutation through the mechanism of natural selection. Or perhaps it is not so clear-cut?

According to Waters (2009), the clarity of the relationship between transmutation (now known as evolution) and natural selection is indeed questionable. Waters suggests that Darwin's attempt to establish a direct connection between the two was not entirely successful, resulting in a significant number of readers (if not the majority) accepting the occurrence of evolution while rejecting natural selection as its mechanism. In this light, we can view *The Origin* as a collection of arguments that offers flexibility, allowing readers to interpret and draw varying conclusions.

The central ideas in *The Origin* revolve around the concept of the tree of life, which illustrates the evolutionary change of species over time. This process involves the extinction of some species while others persist or diverge into multiple descendant species, resembling branches on a tree. Additionally, natural selection is presented as the driving force behind this diversification, acting gradually over generations to favor the survival and adaptation of the fittest individuals, resulting in the modification and specialization of living beings' structures. Waters argues that Darwin faced difficulty in effectively connecting these two central ideas, leading readers to embrace the notion of species change but not fully comprehend or accept the role of natural selection in this process.

Why did this discrepancy arise? Several conjectures can be put forth. Firstly, while the concept of species transmutation was revolutionary, it was not entirely novel, as Lamarck and Chambers had already touched upon it in their works (though Darwin's proposal of a single common ancestor for all living beings distinguished his theory). Secondly, the difficulty might lie in comprehending the gradual and prolonged nature of natural selection, as well as grasping the immense spans of time required for its effects to manifest. Lastly, Darwin's emphasis on other processes, such as the intricate laws governing correlation in the growth of body parts or the influence of use and disuse, might have diverted attention from the centrality of natural selection in explaining species transmutation. Darwin acknowledged and supported these alternative explanations, frequently citing them throughout the text. For instance, he invoked the notion of misuse to account for the presence of rudimentary organs.

Despite being the primary process proposed by Darwin to explain species transmutation, natural selection faced significant criticism. One notable critic was George Mivart (1827–1900), an accomplished anatomist and proponent of transmutation who paradoxically became one of the staunchest opponents of natural selection. In his 1871 book, *On the Genesis of Species*, Mivart dedicates Chapter II to questioning the ability of natural selection to account for the initial stages of useful structures, as the chapter's title suggests. In response, Darwin addressed these arguments in the sixth and final edition of *On the Origin of Species*, published in 1872, devoting an entire chapter to refute Mivart's claims.[4] The dispute between these two scientists has been analyzed by Regner (2006), highlighting that scientific disagreements and debates are intrinsic to the nature of science itself.[5] The theory of natural selection, like any scientific idea, has been subject to scrutiny since its inception, and Darwin, throughout his life, never shied away from engaging in discussions with fellow naturalists, religious figures, and society at large.

We can conclude that Chapter XIV goes beyond being merely an abstract of an abstract. While studies on the rhetorical strategy used in *The Origin*, like Waters (2009), may label it as such, Darwin's concern for the chapter's persuasive effectiveness, as evident in his correspondence with Lyell, suggests that he considered it an important element of conviction. In terms of discourse, Darwin employs strategies such as "*The Origin*'s 'we'" (Depew, 2009) to establish a connection with the reader and "accentuating the negative" (Lennox, 2021) to proactively address potential criticisms of his theory and disarm any potential opponents. After establishing natural selection as the *vera causa* of species transmutation, Darwin speculates on what would drive his opponents to resist the idea and concludes the chapter with further speculations on the potential implications of his theory. Therefore, while it does not introduce new elements to what has already been presented, the significance of Chapter XIV lies in the rhetorical strength that emerges from the novel strategy employed to engage the reader with the argumentative framework meticulously crafted by Darwin throughout the book.

References

Beer, G. (2000). *Darwin's plots: Evolutionary narrative in Darwin, George Eliot and nineteenth-century fiction*. Routledge and Kegan Paul.

Branch, G., Scott, E. C., & Rosenau, J. (2010). Dispatches from the evolution wars: Shifting tactics and expanding battlefields. *Annual Review of Genomics and Human Genetics, 11*, 317–338.

Browne, J. (2006). *Darwin's "Origin of Species": A biography*. Atlantic Books.

Bizzo, N. (2018). Prefácio, revisão técnica e notas. In Darwin, C. (1859/2018). *A Origem das Espécies. Tradução de Daniel Moreira Miranda. Prefácio, revisão técnica e notas de Nelio Bizzo*. Edipro.

[4]In the sixth and final edition of *The Origin* published in 1872, Darwin added a new chapter specifically addressing the objections raised against the theory, primarily by Mivart. These objections had accumulated since the book's initial publication in 1859. As a result, Darwin renumbered the subsequent chapters, causing the chapter previously titled "Recapitulation and Conclusion" to be repositioned from XIV to XV.

[5]This chapter has also highlighted another significant scientific debate, previously mentioned, which revolved around the contrasting perspectives of uniformitarians and catastrophists. Given its historical significance, we argue that this debate can serve as a valuable tool for teaching and examining certain aspects of the Nature of Science. It sheds light on the clash of scientific viewpoints influenced by personal values, even when the researchers involved are analyzing the same body of evidence (Irzik & Nola, 2011; McComas, 2020, pp. 564–5). Thus, it is crucial to acknowledge that science, like any human endeavor, is inherently subjective.

Darwin, C. (1859). *On the origin of species by means of natural selection, or the preservation of favoured races in the struggle for life*. John Murray.
Darwin, C. (1860). *Origin of species* (2nd ed.). John Murray.
Darwin, C. (1862). *On the various contrivances by which British and foreign orchids are fertilized by insects*. John Murray.
Darwin, C. (1868). *The variation of animals and plants under domestication*. John Murray.
Darwin, C. (1871). *The descent of man, and selection in relation to sex*. John Murray.
Darwin Correspondence Project., "Letter no. 2312," https://www.darwinproject.ac.uk/letter/?docId=letters/DCP-LETT-2312.xml.
Darwin Correspondence Project., "Letter no. 2432," https://www.darwinproject.ac.uk/letter/?docId=letters/DCP-LETT-2432.xml.
Darwin Correspondence Project., "Letter no. 2335," https://www.darwinproject.ac.uk/letter/?docId=letters/DCP-LETT-2335.xml.
Darwin Correspondence Project., "Letter no. 2441," https://www.darwinproject.ac.uk/letter/?docId=letters/DCP-LETT-2441.xml.
Darwin Correspondence Project., "Letter no. 2459," https://www.darwinproject.ac.uk/letter/?docId=letters/DCP-LETT-2459.xml.
Darwin Correspondence Project., "Letter no. 2496," https://www.darwinproject.ac.uk/letter/?docId=letters/DCP-LETT-2496.xml.
Darwin Correspondence Project., "Letter no. 2501," https://www.darwinproject.ac.uk/letter/?docId=letters/DCP-LETT-2501.xml.
Depew, D. J. (2009). The rhetoric of the origin of species. In M. Ruse & R. J. Richards (Eds.), *The Cambridge companion to the "origin of species"* (pp. 237–255). Cambridge University Press.
Hodge, J. (2013). Darwin's book: On the origin of species. *Science & Education, 22*, 2267–2294.
Hodge, M. J. S. (1977). The structure and strategy of Darwin's 'long argument'. *British Journal for the History of Science, 10*, 237–246.
Hueda, M. A., & Martins, L. A. P. (2014). As concepções evolutivas de Robert chambers no *vestiges of the natural history of creation* (1844). *Filosofia e História da Biologia, 9*(1), 39–57.
Kampourakis, K., & McComas, W. F. (2010). Charles Darwin and evolution: Illustrating human aspects of science. *Science & Education, 19*, 637–654.
Largent, M. A. (2009). Darwin's analogy between artificial and natural selection in the origin of species. In M. Ruse & R. J. Richards (Eds.), *The Cambridge companion to the "origin of species"* (pp. 14–29). Cambridge University Press.
Lennox, J. G. (2021). Accentuate the negative: Locating possibility in Darwin's 'long argument'. *Studies in History and Philosophy of Science, 87*, 147–157.
McComas, W. F. (2017). Understanding how science work: The nature of science as they foundation for science teaching and learning. *School Science Review, 98*(365), 71–76.
McComas, W. F. (2020). *Nature of science in science instruction. Rationales and strategies*. Springer.
Oliveira, A. W., & Cook, K. L. (Eds.). (2019). *Evolution education and the rise of the creationist movement in Brazil*. Lexington Books.
Paley, W. (1802/1819). *Natural theology* (Collected works: IV). Rivington.
Regner, A. C. K. P. (1995). *A natureza teleológica do princípio darwiniano de seleção natural: A articulação do metafísico e do epistemológico na origem das espécies*. Doctoral thesis. Faculdade de Educação, Universidade Federal do Rio Grande do Sul.
Regner, A. C. K. P. (2006). A polêmica Darwin versus Mivart: uma lição em refutar objeções. *Filosofia e História da Biologia, 1*, 55–89.
Richards, R. A. (2009). Classification in Darwin's origin. In M. Ruse & R. J. Richards (Eds.), *The Cambridge companion to the "origin of species"* (pp. 173–193). Cambridge University Press.
Ruse, M. (1979). *The Darwinian revolution: Science red in tooth and claw*. University of Chicago Press.
Ruse, M. (2009). The origin of the origin. In M. Ruse & R. J. Richards (Eds.), *The Cambridge companion to the "origin of species"* (pp. 1–13). Cambridge University Press.

Ruse, M. (2019). *The Darwinian revolution*. Cambridge University Press.
Secord, J. A. (2000). *Victorian sensation: The extraordinary publication, reception, and secret authorship of vestiges of the natural history of creation*. University of Chicago Press.
Waters, C. K. (2009). The arguments in the origin of species. In J. Hodge & G. Radick (Eds.), *The Cambridge companion to Darwin* (2nd ed., pp. 120–147). Cambridge University Press.
Webster's Complete Dictionary of English Language. (1886). London: George Bell. Retrieved August 5, 2022, from https://archive.org/details/websterscomplete00webs/page/n1/mode/2up?view=theater
White, R. M., Hodge, M. J. S., & Radick, G. (2021). *Darwin's argument by analogy – From artificial to natural selection*. Cambridge University Press.

Gerda Maisa Jensen is a retired teacher from the public school system of the municipality of São Paulo, with 30 years of experience teaching biology in secondary and high school as well as in the special system for Youth and Adults Education (EJA, in the Portuguese acronym, to ages 15 or older who have not regular education completed). She has a master's and a doctorate in the history of science and is affiliated with the Brazilian Association of Philosophy and History of Biology (ABFHiB). Currently, Jensen carries out research in the history of biology and its interfaces with education at the Laboratory of History of Biology and Teaching at the Institute of Biosciences of the University of São Paulo (IB/USP).

Bruno F. Lima teaches natural sciences in the public school system of the municipality of Sao Paulo. Currently, he is developing his M.Sc. Degree project examining the arguments put forth by William Paley, Charles Darwin, and George Mivart regarding the complexity of biological structures, particularly focusing on the vertebrate eye, as a historical case for science education. He is under the supervision of Professor Maria Elice Brzezinski Prestes at the Interunit Graduate Program of Science Education of the University of São Paulo. He is a member of the Laboratory of History of Biology and Teaching at the Institute of Biosciences of the University of Sao Paulo (IB/USP).

Marcelo Monetti Pavani is currently associated with the Laboratory of History of Biology and Teaching at the Institute of Biosciences of the University of Sao Paulo. His present research focuses on the reception of evolutionary ideas in Brazil and their connection with racial theories, eugenics, and Brazilian literature. He works under the supervision of Professor Maria Elice Brzezinski Prestes at the Interunit Graduate Program of Science Education of the University of São Paulo. He teaches biology in secondary schools in the City of São Paulo, Brazil.

Part IV
Epilogue: What Came Next Was Extraordinary

Chapter 24
Continuities and Ruptures: Comparing Darwin's "On the Origin of Species" and the Modern Synthesis

Susana Gisela Lamas

Abstract In his book *On the Origin of the Species,* Charles Darwin provides a great deal of material evidence to support his theory of evolution. Over time, evolution has become accepted as a fact, even though several elements of his work remain controversial to this day. When evolutionary thought is narrated in textbooks and often in scientific publications on evolution, emphasis is placed on the continuities of explanatory mechanisms and principles by passing their ruptures. In his work *Evolution: The Modern Synthesis,* Julian Huxley proposed a Darwinian research program by tackling the problems of evolution, unifying data and theories of genetics with natural selection. From this publication began the development of what is known as the Modern Synthesis, which was carried out with the contribution of researchers from different fields of biology. Its main exponents were Ernest Mayr, Theodosius Dobzhansky, George G. Simpson, and George Ledyard Stebbins. This chapter examines the principles of Darwin's and Modern Synthesis, outlining theoretical continuities and discontinuities and their consequences for modern explanations of biological evolution. It will also consider the effects of these continuities and ruptures on evolutionary biology.

24.1 Introduction

Textbooks generally present the history of evolutionary theory as linear, as if ideas were developed and modified, and their explanatory power increased; that is, they began to explain more and more phenomena. However, the question remains about some elements which are neither explained nor even considered. Continuities of knowledge have been emphasized while ruptures have been hidden. In this chapter, we look at both the continuities and the ruptures of Darwin's *On the Origin of*

S. G. Lamas (✉)
Laboratory of Research on Ontogeny and Adaptation (LINOA), National University of La Plata, La Plata, Argentina
e-mail: sglamas@fcnym.unlp.edu.ar

M. Elice Brzezinski Prestes (ed.), *Understanding Evolution in Darwin's "Origin"*, History, Philosophy and Theory of the Life Sciences 34,
https://doi.org/10.1007/978-3-031-40165-7_24

Species and compare them to the evolutionary explanations of Huxley's *Modern Synthesis*.

24.2 Some of Darwin's Influences

The work of Darwin is vast, and here we will focus on just some issues developed in *On the Origin of Species*. Darwin recognized the influence of different authors, one of whom was Charles Lyell, who (following the theoretical line of James Hutton) suggested that the earth was formed gradually based on the same geological phenomena, such as erosions, earthquakes, volcanic processes, floods, etc. This framework, actualism, is sustained in geological studies to the present day. It contrasts with catastrophism, in which the geological structure of the earth with its layers and strata was formed due to great catastrophes over a relatively short period of time.

Another author who influenced Darwin was Jean Baptiste Pierre Antoine de Monet, Knight of Lamarck. He considered that the earth was formed gradually and that organisms became modified due to changes in their environment. Lamarck asserted that if an acquired character is present in both biological parents and helps individual needs, such as the need to eat, it is inherited. Lamarck was not the first to suggest this: throughout the sixth, seventh, and eighth centuries, authors claimed that inheritance brings modification (Martins, 2015). This phenomenon was called the *inheritance of acquired characters*. According to Lillian Martins, when Lamarck wrote his work, this idea was accepted as a fact, so he did not develop it as extensively as other parts of his theory. Accidental changes, for example, the loss of a limb due to an accident, were not inherited; only those changes related to the needs of the organism over a long period of time were inherited.

Another of his laws was use and disuse. According to Lamarck (1809), circumstances decide if an organism uses an organ frequently or whether the organ is relatively unused. The more used organs will be maintained and developed, while those less used will tend to disappear (always in the adult phase and not during the organism's early development). Darwin took up both these laws.

Catastrophism has not only been applied to geology but also biology and is related to fixist theories. The existence of extinct species in fossils is explained by independent acts of creation and catastrophes leading to their extinction. These theories disregard the possibility of transformation or changes in species. For Lamarck, it was necessary to explain not only mechanisms of transformation but also those of adaption.[1] Mayr (1982) states:

> Lamarck never explicitly articulated a concept of adaptation, but the entire causal chain of evolution postulated by him inevitably had to result in adaptation. Since the evolutionary force described by him was not teleological but materialistic, it produced adaptation by

[1] According to Lilian Martins (2002), Lamarck at first supported fixist theorists, but later changed his ideas.

> natural means. (...). For Lamarck adaptation was the inevitable end product of the physiological processes (combined with an inheritance of acquired characters) necessitated by the needs of organisms to cope with the changes in their environment. I can see no other way to designate his theory of evolution than as adaptive evolution. The acquisition of new organs and of new faculties was clearly an adaptive process. Accepting his premises, Lamarck's theory was as legitimate a theory of adaptation as that of Darwin. Unfortunately, these premises turned out to be invalid. (Mayr, 1982, pp. 358–359)

A fundamental experience in Darwin's life was his journey on board the Beagle for 5 years, traveling through the current territories of Brazil, Uruguay, Chile, Peru, Galapagos Islands, Australia, and South Africa, collecting and observing a vast diversity of organisms. Several years later, while writing *On the Origin of Species,* he asks how such a variety among species could have originated and why they are so well adapted to their environments. Answering these two key questions will allow Darwin to provide a materialistic explanation of the change, that is, based on natural processes.

24.3 Darwin's Proposal

The works of Charles Darwin and especially *On the Origin of Species*, first published in 1859, are considered the beginning of current evolutionary theories. Nevertheless, in the first edition of his book, Darwin never used the Word "evolution" to refer to the processes he was describing. According to the Stanford Encyclopedia of Philosophy, until the mid-nineteenth century, the term "evolution" referred to the development of an individual embryo; from 1860, it began to describe species changes. Darwin used the word for the first time in his book *Descent of Man,* published in 1871.

In this work, we will use the 1872 edition since here he responds to some criticisms. Darwin analyses a significant quantity of data in *On the Origins of Species* from his own observations and other scientists from different disciplines (zoology, botany, geology, etc.), as well as cultivators of plants, breeders of doves, dogs, horses, etc. He also draws on references from his other works to provide more examples of cases. One of the main questions he proposes in his work is about the origin of the variability, about the gradual modifications of individuals until a new specie is formed. These are two questions: how variations originate and why these traits become fixed in the population. We will briefly analyze both questions.

Chapter V of his book presents the Laws of Variation. He states that changed conditions, use and disuse, correlated variation and compensation, and economy of growth, among others, are the different causes of variation; nonetheless, he recognizes that they are not sufficient to explain the whole variability.

Darwin asserts that *changed conditions* could have definite or indefinite results. He, therefore, recognizes that there is substantial flexibility in organisms and various responses to environmental conditions. More flexibility is reflected in more variety between organisms. He affirms:

> Changed conditions act in two ways, directly on the whole organism or on certain parts alone, and indirectly through the reproductive system. In all cases there are two factors, the nature of the organism, which is by far the more important of the two, and the nature of the conditions. The direct action of changed conditions leads to definite or indefinite results. In the latter case the organisation seems to become plastic, and we have extensive fluctuating variability. In the former case the nature of the organism is such that it yields readily, when subjected to certain conditions, and all, or nearly all, the individuals become modified in the same way. (Darwin, 1872/2009, p. 106)

Darwin quoted cases cited by other authors. For example, a bird of the same specie has brighter colors if it lives on the coast than on an island. Likewise, a marine shellfish will have brighter colors depending on whether it lives in a more or less saline environment or shallow or deeper waters. However, he also knows that some species show no variety even when individuals live in different environments. He, therefore, concludes that there is no reliable explanation for the presence of variations:

> Again, innumerable instances are known to every naturalist, of species keeping true, or not varying at all, although living under the most opposite climates. Such considerations as these incline me to lay less weight on the direct action of the surrounding conditions, than on a tendency to vary, due to causes of which we are quite ignorant. (Darwin, 1872/2009, p. 107)

Here he is admitting a tendency of organisms to change, although he cannot explain their mechanism.

Darwin gives examples of use *and disuse*, such as:

> The ostrich indeed inhabits continents and is exposed to danger from which it cannot escape by flight, but it can defend itself, by kicking its enemies, as efficiently as many quadrupeds. We may believe that the progenitor of the ostrich genus had habits like those of the bustard, and that, as the size and weight of its body were increased during successive generations, its legs were used more and its wings less, until they became incapable of flight. (Darwin, 1872/2009, p. 108)

In this case, a feature such as an inability to fly is explained by the lack of use of wings and also due to body weight. The legs, on the other hand, are used for defense. Therefore, organisms become differentiated from their ancestors by developing a part of the body that is more used and gradually stop using another part which in turn leads to a change in abilities, such as flying, and in some cases, become atrophied.

One key point in his argument is that organs disappear and remain as vestiges showing disuse; organs that are present but do not have any function go clearly against the fixism and creationist theories in which organisms are perfectly adapted to their environment. These cases of maladaptation were of utmost importance to his theory.

> It appears probable that disuse has been the main agent in rendering organs rudimentary. It would at first lead by slow steps to the more and more complete reduction of a part, until at last it became rudimentary – as in the case of the eyes of animals inhabiting dark caverns, and of the wings of birds inhabiting oceanic islands, which have seldom been forced by beasts of prey to take flight and have ultimately lost the power of flying. Again, an organ, useful under certain conditions, might become injurious under others, [. . .]; and in this case natural selection will have aided in reducing the organ, until it was rendered harmless and rudimentary. (Darwin, 1872/2009, pp. 400–401)

The ways in which Darwin applies the concept of use and disuse differ from those of Lamarck in that, for Darwin, it is possible to inherit accidental characters (e.g., after surgery). A feature could also be inherited even if it is in only one of the parents (Martins, 2015).

Another principle used to explain variability is *correlated variation.* He shows the importance that a change could have on the structure of the whole organism:

> It is believed by some authors that with birds the diversity in the shape of the pelvis causes the remarkable diversity in the shape of the kidneys. Others believe that the shape of the pelvis in the human mother influences by pressure the shape of the head of the child. In snakes, according to Schlegel, the shape of the body and the manner of swallowing determine the position and form of several of the most important viscera. (Darwin, 1872/2009, p. 115)

Some traits will result from modifications to other parts of the organism's structure, showing an integrative vision. However, this integrative conception was one of the elements which were omitted by Modern Synthesis.

The *law of compensation or balancement of growth* claims that there is a natural tendency to economize. Darwin explains it as follows:

> The elder Geoffrey and Goethe propounded, at about the same time, as Goethe expressed it, "in order to spend on one side, nature is forced to economise on the other side." I think this holds true to a certain extent with our domestic productions: if nourishment flows to one part or organ in excess, it rarely flows, at least in excess, to another part; thus it is difficult to get a cow to give much milk and to fatten readily. (Darwin, 1872/2009, p. 117)

As we stated, we can analyze two issues of Darwin's theory: one is the origin of variability, and the other is the mechanism by which these slight variations develop until a new specie is formed (i.e., the speciation mechanism). Regarding *On the Origin*, Darwin admits that his Laws of Variation are insufficient to explain the reasons for variability. Nevertheless, he does not doubt that variability exists; for him, it is an evident fact. The other problem is how species differentiate; natural selection is the principle used to explain why some features remain in the population.

> On the whole, we may conclude that habit, or use and disuse, have, in some cases, played a considerable part in the modification of the constitution and structure; but that the effects have often been largely combined with, and sometimes overmastered by, the natural selection of innate variations. (Darwin, 1872/2009, p. 114)

Darwin observed how animal breeders and farmers select organisms with some traits and allow them to reproduce, obtaining offspring with these slight variations that, over generations, will produce descendants with the desired feature. So likewise, in the wild, organisms present variations that are advantageous to themselves. Therefore these variations will be selected. For this reason, chapter IV is called Natural Selection or the Survival of the Fittest:

> This preservation of favourable individual differences and variations, and the destruction of those which are injurious, I have called Natural Selection, or the Survival of the Fittest. Variations neither useful nor injurious would not be affected by natural selection, and would be left either a fluctuating element, as perhaps we see in certain polymorphic species, or would ultimately become fixed, owing to the nature of the organism and the nature of the conditions. Several writers have misapprehended or objected to the term Natural Selection.

> Some have even imagined that natural selection induces variability, whereas it implies only the preservation of such variations as arise and are beneficial to the being under its conditions of life. No one objects to agriculturists speaking of the potent effects of man's selection; and in this case the individual differences given by nature, which man for some object selects, must of necessity first occur. (Darwin, 1872/2009, p. 63)

This is one of the most important contributions of Darwin since he shows how the mechanism of natural selection could explain the variability in organisms with a common origin. This is crucial for Darwin because it not only allows him to justify that species change but also explains how two different species can have a common origin. He introduces a model in the form of a tree to understand the relations between different species. He gives a hypothetical example: if in a population of bears, some of them were more agile and faster than others, and if in a season of the year there were less prey than usual, then the faster and more agile bears would survive the others. Therefore, faster and more agile bears would generate more offspring, and only fast and agile bears would remain after some generations. The struggle for survival is implicit in the survival of the fittest. Darwin's model explains the changes in species based on variability, the inheritance of these variations, and natural selection.[2]

Darwin's proposal differs from Lamarck's because the latter advocates a linear model with a tendency to perfection, but this is not supported in Darwin's model (Martins, 2002).

One of Darwin's major problems was explaining the mechanism of inheritance. While inheritance was evident, he could not explain how it was done. He suggested pangenesis, a principle by which fluids pass from parents to offspring, but he was dissatisfied with this explanation and admitted his ignorance of the hereditary mechanism. This large gap in his theory has only recently been resolved with the development of genetics.

Finally, before considering the Modern Synthesis, we will underline the last sentence of the introduction of *On the Origin of Species*, and we will look at this later: "I am convinced *that Natural Selection* has been the *main but not exclusive means of modification*" [emphasis added] (Darwin, 1872/2009, p. 4).

24.4 The Modern Synthesis

At the end of the nineteenth and beginning of the twentieth centuries, genetics began to develop and play an essential role in evolutionary theories. For example, the mutation theory affirms that new varieties and species are formed by mutations, rejecting the gradual genetic changes. Thus, a gap is formed between experimental geneticists and naturalists since their languages, problems, and methodologies are

[2] According to Mayr (1982), Darwin also proposed, although not explicitly, geographical isolation as another mechanism of speciation.

very different. According to Mayr (1982), the ideas of Huxley's book *Evolution: The Modern Synthesis* provided a bridge between them, arriving at the following agreements:

> (1) that evolution is gradual, being explicatory in terms of small genetic changes and recombination and in terms of the ordering of this genetic variation by natural selection; and (2) that by introducing the population concept, by considering species as reproductively isolated aggregates of populations, and by analyzing the effect of ecological factors (niche occupation, competition, adaptive radiation) on diversity and on the origin of higher taxa, one can explain all evolutionary phenomena in a manner that is consistent both with the known genetic mechanisms and with the observational evidence of the naturalists. Julian Huxley (1943) designated the achievement of consensus on these points as the evolutionary synthesis. (Mayr, 1982, p. 567)

Huxley states that at the end of the nineteenth and the beginning of the twentieth centuries, Darwinism presented an "eclipse" due to the development of Mendelian genetics and fields of physiology. Indeed, in these areas of knowledge, changes could be explained without resorting to natural selection. This metaphor is interesting because an eclipse is a transitory concealed state, and this is precisely what Huxley intended with his work: that Darwin shines again. In his own words, he attempts to "harmonize" facts and investigative tools of areas of biology (such as Mendelian genetics, embryology, ecology, systematics, paleontology, and compared anatomy) with nonbiological disciplines (mathematical analysis, geography, and geology). He calls this *Modern Synthesis* because it seeks to unify using an evolutionist perspective in which natural selection serves as an explanatory principle or general mechanism for change. Furthermore, the development of genetics provides the possibility to evaluate this mechanism quantitatively.

It should be noted that Julian Huxley is the grandson of Thomas Huxley, known as "Darwin's bulldog" for his defense of Darwin's theory of evolution. In his work, Julian Huxley recognizes his interest that Darwin's theory reappears. On this point, Smocovitis (1996) writes:

> The unification of biology and the emergence of evolutionary biology took place just as the centenary of the publication of Darwin's Origin was approaching. Gathering to re-examine and re-assess the work of this "great man of science," evolutionary biologists and historians would begin to contribute to the burgeoning literature of Darwin studies. Re-reading the present into the past, they reinvented Darwin and Darwinism as neo- Darwinism, and reinterpreted his "theory of descent with modification" as evolution by means of natural selection. Darwin was to be reconstructed once again as the "founding father" of the discipline of evolutionary biology. Yet though Darwin was to be repeatedly hailed as the Newton of biology, it was the "modern synthesis" that would function as the biological analogue of the "Newtonian synthesis" in the grand narrative of the history of science. (Smocovitis, 1996, pp. 56–57)

The Modern Synthesis is presented as an evolutionary theory "inherited" from Darwin, with natural selection as its unifying principle and population genetics as the discipline which explains evolution through mathematical models. The way that it is presented emphasizes the continuities and omits the ruptures. Nevertheless, the

Modern Synthesis will become a perspective or a general frame from which different ideas develop. For example, Mayr (1982) states:

> Dobzhansky, Rensch, Mayr, Huxley, Simpson, and Stebbins, among others, showed that the major evolutionary phenomena such as speciation, evolutionary trends, the origin of evolutionary novelties, and the entire systematic hierarchy could be explained in terms of the genetic theory as matured during the 1920s and 30s. Except for shifts in emphasis and for a far more precise analysis of all the various mechanisms, the synthetic theory of evolution is the paradigm of today. (Mayr, 1982, p. 118)

Natural selection is understood as differential reproduction according to adaptive advantage. Evolutionary processes and changes are restricted to gene frequencies and will be explained by genetic drift, gene flow, random factors, and mutations. In this way, the variations Darwin referred to remain restricted to a genetic level. Genotype and phenotype are differentiated, and genes are seen as responsible for producing phenotype characters.

On speciation mechanisms:

> Mayr and the other founders of the modern synthesis were convinced that speciation occurred only through an initial phase of geographical isolation breaking a once homogeneous population into sub-populations which could begin to evolve in different directions according to the demands of their local environment [...]. (Bowler, 2003, p. 337)

Mayr calls this mechanism *allopatric speciation*. Dobzhansky (1951) recognizes that there could be other nongenetic factors of speciation, which he calls *isolating mechanisms*. Hence, evolution is only understood at a genetic level. So, this way of conceiving evolution is reductionist. Mayr (2001) claims that synthetic theory holds:

> Evolution is change in the properties of populations of organisms over time. In other words, the population is the so-called unit of evolution. Genes, individuals, and species also play a role, but it is the change in populations that characterizes organic evolution. (Mayr, 2001, p. 9)

Finally, Weissmann's Barrier was another of the ideas accepted by synthetic theory. Weissman suggests that hereditary information moves only from germline cells to somatic cells and concludes from his experimental research (he cut off mouse tails for numerous generations and the offspring fully retained tails) that extragenetic inheritance does not exist.[3]

Below we will analyze the continuities and ruptures in the works of Darwin and the Modern Synthesis and discuss "the cost" of these ruptures for evolutionary thought.

[3]August Weismann admitted that these experiments did not allow him to oppose Lamarckism, because it is not a functional adaptation but a sudden and accidental mutilation (see Gould, 2002, p. 2001). However, both Weismann and authors of Modern Synthesis had cited these experiments as proof that refute the laws of Lamarck.

24.5 Ruptures and Continuities

In the section of this work where we analyzed the ideas of Darwin, we looked at two different questions: (1) how variation appears, that is, its origin, and (2) why those traits remain fixed in the population.

We have seen that the most important mechanism of change for Darwin was a natural selection which proceeds gradually and results over long periods of time in the diversification of the species. Therefore, large changes (macroevolution) can be understood based on small changes (microevolution). Modern Synthesis takes up these explanations of Darwin, resulting in the following agreements:

The principles for studying evolution are: (1) natural selection as the change mechanism; (2) the gradualism of the change; (3) a continuum between micro and macroevolution.

However, there are also differences between these theories related to the rejection of two assumptions: (1) the inheritance of acquired characters and (2) the use and disuse to explain the development or reduction of an organ. When synthetic biologists mention the elements of Darwin's theory that they reject, they always relate them to Lamarckian assumptions. But we will see that this is not altogether true.

Synthetic biologists also ignored other ideas developed by Darwin. As we saw above, in *On the Origin,* he reveals the Laws of Variation, some of which were indeed influenced by Lamarck. Still, others were unrelated, such as correlated variation and compensation and the economy of growth. The correlated variation is related to the body plan and embryological questions. Synthetic biologists did not reject them, but these were research fields that they did not develop.[4] Thus, we could conjecture that they were not considered necessary to explain evolution. The same process occurred with physiology which is present in Darwin's work but neglected by Modern Synthesis.

The differences can be seen in the following schematic way:

	Darwin	Synthetic theory
Inheritance of acquired characters	√	X
Use and disuse	√	X
Correlated variation and compensation	√	X
Embryology is an essential discipline for understanding evolutionary processes	√	X
Physiology is an essential discipline for understanding evolutionary processes	√	X
Evolution is explained based on a level of organization (genetic)	X	√
Speciation occurred only through an initial phase of geographical isolation	X	√

[4] Gould and Lewontin (1979) criticize the Adaptationist Program and its atomization of organisms.

We can conclude that Synthetic theory was a "hardening" of Darwin's theory since some problems and disciplines were left out of the discussion on evolution. Fortunately, these fields of study continue to develop.

24.6 Conclusion

In much of the specific bibliography and textbooks, the Modern Synthesis is considered the "heir" of Darwin, in which only the ideas of Lamarck are rejected. Nevertheless, other essential and profound differences are found when both works are closely examined. Although Huxley denied this, several topics and disciplines were left out (1943). It is interesting to observe that in current research, precisely those neglected areas are contributing most to evolutionary knowledge at present.

Some examples include Evo-devo, which shows the developmental processes that produce the features in every generation of an evolving lineage (Raff, 2000). Evo-devo is nothing more than developmental biology (embryology) related to evolution. Waddington's ideas on epigenetics (1942) state that there are "canalization" processes of development, and very often, genetic and phenotypic variations are not coupled (see also Jablonka & Lamb, 1995). The niche construction theory recognizes organisms' influences on their environment, modifying selection pressures (Odling-Smee, 1995; Odling-Smee et al., 1996, 2003).

There have also been some crucial critics of the position that Darwin and synthetic theory were in agreement. For example, the theory of Punctuated Equilibrium by Gould and Eldredge (1993) denies that large changes (macroevolution) are due to small changes (microevolution). The Neutral Theory of Molecular Evolution by Kimura (1983) states that most genetic differences between species and polymorphisms within species are selectively neutral. Moreover, the classical view of a gene as a *discrete element* in the genome has been shaken by the Genome Project (Gerstein et al., 2007; Gingeras, 2007).

To summarize, the synthetic theory developed some aspects of Darwin's theory, neglecting others that are fundamental to understanding evolutionary phenomena. Consequently, different disciplines are built upon a reductionist view of evolutionary processes. This picture is changing thanks to the growth of new disciplinary fields and technological developments. Unfortunately, these changes are not reflected in most textbooks. To tell the history of a discipline is also a fundamental issue in the teaching of science since they show us not only the answers but also the key questions. After more than 160 years since the first edition of *On the Origin of Species*, questions about how variation appears and why it becomes fixed in the population are still pertinent. Unfortunately, a consensus has not been agreed upon in the scientific community. Perhaps the problem is the attempt to establish a limited number of processes to explain to account for such wide-ranging evolutive phenomena. I think the best way to conclude this work is to cite the late great Richard Lewontin (2002):

> What is already known about evolution shows us that there are no universal rules and even what appear to be regularities have many informative exceptions. Evolution is a loose and complex process, the result of a number of interacting, individually weak forces with many alternative outcomes, and at all times contingent on previous history. The best answer to any question about evolution is the lawyer's answer to any general question about the law: "It depends on the jurisdiction." That is why the program of evolutionary investigation never comes to an end – and, so often, never to a conclusion. (Lewontin, 2002, p. 17)

References

Bowler, P. J. (2003). *Evolution: The history of an idea*. University of California Press.

Darwin, C. R. (1872/2009). *On the origin of species* (6th ed.). Cambridge University Press.

Dobzhansky, T. (1951). *Genetics and the origin of species (no. 11)*. Columbia University Press.

Gerstein, M. B., Rozowsky, J. S., Zheng, D., Du, J., Korbel, J. O., Emanuelsson, O., Zhang, Z. D., Weissman, S., & Snyder, M. (2007). What is a gene, post-ENCODE? History and updated definition. *Genome Research, 17*, 669–681.

Gingeras, T. R. (2007). Origin of phenotypes: Genes and transcripts. *Genome Research, 17*, 682–690.

Gould, S. J. (2002). *The structure of evolutionary theory*. The Belknap Press of Harvard University Press Cambridge.

Gould, S. J., & Eldredge, N. (1993). Punctuated equilibrium comes of age. *Nature, 366*, 223–227.

Gould, S. J., & Lewontin, R. C. (1979). The spandrels of San Marco and the Panglossian paradigm: A critique of the Adaptationist programme. *Proceedings of the Royal Society of London, Series B, 205*(1161), 581–598.

Huxley, J. (1943). *The modern synthesis*. George Allen & Unwin.

Jablonka, E., & Lamb, M. J. (1995). *Epigenetic inheritance and evolution: The Lamarckian dimension*. Oxford University Press.

Kimura, M. (1983). *The neutral theory of molecular evolution*. Cambridge University Press.

Lamarck, J. B. (1809). *Philosophie zoologique*. Dentu.

Lewontin, R. C. (2002). Directions in evolutionary biology. *Annual Review of Genetics, 36*(1), 1–18.

Martins, L. A. C. P. (2002). Nos tempos de Lamarck: o que ele realmente pensava sobre evolução orgânica. Programa de Estudos Pós-Graduados em História da Ciência, Pontifícia Universidade Católica de São Paulo.

Martins, L. A. C. P. (2015). A herança de caracteres adquiridos nas teorias "evolutivas" do século XIX, duas possibilidades: Lamarck e Darwin. *Filosofia e História da Biologia, 10*(1), 67–84.

Mayr, E. (1982). *The growth of biological thought: Diversity, evolution, and inheritance*. Harvard University Press.

Mayr, E. (2001). *What evolution is*. Phoenix.

Odling-Smee, F. J. (1995). Niche construction, genetic evolution and cultural change. *Behavioural Processes, 35*(1–3), 195–205.

Odling-Smee, F. J., Laland, K., & Feldman, M. W. (1996). Niche construction. *American Naturalist, 147*, 641–648.

Odling-Smee, F. J., Laland, K., & Feldman, M. W. (2003). *Niche construction: The neglected process in evolution*. Princeton University Press.

Raff, R. A. (2000). Evo-devo: The evolution of a new discipline. *Nature Reviews. Genetics, 1*(1), 74–79.

Smocovitis, V. B. (1996). *Unifying biology. The evolutionary synthesis and evolutionary biology*. Princeton University Press.

Waddington, C. H. (1942). Canalization of development and the inheritance of acquired characters. *Nature, 3811*, 563–565.

Susana Gisela Lamas is currently a tenured professor at the Faculty of Natural Sciences at the National University of La Plata, Argentina. Her main field of research and teaching is the epistemological and methodological analysis of biological evolutionary theories. She has directed several research projects. Dr. Lamas is a member of several academic societies and has presented her work at numerous international congresses and publications.

Chapter 25
From the Modern Synthesis to the Other (Extended, Super, Postmodern...) Syntheses

Thierry Hoquet

Abstract Biology has always been in search of "syntheses." Darwin's *Origin of Species* (Darwin, On the origin of species by means of natural selection, or the preservation of favoured races in the struggle for life. John Murray, London, 1859) gave maybe the first attempt to reconcile and think together various fields of biology such as biogeography, embryology, systematics, and paleontology. In the 1930s and 1940s, the Modern Synthesis emerged, based on the change of frequency of genes in a population by means of natural selection. The Synthesis unified different biological disciplines (Genetics, Cytology, Embryology, Systematics, Botany, Paleontology, Morphology) and emerged in different countries (USA, Britain, Germany) (Mayr and Provine, The evolutionary synthesis: Perspectives on the unification of biology. Harvard University Press, Cambridge, MA/London, 1980). However, as the Synthesis was ripening into an orthodox view on the process of organic evolution, several have complained of its "narrowing" and even of its "hardening" (Gould, Dimensions of Darwinism: Themes & counterthemes in twentieth-century evolutionary theory. Cambridge University Press/Éditions de la Maison des Sciences de l'Homme, Cambridge, New Rochelle/Paris, 1983). Moreover, several of its features were repeatedly challenged: especially the gradual approach to evolution and the use of microevolution as a proxy for macroevolution have been under fire. Major challenges include the neutralist view of mutation (Kimura) and the question of Punctuated Equilibria (Gould and Eldredge). More recently, new experimental data has complemented our views of the development of organisms (evo-devo) and the inheritance of characters (epigenetics). Some claim that the Modern Synthesis Theory of evolution should be rejected or simply revised or extended in the face of new biological data.

T. Hoquet (✉)
University Paris-Nanterre, Paris, France
e-mail: t.hoquet@parisnanterre.fr

M. Elice Brzezinski Prestes (ed.), *Understanding Evolution in Darwin's "Origin"*, History, Philosophy and Theory of the Life Sciences 34,
https://doi.org/10.1007/978-3-031-40165-7_25

25.1 Introduction

The Modern evolution Synthesis (MS for short) is characterized by some general principles: evolution is a gradual process, primarily driven by natural selection, operating on random variation. The Synthesis also has an institutional aspect: as a "treaty" (Depew & Weber, 1988) between various disciplines (such as genetics, paleontology, or evolution), uniting the field of biological studies. A crucial tenet of the MS bears on the idea of compatibility between microevolution and macroevolution, in other terms: the belief that the results of population genetics do not conflict with the patterns exhibited by the paleontological fossil records, with the hierarchical structure of the Linnaean classification, or with the data on competition, symbiosis, and adaptation as revealed by the ecological sciences. It may be tempting to consider the MS as a classic example of a theoretical framework: what the historian and philosopher of science Thomas Kuhn (1962/1970) called a "paradigm" or a "disciplinary matrix," or what Imre Lakatos (1978) called a "research program."

But the epistemological challenge of defining what counts as a "paradigm" and on what criteria often leads to the different question of identifying "paradigm shifts" or "scientific revolutions." It also constitutes a personal challenge for biologists, imagining themselves as more than just "workers in the factory" or simply in search of "career niches" (Kitcher, 2000).

Richard Dawkins' *The Extended Phenotype* (1982) is one example of a call to "extend" the scope of research in biology, promoting the idea that the information included in the *genotype* also codes for the behavior and the transformation of the milieu – that it contains instructions for constructing structures beyond the organism. Similarly, biologists' – or, for that matter, philosophers' – discontent with the current state of biological affairs is commonplace. Calls for a "new evolutionary paradigm" can even become mainstream.

The Darwinian theory of evolution has always been a target for critics (e.g., Mivart, 1871). The Modern Synthesis – mostly an Anglo-American creation – was long to be accepted in some countries, such as France, where it was met with a long-standing resistance. Leading zoologists such as Grassé (1973, 1978), still following the footsteps of Bergson's *Creative Evolution* (1907), were convinced that evolution is not explained by the gradual accumulation of mutations (on the Synthesis in France, see, e.g., Loison, 2020). Microevolutionary changes in gene frequencies were deemed inadequate to account for big evolutionary events such as the transition from reptiles to mammals or fish to amphibians or the origins of tetrapods.

So, especially from a French vantage point, for which the right to teach the MS in biological curricula has been a difficult conquest, it is especially striking to realize that, as a matter of fact, challenges to the MS have been, regularly and by waves, hitting the headlines of biological journals even in the Anglo-American world of scientific publications. If not straightaway critical of the MS, a non-neglectable part of biological literature (both lay and professional) is tainted with a shared feeling of "unfinished" business – be it in the 1980s (Eldredge, 1985; Reid, 1985) or more recently (Laland, 2017).

The birth date of the MS may be arbitrarily fixed around 1946–1947 with the founding of the Society for the Study of Evolution (1946), the publication of the journal *Evolution*, and the Princeton conference of 1947 (see Mayr & Provine, 1980, Smocovitis, 1996). The critics started to pile up about 40 years later, with several attacks by Gould (see especially 1980, 1983, and Gould & Lewontin, 1979), climaxing with a call for an "evolutionary dis-synthesis" (Antonovics, 1987). Some biologists imagined themselves quite archetypically as Hercules at the "crossroads" (Depew & Weber, 1985), drawing new "paths into the future" of biology (Pollard, 1984) or calling for a "new evolutionary paradigm" (Ho & Saunders, 1984) (for a review of those books, see Burian, 1988).

Again, in the Summer of 2008, for the sixtieth birthday of the MS, 16 biologists and theoreticians of biology convened in Altenberg (Austria) at the initiative of Stony Brook evolutionary ecologist Massimo Pigliucci. Among them were several proponents of an "Extended evolutionary Synthesis" (ES) or even of a "Postmodern evolution," as some observers then claimed (Whitfield, 2008).

However, the idea of moving beyond the MS, a theory that has held sway since the middle of the twentieth century, might be perceived as intellectual relativism, suggesting that several evolutionary theories are competing, all equivalent in their pretentions. The risk is well perceived by promoters of the ES and is fully exploited by longtime radical critics of the Darwinian theory (e.g., Mazur, 2010).

As a matter of fact, the debate on the current state of the theory of evolution is complex, combining several fields of conceptual and empirical biology, but also the history of science and philosophy of biology. While some biologists have constantly leaned toward reforming the conceptual framework at hand, in view of recent developments in science and empirical discoveries, on the other hand, others have often resisted the urge to throw the baby with the bathwater. They have shown that the MS has always been broader and swifter than one might have thought. In the remnant of this chapter, I will review several issues at stake with the Modern Synthesis and its several challenges during its history (development, inheritance, variation, ecology. . .). One of the issues at stake is obviously to decide whether the MS is only "imperfect," "incomplete," or whether it is fully "incorrect"; whether it is merely an "unfinished" business or a dead end that has to be fully dismantled (compare for instance Eldredge, 1985 and Antonovics, 1987).

25.2 The MS as the Second Synthesis

The word "synthesis" primarily refers to a process of "unification" (Mayr & Provine, 1980). We speak of "Synthesis" in order to emphasize an attempt to integrate various empirical fields and provide a unified theoretical framework. But the "Modern" synthesis is not the first of its kind. As was sometimes emphasized, it is a "Darwinian" synthesis (Huxley, 1942) (or even "neo-Darwinian"), coming only second to Darwin's endeavors. The evolutionary theory started with Darwin's own attempt to provide his field of research with a synthesis: *On the Origin of Species* (1859). This book evinced the hidden bonds between the works of paleontologists, systematists, comparative anatomists, botanical geographers, embryologists, hybridists, breeders, etc. The *Origin* is often read as a monument and milestone of biological research, uniting the great domains of natural history. As noted by W.K. Brooks (1883), Darwin "has marshalled an overwhelming array of facts with masterly skill." Two facts should be noted here. First, the *Origin* was itself an evolving synthesis, and Darwin kept on extending it throughout his life: from 1859 to 1872, he issued no less than six different editions, varying greatly in size and contents (Peckham, 1959/2009; Hoquet, 2013). Second, Darwin hoped to overcome many objections that his predecessors had advanced against the evolutionary interpretation of life; but his own theory was met with sharp criticisms. The most prominent theoretical problem of this period bore on the theory of variation and inheritance. Darwin believed in pangenesis, a theory of heredity that was not particulate and included "Lamarckian" features such as "blending inheritance" and inheritance of acquired characters. In other terms, Darwin's theory of inheritance was not "Mendelian" (Bowler, 1983, 1988). After Darwin's death, A.R. Wallace provided his fellow Darwinians with another attempt at synthesis with his *Darwinism* (1889), triggering George Romanes's (1892–1897) response that Wallace and others (such as Weismann's germplasm theory) were, in fact, "Neo" or "Ultra" Darwinists as they entirely rejected the role of Lamarckian factors in evolution. Similarly, the MS is Darwinian in character. In Huxley's words:

> The Darwinism thus reborn is a modified Darwinism since it must operate with facts unknown to Darwin; but it is still Darwinism in the sense that it aims at giving a naturalistic interpretation of evolution and that its upholders, while constantly striving for more facts and more experimental results, do not, like some cautious spirits, reject the method of deduction. (Huxley, 1942, p. 27)

But this "Darwinian" character has remained contentious. Lancelot Hogben (1931), for one, strongly contended that "the difference between Darwin's hypothesis and the view which modern geneticists imply when they speak of natural selection might seem to be so great as to merit an alternative term for Natural selection." This clearly shows that, as biologists keep discovering new facts and devising new tools, science has a history, so its past often looks like its infancy. For instance, the advent of the MS was undoubtedly marked by new mathematical tools. Biological concepts and theories need to be constantly refined to accommodate new features of life that were previously unknown. Scientists do the best they can with the evidence available in their day. The only relevant question is whether the old theories still provide useful foundations or whether they crumble or disintegrate in the face of the new findings.

25.3 Key Aspects of the MS

The Modern Synthesis is itself a complex historical object and even "a moving target" (as Burian put it, 1988). The MS developed over several decades (Smocovitis, 1996; Bertoldi, 2020). Hence the idea that the MS took several shapes during its history. The term "Synthesis" is also confusing: is the MS the unification of the various branches of biology or the skillful reconciliation or even fusion of Darwin's natural selection and Mendel's theory of heredity (Gayon, 1992, 1998)? The MS was made possible by a first generation of biologists such as Ronald Fisher (1930), J.B.S. Haldane (1932), and Sewall Wright (1931), who built the tools permitting the mathematical description of the genetic makeup of populations and how it changes. But the Modern (evolutionary) Synthesis per se – a phrase initially introduced by Julian Huxley (1942) – describes the conceptual and theoretical framework that has defined evolutionary theory since the 1940s. The MS was primarily concerned with identifying the "creative" power of evolution: natural selection is not only a sieve or a mechanism that "prunes" the tree of life but rather a composer of symphonies. As Huxley puts it, "neither mutation nor selection alone is *creative* of anything important in evolution; but the two in conjunction are *creative*. [. . .] The two processes are complementary. Their interplay is as indispensable to evolution as is that of hydrogen and oxygen to water" (1942, pp. 28–29, emphasis added). The MS was initially understood as the work of several "bridge builders," who realized the integration of several strands or traditional subdisciplines of biological thought among which genetics (under the guise of the genetics of natural populations – Dobzhansky, 1937), systematics (Mayr, 1942), paleontology (Simpson, 1944), cytology (White, 1945), morphology (Rensch, 1947), or botany (Stebbins, 1950). This "bridging" of disciplines was due to "linking [them] to the core theory forged during the first phase" (Gould, 2002, p. 504). Each of these biologists made his own contribution to the theory: Dobzhansky worked with natural populations and contributed to the foundation of population genetics; Mayr reflected on the phenomenon of "allopatric speciation" and defined the "biological species concept," etc. Another significant contribution to this "Neo-Darwinian" Synthesis was David Lack's reinterpretation of Darwin's finches and the speciation process in the Galapagos islands (Lack, 1947), focusing on genetic variation, geographic isolation, and natural selection as the principal mechanisms of evolutionary change.

Another topic of special relevance here is the question of the "hardening of the Synthesis" (Gould, 1983). Some theoretical alternatives were present at the beginning of the MS (e.g., on the rhythm and patterns of macroevolution in the paleontological records) but were later dismissed without sufficient examination of scientific evidence. Some aspects of the evolutionary and biological phenomena were initially thought to be explained by the MS, then later were considered insufficiently explained by the theory. Hence the will to extend the MS might well be better understood as a way to bring the theory back to its initial openness or softness, or in a Gouldian fashion, to its "pre-hardening" state: to maintain it as an *empirical* rather than to consider it as an a priori doctrine. The role of G.G. Simpson in the MS is here especially discussed in the question: which patterns of speciation?

The fact that the MS has been constantly and gradually updating its thinking and its core tenets makes the MS a moving target. Hence it may be difficult to determine precisely what is and is not the "MS."

25.4 Key Aspects of the MS

In the received view, the MS was, as a theory, the combination of several ingredients.[1] First and foremost comes the focus on population biology. Quantification and modelling are emphasized in the MS, while the fossil record was practically useless to the theory. Another prominent feature of the MS is the extrapolation from micro to macroevolution; or macroevolution as a mere extrapolation of genetic events at the microscale. Dobzhansky famously explained this point with the following analogy:

> Experience seems to show, however, that there is no way toward an understanding of the mechanisms of macro-evolutionary changes, which require time on a geological scale, other than through a full comprehension of the micro-evolutionary processes observable within the span of a human lifetime and often controlled by man's will. For this reason, we are compelled at the present level of knowledge reluctantly to put a sign of equality between the mechanisms of macro- and micro-evolution, and, proceeding on this assumption, to push our investigations as far ahead as this working hypothesis will permit. (Dobzhansky, 1937, p. 12)

The MS is based on the general assumption that the genetic bases of interspecific differences can be approached from the genetic bases of within-species variation. What Dobzhansky describes as a methodological "assumption" and even an obligation ("we are compelled") may also be seen as an arbitrary "extrapolation" by which "macroevolution can be reduced to microevolution" (Gilbert et al., 1996, p. 358). But this methodological proxy brought immediate results as microevolutionary changes could be easily perceived in the field or the lab, while macroevolutionary changes could not be observed over a lifetime. This kind of statement allowed Kettlewell (1959) to say that his observations on "industrial melanism" in pepper moths were "Darwin's missing evidence." Insensibly, with such a definition, the MS moves away from fossil morphology and embryonic structures.

Other prominent theoretical aspects of the MS were: evolution by the accumulation of small genetic changes, guided by Darwin's natural selection (including sexual selection), leading to gradual change, patterns of classification based strictly on genealogical lineages, and biogeography explained by divergent adaptation. Douglas Futuyma summarized those central tenets of the MS in the following words:

> The major tenets of the evolutionary synthesis, then, were that populations contain genetic variation that arises by random (i.e., not adaptively directed) mutation and recombination; that populations evolve by changes in gene frequency brought about by random genetic drift, gene flow, and especially natural selection; that most adaptive genetic variants have individually slight phenotypic effects so that phenotypic changes are gradual (although some alleles with discrete effects may be advantageous, as in certain color polymorphisms); that diversification comes about by speciation, which normally entails the gradual evolution of reproductive isolation among populations; and that these processes, continued for sufficiently long, give rise to changes of such great magnitude as to warrant the designation of higher taxonomic levels (genera, families, and so forth). (Futuyma, 1986, p. 12)

[1] I will leave aside here the metaphysical aspects of the MS. For a critical outlook, see e.g., Antonovics (1987) and for an overview, Delisle (2009).

In this important textbook account,[2] one must notice the phrase: "populations evolve by changes in gene frequency," as redefining evolution as "changes in gene frequency" is anything but obvious. We should also note the emphasis on the randomness of mutations and the centrality of population genetics. The definition of "randomness" is very precise; it means: not oriented by the requirements of phenotypic adaptation.

However, we should always keep in mind that there were clear lines of disagreement between the various participants of the MS. Here, we may distinguish between the views of Dobzhansky, for whom "evolution is a change in the genetic composition of populations" (1937, p. 11) and those supported by Mayr (1942), for whom evolutionary biology deals with highly complex realities (not genes, not changes in allele frequencies). Hence, we would have to distinguish between "populations" that are reproducing through the ecology of the organisms and "population genetics," barely concerned with changes in allele frequency; and between two different concepts of natural selection: as an ecological process leading to adaptation and speciation; and as a quantifiable bias in the transmission of alleles between generations (Huneman, 2014; Welch, 2017).

Also, against the received view that the MS emerged from the mathematical school of Fisher, Haldane and Wright (e.g., Crow & Kimura, 1956), Mayr bluntly asked what their contribution to evolutionary theory was. He also opposed a physiological view of the biological population to the abstract approach he called "beanbag genetics" (conceiving of the population as a bag of colored genes) – triggering a violent response by Haldane (1964). Hence what is often taken as a "major tenet" or a characteristic of the MS (such as "beanbag genetics") was often an object of fierce controversy among the main proponents of the MS.

25.5 Four Challenges to the MS

Throughout its history, the MS was met with several challenges that were often considered, at first, fatal blows. The most famous are the following ones: neutralism, punctuated equilibria, evo-devo, eco-evo-devo.

25.5.1 Neutralism

The ultimate source of genetic variability in a population is mutation. Mutations are defined as substitutions, gains and losses in DNA base arrangements. Besides, in the MS, evolution is defined as a "change in the genetic composition of populations"

[2] Textbook exposés of the MS were scarce. Antonovics claimed in 1987: "Even now it is hard to find a good text on evolutionary biology (I think of Futuyma 1979 as an exception)."

(Dobzhansky, 1937). Hence the importance of mutations in the mechanism of evolution. But in 1968, the Japanese geneticist Kimura Motoo, a longtime specialist of genetic changes in natural populations, showed that neutral (or nearly neutral) mutations were being produced in each generation at a much higher rate than was thought before (Kimura, 1968a, b). This newly acknowledged result regarding the genetic structure of biological populations led to a shift of emphasis from natural selection to random genetic drift in finite populations. Kimura was clearly targeting the MS, especially in Ernst Mayr's work: Mayr had acknowledged the role of genetic drift in the case of the founder principle but denied the importance of genetic drift due to finite population numbers (Kimura, 1968a, p. 626). This is the base of the so-called "neutralist challenge" to the MS, which was promoted in the scientific literature as "non-Darwinian evolution" (King & Jukes, 1969). The "neutralist" challenge to the MS suggests that speciation events and gene evolution may not be equated (Kimura, 1983).[3]

25.5.2 Punctuated Equilibria

Punctuated equilibrium was another challenge to the Modern Synthesis brought about by two paleontologists, Eldredge and Gould (1972). In the *Origin of species* (1859), Darwin famously emphasized the "imperfection of the fossil record" (chapter IX) while scrutinizing the "geological succession of organic beings" (chapter X). The incompleteness of the fossil record and possible gaps in it were a recurrent theme in paleontology, sometimes used to support the claim that the spontaneous generation of forms was a fact. Eldredge and Gould targeted the MS as a theoretical framework ending up "color [ing] perception to such a degree that new notions seldom arise from facts collected under the influence of old pictures of the world" (Eldredge & Gould, 1972, p. 83). The authors targeted "phyletic gradualism": the idea that an unbroken fossil series linked two paleontological species by insensible gradations (apart from the occurrence of "gaps" or "imperfections" in the archive). Instead of gradualism, Eldredge and Gould promoted "punctuated equilibria": the idea of homeostatic equilibria only occasionally disturbed by rapid episodes of speciation. Hence the pattern observed in the evolution is that of a longtime stasis punctuated by sudden shifts in phenotype. Punctuated equilibria have also nourished anti-Darwinian stances on the ground that the MS implied that macroevolutionary processes (speciation and morphological diversification) were gradual. However, Stebbins and Ayala (1981) have shown that microevolutionary principles were compatible with both gradualism and punctualism. Claims have been made that Gould et al. were attacking a straw man, especially a definition of "gradualism" that did not apply to the MS (Charlesworth et al., 1982; Carroll, 2004 and, for a different perspective, Stidd, 1985).

[3]For a presentation of neutralism and Kimura's ambiguities, see for instance Tanghe et al. (2018, note 15, p. 9 of 21).

25.5.3 Evo-Devo

However, the main question that comes to the fore in discussions on extending the MS is certainly the return of development (Gilbert et al., 1996). Embryonic development raises the problem of morphology and questions about the concepts of *form*, *shape* or *formation* (Raff, 1996). Development was a central feature of early evolutionary theories, such as that of Haeckel, for whom, most notoriously, phylogeny recapitulates ontogeny (Gould, 1977; Barnes, 2014). Hence morphogenesis may be said to be central in what Gilbert et al. call "some 'Unmodern Synthesis [. . .] the notion that evolution was caused by changes in development" (1996, p. 358) – a "synthesis" that was later replaced by the MS which treated morphology as a black box incidental to core evolutionary processes. Developmental biology was left out of the MS also because of its alleged "failure [. . .] to assimilate Darwinism" (Ghiselin, 1980). A concern for the evolution of morphological diversity only re-emerged later within the so-called "evo-devo" approach (Carroll, 2008). This can be rephrased as: the MS does not capture multiple levels of biological organization (Love, 2006).

One of the reasons for leaving development outside of the MS is Mayr's distinction between proximate and ultimate causes (1961, 1993). For Mayr, the genetic bases of development count as proximate causes, whereas the evolutionary processes are deemed as ultimate causes.

But form development is intrinsically part of evolution as evolution is "transformation," understood literally as a *change of form*. The discovery of the homeobox (McGinnis et al., 1984; Scott & Weiner, 1984), a network of genes involved in the regulation of patterns of anatomical development (morphogenesis), was especially important, together with the fact that the coding sequence was strikingly well-conserved in many other animals, including vertebrates (e.g., Duboule & Dollé, 1989). In the 1990s, new research in the field of evo-devo contributed to putting to the fore the question of form and the related issue of development. However, the revolutionary impact of evo-devo is subject to a great variety of appreciations (see a review in Carroll, 2008). Especially, stating that "form evolves largely by altering the expression of functionally conserved proteins" (Carroll, 2008, p. 34) may be the foundation of a "genetic theory of a morphological evolution." Still, it does not by any means equate with a genetic theory of the origin of adaptation or that molecular changes contribute substantially to organismal adaptation (contra Hoekstra & Coyne, 2007).

25.5.4 Evo-Eco

A new dimension was later added to the development puzzle: the ecological dimension. Traditionally, the environment was thought to be an external passive element, posing problems or riddles that organisms need to solve. However, recent research such as "niche construction" challenged this basic organism/environment divide.

Niche-constructionists and proponents of an Evo-Eco Synthesis (EES, standing also for "extended evolutionary synthesis") claimed that the MS was largely gene-centered and that it basically ignored the ecological dimension. Against that, the EES proponents chose to focus on the organism. For instance, Laland et al. (2014, p. 161) argue that "important drivers of evolution, ones that cannot be reduced to genes, must be woven into the very fabric of evolutionary theory" (see also Sultan 2007 and Scott-Phillips et al., 2014). But the claims that "niche construction" theory leads to an EES have been contested (see especially Gupta et al., 2017a, and the replies ensuing: Feldman et al., 2017 and Gupta et al., 2017b).

25.6 Conceptual Challenges to the MS

Huneman (2014) suggested that most issues bearing on the MS could be reconstructed along two orthogonal lines: a debate on whether natural selection is the main agent operating in evolutionary processes (which would imply a trend toward optimization); and a debate on whether evolution is mostly genetic or on the contrary whether biology should acknowledge the causal role of organisms. But many more conceptual aspects of the MS are discussed and debated.

Critics of the MS (and correlate calls towards an EES or more simply "Extended Synthesis" — ES— such as Pigliucci, 2009) are generally motivated by the general feeling that biology has changed. New data piling up in biology seems to deeply and radically outdate old-fashioned theories. Several topics are commonly called in what may sometimes look like a "laundry-list" (Welch, 2017): modularity, evolvability, epigenetic inheritance, complexity theory, emergence, robustness, developmental (or phenotypic) plasticity, genetic accommodation, etc.

Various conceptual and theoretical aspects of the MS have also been challenged: their concept of the gene and their notion of the organism/environment relationship. Another issue at stake is the question of inheritance. For a long time, the MS was associated with a complete rejection of Lamarckism and the inheritance of acquired characters. However, more recently, many have thought that the question of inheritance might not be definitely settled and that other mechanisms (in addition to the "Mendelian" or genetic ones) should be taken into account.

The way the MS conceptualized the relation between genotypes and phenotypes was also a matter of debate (Pigliucci, 2010).

The MS was targeted for being population genetics, mainly dealing with the changes in the ratios of alleles in a population rather than with the actual behavior of animals or the development of their form. The MS was described as a highly *genetic* and even *genocentric* theoretical framework. The common metaphors of a "genetic blueprint" or a "genetic program" have been called "instructive" and criticized for being insufficiently "Darwinian" (Kupiec, 2009). More recently, the advent of genomics and adjunct disciplines, such as metabolomics, has considerably changed our understanding of the relationship between genotype and phenotype. In the genocentric view of the MS, it may be crucial to decode the genes as they are a key to the production of the organism, its traits, and its biology. But notions like epistasis have long ago challenged the idea that one gene determines or codes for one trait. Epistasis implies the idea of multi-gene causality but in a limited fashion. In the postgenomic view, epistasis is no longer the exception but the rule. Emphasis is on organismic complexity, and the action of multiple genes causally explains the organism. The focus is no longer on the effects of individual genes but the network of genes and gene-products (Cork & Purugganan, 2004).

But natural selection is not only viewed as dynamics of alleles: it can also be perceived in the style of economics as an optimizing process (Huneman, 2014). The idea of an EES is also linked to a persistent interrogation of adaptation (and "adaptationism," as Gould & Lewontin, 1979 put it). EES aims to reform the mathematical theory that supports the MS. For instance, Sewall Wright (1931) introduced the idea of the "adaptive landscape" in population genetics theory. An "adaptive landscape" has peaks and valleys. Wright's model generated research on how a population may effect a "peak shift" while other populations are stuck in a suboptimal area (see, among others, the critics by Gavrilets, 2004, and comments by Pigliucci, 2008).

The "Tree of Life" concept is that of a branching graph connecting species in a way that reflects their common ancestry. This view has also been contested based on empirical data mainly coming from the bacterial world. The "Tree of Life" view of phylogeny has recently been challenged by new data on the ubiquity of lateral or horizontal gene transfer in bacteria (e.g., Woese, 2004). Other phenomena are symbiogenesis and differential gene lineage sorting (Rose & Oakley, 2007). Koonin (2011) has been especially vocal in claiming that the evolution of life cannot be presented as a tree anymore and that it constitutes a serious upheaval to the MS.

25.7 "Extended," "Postmodern" or "Super" Syntheses: Does the MS Need "Softening"?

Since Antonovics (1987) called for a radical evolutionary "Dis-Synthesis," a dismantling of the MS and a return to a pre-Synthesis state of science, the MS has been constantly under fire.

Koonin (2009) called for a "Postmodern Synthesis." The term "postmodern" may be equated with the acceptance of explanatory pluralism (Doolittle 2007), but it is also associated with strong connotations and leads to hermeneutical problems (Rose & Oakley, 2007, p. 13 of 17; Whitfield, 2008). Other words such as "Supersynthesis" (Ricqlès & Padian, 2009) were used in order to "soften" the MS (I take here "softening" to be the opposite of the "hardening" earlier denounced by Gould, 1983). By "softening," I mean to include other elements of the Darwinian legacy: laws of the correlation of growth, the question of ontogeny, geological and fossil records incomplete but informative; classification based strictly on genealogy; morphology explained by ontogeny and selection. But new data may have two different sets of effects: (1) to prove the mere insufficiency of the MS; leading, for instance, to Pigliucci's (2007) claims that "one would not expect the original synthesis to be able to address directly the wealth of information emerging from genomics [...] and the other new '-omics' sciences"; or (2) to refute the basic tenets of the MS, calling for "the overcoming of gradualism, externalism, and gene centrism" in a move perceived as the "general hallmarks of the Extended Synthesis" (Pigliucci & Muller, 2010, p. 14). As Pigliucci and Muller (2010) explained, there are various ways to expand or extend the MS. Some maintain that the MS and its central tenets are still valid, while others lean toward major revisions or want to relax some of its assumptions; others want to tear the whole building down and start a new theory from scratch. The question is whether "extended" versions merely introduce "augmentations" of the basic MS structure or whether they demand (or lead to) a new Synthesis set on completely new bases. Are new elements already included in the MS or merely consistent with the general MS structure? Or do they require a broad reevaluation of the standard theory?

How many biological syntheses are there, or have there been? How many synthetical events or episodes or "re-integration" have taken place in the history of biology? Darwin's *Origin* and the Modern Synthesis were undoubtedly important events of "synthesis." But Doolittle (2007) suggests that several theoretical events happened since the 1940s: a "Molecular Synthesis" (pre-genomic molecular biology); a "Genomic Synthesis" (also called "systems biology"), dealing with complex networks of biological regulation, and interested in organisms as dynamical systems; a "Metagenomic Revolution," deconstructing the belief that organisms must be clustered in species.

25.8 Conclusion

Debates on whether an ES is currently replacing the pattern of a MS might be usefully used with students to see how science and philosophy interact fruitfully (Pigliucci & Finkelman, 2014). While it is clear that most biologists conduct their research without examining their philosophical implications, it can also be argued that scientific work is heavily loaded with epistemological questions. In the case of the MS, some important issues are: what kind of mechanism or processes can be accurately called "Darwinian"? What counts as a "paradigm shift" or a "scientific revolution"? On the other hand, many claims that a new "Synthesis" is afoot, often hiding hopes for a more "holistic" biology.

The idea of a theoretical move from MS to ES challenges our view of the history of science as a continuously developing thread. The question is to understand how new discoveries can lead to conceptual breakthroughs. As a matter of fact, science is perpetually changing: think of the emergence of pattern cladism, Hennigian phylogenetics, or Croizatian vicariance biogeography. Another issue is whether or not the developments of evolutionary biology can be viewed as a refutation of Darwin or whether they stay on the same trail although contradicting some of his initial intuitions (such as gradualism) (Koonin, 2009).

Recently, Welch (2017) wondered why evolutionary biology seems to be plagued by a "steady stream of claims that the field needs urgent reform." Two hypotheses seem to prevail: either evolutionary theory is seriously deficient and incapable of incorporating new ideas; or rather, the field seems prone to attracting discontent. Welch chose the latter explanation.

There are also constant attempts to distinguish old-time critics and new ones (the latter being sounder and stronger than the former).[4] But the constancy of critics that show up is also a striking feature of evolutionary debates, such as debates on the structure of the "eye" (Hoquet, 2020). It seems important here to note that discussing the MS and its replacement by other kinds of extended syntheses, calling for the advent of a "new biology" (Woese, 2004; Rose & Oakley, 2007), is not a biological issue. It involves metanarratives that cannot be judged to be valid or invalid (Doolittle, 2007). Another metabiological issue would be to ask whether, in alternance with movements of integration, there also occurs opposite events of splintering or fractioning – events that "disintegrate" biology.

As a historian and a philosopher of biology, I may be prone to "neophilia," the temptation to identify groundbreaking events and paradigm shifts. Nevertheless, I believe that science needs theories to organize facts and the power of general explanations to interpret them and not be confined to mere piles of case studies. Moreover, I fully agree with Tanghe et al. (2018, p. 16 of 21) that "the MS has not been, and should not be, replaced with a heterogeneous and amorphous explanatory toolkit, nor with an EES, but rather with an improved version of itself." I do not think that biology would gain by replacing a state of imperfect (modern) synthesis with a state of intellectual (postmodern) disarray.

References

Antonovics, J. (1987). The evolutionary dys-synthesis. *American Naturalist, 129*, 321–331.

Barnes, M. E. (2014). Ernst Haeckel's Biogenetic Law (1866). *Embryo Project Encyclopedia (2014-05-03).* http://embryo.asu.edu/handle/10776/7825

Bergson, H. (1907). *L'Évolution créatrice*. Alcan.

Bertoldi, N. (2020). *La formalisation de la théorie de l'évolution: Une approche historique et philosophique*. Université Paris I Panthéon-Sorbonne: PhD thesis.

Bowler, P. J. (1983). *The eclipse of Darwinism: Anti-Darwinian evolution theories in the decades around 1900*. John Hopkins Press.

[4] See for instance, Pigliucci, 2010, p. 560: "These findings of course do not mark a sharp boundary between micro- and macro-evolution. We are not talking about Goldschmidt (1940)-type 'hopeful monsters'. Nonetheless, criticism of the simplistic view of macro-evolution that has characterized the Modern Synthesis, and that did inspire Goldschmidt (1940) to write *The Material Basis of Evolutionary Change* is finally beginning to be vindicated by new knowledge about those very material bases".

Bowler, P. J. (1988). *The non-Darwinian revolution: Reinterpreting a historical myth*. John Hopkins University Press.
Brooks, W. K. (1883). *The law of heredity: A study of the cause of variation, and the origin of living organisms*. John Murphy.
Burian, R. (1988). Challenges to the evolutionary synthesis. *Evolutionary Biology, 23*, 247–269.
Carroll, J. (2004). Modern Darwinism and the pseudo-revolutions of Stephen Jay Gould. In J. Carroll (Ed.), *Literary Darwinism: Evolution, human nature, and literature* (pp. 227–246). Routledge.
Carroll, S. B. (2008). Evo-devo and an expanding evolutionary synthesis: A genetic theory of morphological evolution. *Cell, 134*, 25–36.
Charlesworth, B., Lande, R., & Slatkin, M. (1982). A Neo-Darwinian commentary on macroevolution. *Evolution, 36*, 474–498.
Cork, J. M., & Purugganan, M. D. (2004). The evolution of molecular genetic pathways and networks. *BioEssays, 26*, 479–484.
Crow, J., & Kimura, M. (1956). *Some genetic problems in natural populations*. In Proceedings of the third Berkeley symposium on mathematical statistics and probability, Volume 4: Contributions to biology and problems of health (pp. 1–23). University of California Press.
Darwin, C. (1859). *On the origin of species by means of natural selection, or the preservation of favoured races in the struggle for life*. John Murray.
Dawkins, R. (1982). *The extended phenotype: The gene as the unit of selection*. W. H. Freeman.
De Ricqlès, A., & Padian, K. (2009). Is there a role for the Modern Synthesis in the Super-Synthesis? (Quelques apports à la théorie de l'Évolution, de la "Synthèse orthodoxe" à la "Super synthèse évo-dévo" 1970-2009 : un point de vue). *Comptes rendus-Palevol, 8*(2–3), 341–364.
Delisle, R. G. (2009). *Les philosophies du néo-darwinisme: Conceptions divergentes sur l'homme et sur le sens de l'évolution*. Presses universitaires de France.
Depew, D. J., & Weber, B. H. (Eds.). (1985). *Evolution at a crossroads: The new biology and the new philosophy of science*. MIT Press.
Depew, D. J., & Weber, B. H. (1988). Consequences of nonequilibrium thermodynamics for the Darwinian tradition. In B. H. Weber, D. J. Depew, & J. D. Smith (Eds.), *Entropy, information, and evolution: New perspectives on physical and biological evolution*. MIT Press.
Dobzhansky, T. (1937). *Genetics and the origin of species*. Columbia University Press.
Doolittle, F. W. (2007). Review appended to Rose and Oakley (2007).
Duboule, D., & Dollé, P. (1989). The structural and functional organization of the murine HOX gene family resembles that of drosophila homeotic genes. *EMBO Journal, 8*, 1497–1505.
Eldredge, N. (1985). *Unfinished synthesis: Biological hierarchies and modern evolutionary theory*. Oxford University Press.
Eldredge, N., & Gould, S. J. (1972). Punctuated equilibria: An alternative to phyletic gradualism. In T. J. M. Schopf (Ed.), *Models in paleobiology* (pp. 82–115). Freeman, Cooper.
Feldman, M. W., Odling-Smee, J., & Laland, K. N. (2017). Why Gupta et al.'s critique of niche construction is off target. *Journal of Genetics, 96*(3), 505–508.
Fisher, R. A. (1930). *The genetical theory of natural selection*. Oxford University Press.
Futuyma, D. J. (1986). *Evolutionary biology*. Sinauer.
Gavrilets, S. (2004). *Fitness landscapes and the origin of species*. Princeton University Press.
Gayon, J. (1992). *Darwin et l'après-Darwin: Une histoire de l'hypothèse de sélection naturelle*. Kimé.
Gayon, J. (1998). *Darwinism's struggle for survival: Heredity and the hypothesis of natural selection*. University Press.
Ghiselin, M. T. (1980). The failure of morphology to assimilate Darwinism. In E. Mayr & W. B. Provine (Eds.), *The evolutionary synthesis: Perspectives on the unification of biology* (pp. 180–193). Harvard University Press.
Gilbert, S. F., Opitz, J. M., & Raff, R. A. (1996). Resynthesizing evolutionary and developmental biology. *Developmental Biology, 173*, 357–372.

Goldschmidt, R. (1940). *The material basis of evolution*. Yale University Press/H. Milford/Oxford University Press.

Gould, S. J. (1977). *Ontogeny and phylogeny*. Harvard University Press.

Gould, S. J. (1980). Is a new and general theory of evolution emerging? *Paleobiology, 6*, 119–130.

Gould, S. J. (1983). The hardening of the modern synthesis. In M. Grene (Ed.), *Dimensions of Darwinism: Themes & counterthemes in twentieth-century evolutionary theory* (pp. 71–93). Cambridge University Press/Éditions de la Maison des Sciences de l'Homme.

Gould, S. J. (2002). *The structure of evolutionary theory*. Belknap Press of Harvard University Press.

Gould, S. J., & Lewontin, R. C. (1979). The spandrels of san Marco and the Panglossian paradigm: A critique of the adaptationist programme. *Proceedings of the Royal Society of London, Series B, 205*(1161), 581–598.

Grassé, P.-P. (1973). *L'évolution du vivant: Matériaux pour une nouvelle théorie transformiste*. A. Michel.

Grassé, P.-P. (1978). *Biologie moléculaire, mutagenèse et évolution*. Masson.

Gupta, M., Prasad, N. G., Dey, S., Amitabh, J., & Vidya, T. N. C. (2017a). Niche construction in evolutionary theory: The construction of an academic niche? *Journal of Genetics, 96*(3), 491–504.

Gupta, M., Prasad, N. G., Dey, S., Amitabh, J., Vidya, T. N., Feldman, C., et al. (2017b). Do protest too much, we think. *Journal of Genetics, 96*(3), 509–511.

Haldane, J. B. S. (1932). *The causes of evolution*. Longmans Green.

Haldane, J. B. S. (1964). A defense of beanbag genetics. *Perspectives in Biology and Medicine, 7*, 343–359.

Ho, M.-W., & Saunders, P. T. (Eds.). (1984). *Beyond neo-Darwinism: An introduction to the new evolutionary paradigm*. Academic Press.

Hoekstra, H. E., & Coyne, J. A. (2007). The locus of evolution: Evo devo and the genetics of adaptation. *Evolution, 61*, 995–1016.

Hogben, L. T. (1931). *Genetic principles in medicine and social science*. Williams & Norgate.

Hoquet, T. (2013). The evolution of Darwin's *origin* (1859-1872). In M. Ruse (Ed.), *The Cambridge encyclopedia of Darwin and evolutionary thought* (pp. 158–164). Cambridge University Press.

Hoquet, T. (2020). L'œil de Méduse. *Critique, 874*, 248–261.

Huneman, P. (2014). A pluralist framework to address challenges to the modern synthesis in evolutionary theory. *Biological Theory, 9*(2), 163–177.

Huxley, J. (1942). *Evolution: The modern synthesis*. Princeton University Press.

Kettlewell, H. B. D. (1959). Darwin's missing evidence. *Scientific American, 200*(3), 48–53.

Kimura, M. (1968a). Evolutionary rate at the molecular level. *Nature, 217*, 624–626.

Kimura, M. (1968b). Genetic variability maintained in a finite population due to mutational production of neutral and nearly neutral isoalleles. *Genetics Research, 11*, 247–269.

Kimura, M. (1983). *The neutral theory of molecular evolution*. Cambridge University Press.

King, J. L., & Jukes, T. H. (1969). Non-Darwinian evolution. *Science, 164*, 788–798.

Kitcher, P. (2000). Patterns of scientific controversies. In P. Machamer, M. Pera, & A. Baltas (Eds.), *Scientific controversies: Philosophical and historical perspectives*. Oxford University Press.

Koonin, E. V. (2009). Towards a postmodern synthesis of evolutionary biology. *Cell Cycle, 8*(6), 799–800.

Koonin, E. V. (2011). *The logic of chance: The nature and origin of biological evolution*. FT Press.

Kuhn, T. S. (1962/1970). *The structure of scientific revolutions: New augmented edition*. Chicago University Press.

Kupiec, J.-J. (2009). *The origin of individuals*. World Scientific.

Lack, D. (1947). *Darwin's finches*. Cambridge University Press.

Lakatos, I. (1978). The methodology of scientific research programmes. In J. Worrall & G. Currie (Eds.), *Philosophical papers. Vol. 1*. Cambridge University Press.

Laland, K. N. (2017). *Darwin's unfinished symphony: How culture made the human mind*. Princeton University Press.

Laland, K. N., Uller, T., Feldman, M., Sterelny, K., Müller, G. B., Moczek, A., Jablonka, E., Odling-Smee, J., et al. (2014). Does evolutionary theory need a rethink? *Nature, 514*, 161–164.

Loison, L. (2020). Enseigner la théorie synthétique de l'évolution au Collège de France: Les cours d'Ernst Mayr dans la chaire de François Jacob (1978). In A. Compagnon & C. Surprenant (Eds.), *Darwin au Collège de France, XIXe-XXIe siècles* (pp. 159–171). Paris.

Love, A. C. (2006). Evolutionary morphology and Evo-devo: Hierarchy and novelty. *Theory in Biosciences, 124*, 317–333.

Mayr, E. (1942). *Systematics and the origin of species: From the viewpoint of a zoologist*. Columbia University Press.

Mayr, E. (1961). Cause and effect in biology. *Science, 131*, 1501–1506.

Mayr, E. (1993). Proximate and ultimate causation. *Biology and Philosophy, 8*, 95–98.

Mayr, E., & Provine, W. B. (Eds.). (1980). *The evolutionary synthesis: Perspectives on the unification of biology*. Harvard University Press.

Mazur, S. (2010). *The Altenberg 16: An exposé of the evolution industry*. North Atlantic Books.

McGinnis, W., Garber, R. L., Wirz, J., Kuroiwa, A., & Gehring, W. J. (1984). A homologous protein-coding sequence in drosophila homeotic genes and its conservation in other metazoans. *Cell, 37*, 403–408.

Peckham, M. (Ed.). (1959/2009). *The origin of species: A variorum text – Edited by Morse Peckham*. University of Pennsylvania Press.

Pigliucci, M. (2007). Do we need an extended evolutionary synthesis? *Evolution, 61*, 2743–2749.

Pigliucci, M. (2008). Sewall Wright's adaptive landscapes: 1932 vs. 1988. *Biology and Philosophy, 23*, 591–603.

Pigliucci, M. (2009). An extended synthesis for evolutionary biology: The year in evolutionary biology 2009. *Annals of the New York Academy of Sciences, 1168*, 218–228.

Pigliucci, M. (2010). Genotype-phenotype mapping and the end of the "genes as blueprint" metaphor. *Philosophical Transactions of the Royal Society of London. Series B, Biological Sciences, 365*(1540), 557–566.

Pigliucci, M., & Finkelman, L. (2014). The extended (evolutionary) synthesis debate: Where science meets philosophy. *Bioscience, 64*(6), 511–516.

Pigliucci, M., & Muller, G. B. (2010). Elements of an extended evolutionary synthesis. In M. Pigliucci & G. B. Muller (Eds.), *Evolution: The extended synthesis*. MIT Press.

Pollard, J. W. (Ed.). (1984). *Evolutionary theory: Paths into the future*. Wiley.

Raff, R. A. (1996). *Shape of life: Genes, development, and the evolution of animal form*. University of Chicago Press.

Reid, R. G. B. (1985). *Evolutionary theory: The unfinished synthesis*. Cornell University Press.

Rensch, B. (1947). *Neuere Probleme der Abstammungslehre: Die Transspezifische Evolution*. Ferdinand Enke.

Romanes, G. J. (1892–1897). *Darwin and after Darwin, an exposition of the Darwinian theory and a discussion of post-Darwinian questions* (3 vols). Longmans.

Rose, M. R., & Oakley, T. H. (2007). The new biology: Beyond the modern synthesis. *Biology Direct, 2*, 30.

Scott, M. P., & Weiner, A. J. (1984). Structural relationships among genes that control development: Sequence homology between the Antennapedia, Ultrabithorax, and fushi tarazu loci of drosophila. *PNAS, 81*, 4115–4119.

Scott-Phillips, T. C., Laland, K. N., Shuker, D. M., Dickins, T. E., & West, S. A. (2014). The niche construction perspective: A critical appraisal. *Evolution, 68*(5), 1231–1243.

Simpson, G. E. G. (1944). *Tempo and mode in evolution*. Columbia University Press.

Smocovitis, V. B. (1996). *Unifying biology: The evolutionary synthesis and evolutionary biology*. Princeton University Press.

St. Mivart, G. J. (1871). *On the genesis of species*. Macmillan.

Stebbins, G. L. (1950). *Variations and evolution in plants*. Columbia University Press.

Stebbins, G. L., & Ayala, F. J. (1981). Is a new evolutionary synthesis necessary? *Science, 213*, 967–971.
Stidd, B. M. (1985). Are punctuationists wrong about the modern synthesis? *Philosophy of Science, 52*, 98–109.
Sultan, S. E. (2007). Development in context: The timely emergence of eco-devo. *TREE, 22*, 575–582.
Tanghe, K. B., De Tiège, A., Pauwels, L., Blancke, S., & Braeckman, J. (2018). What's wrong with the modern evolutionary synthesis? A critical reply to Welch (2017). *Biology and Philosophy, 33*, 23.
Wallace, A. R. (1889). *Darwinism: An exposition of the theory of natural selection with some of its applications*. Macmillan.
Welch, J. J. (2017). What's wrong with evolutionary biology? *Biology and Philosophy, 32*(2), 263–279.
White, M. J. D. (1945). *Animal cytology and evolution*. Cambridge University Press.
Whitfield, J. (2008). Postmodern evolution? *Nature, 455*, 281–284.
Woese, C. (2004). A new biology for a new century. *Microbiology and Molecular Biology Reviews*, 173–186.
Wright, S. (1931). Evolution in Mendelian populations. *Genetics, 16*, 97–159.

Thierry Hoquet is a Professor of Philosophy of Science in the Department of Philosophy, at Paris Nanterre University, France. His publications focus on different topics such as: (1) eighteenth-century and nineteenth-century natural history (Buffon, Linnaeus, Darwin and the history of Darwinism); (2) the concept of sex in biology, especially with regard to sexual selection, and the sex/gender dualism; (3) the role of words and metaphors in science. His publications in English include *Revisiting the* Origin of Species*: The Other Darwins* (Oxford, Routledge, 2018), and an edited volume: *Current Perspectives on Sexual Selection: What's left after Darwin?* (Dordrecht, Springer, 2015). In journals, see also "Beyond Coy Females and Eager Males: The Evolution of Darwin's Sexual selection" in *Challenging Popular Myths of Sex, Gender and Biology*, 2013; "Bateman's Data: Inconsistent with 'Bateman's Principles'", *Ecology and Evolution*, vol. 10 (2020) (with W.C. Bridges & P.A. Gowaty); "Pluralizing Darwin: Making Counter-Factual History of Science Significant," *Journal for General Philosophy of Science* (2021).

Correction to: Origin's Chapter VIII: Darwin for and Against Hybridism

P. Lorenzano

Correction to:
Chapter 19 in: M. Elice Brzezinski Prestes (ed.), *Understanding Evolution in Darwin's "Origin"*, History, Philosophy and Theory of the Life Sciences 34, https://doi.org/10.1007/978-3-031-40165-7_19

Due to an unintentional oversight, the printed version of this chapter did not include some of the bibliographical references used in its preparation, as well as some clarifications that should have appeared on chapter along with the final paragraph. The same has been included in this corrected version.

Missing bibliography which should be listed in the References (pp. 317–320):

Beatty, J. (1985). Speaking of species: Darwin's strategy. In D. Kohn (Ed.), *The Darwinian heritage* (pp. 265–282). Princeton University Press.

Darwin, C. (1872). On the origin of species by means of natural selection, or the preservation of favoured races in the struggle for life (6th ed.). John Murray.

Darwin, C. (1876). The effects of cross and self fertilisation in the vegetable kingdom. John Murray.

Ginnobili, S. (2013). On the origin of «that thing you call "species"». In P. Lorenzano, L. A.-C. P. Martins, & A. C. Regner (Eds.), *History and philosophy of life sciences in the South Cone* (pp. 173–184). College Publications.

Olby, R. (2009). Variation and inheritance. In M. Ruse & R. Richards (Eds.), *The Cambridge companion to the "Origin of species"* (pp. 30–46). Cambridge University Press.

The updated version of this chapter can be found at
https://doi.org/10.1007/978-3-031-40165-7_19

M. Elice Brzezinski Prestes (ed.), *Understanding Evolution in Darwin's "Origin"*,
History, Philosophy and Theory of the Life Sciences 34,
https://doi.org/10.1007/978-3-031-40165-7_26

Reznick, D. N. (2009). *The Origin then and now: An interpretive guide to the Origin of species*. Princeton University Press.

Secord, J. A. (1985). Darwin and the breeders: A social history. In D. Kohn (Ed.), *The Darwinian heritage* (pp. 519–542). Princeton University Press.

Smocovitis, V. B. (2009). Darwin's botany in the *Origin of species*. In M. Ruse & R. Richards (Eds.), *The Cambridge companion to the "Origin of species"* (pp. 216–236). Cambridge University Press.

On p. 309:

On the general place of Chapter VIII in Darwin's Origin, reference should be made to the work of Olby (2009), Reznick (2009), and Smocovitis (2009).

> Chapter VIII is thus part of the set of Chapters VI, VII, VIII – in which Darwin gives answers to the possible difficulties and objections that could be raised to his theory (Olby, 2009; Reznick, 2009; Smocovitis, 2009).

On p. 309:

The problem related to the "definition" of the species should be complemented by the reference to Reznick (2009) and Beatty (1985), and to the discussion of the latter by Ginnobili (2013).

> One of these questions involves issues around the "definition" of the concept of species, that is, of what a species is (Reznick, 2009; Beatty, 1985). However, it is worth noting that in this context the term "definition" is used in a rather loose way of giving operational criteria, methods, or ways to distinguish species from varieties, i.e., to determine the extension of the concept of species, meaning how it can be determined what are the objects that fall under the concept of species, and how a species differs and is related to varieties, and not in the strict logical sense of providing necessary and sufficient conditions for the application of the concept of species (Ginnobili, 2013).

On p. 309:

To the mention of Darwin's discussion of the role that sterility plays in the so-called species problem, reference to Olby (2009) should be included.

> As a part of his general argument against the view that "species were immutable productions, and had been separately created" and in favour of the conception that "species undergo modification, and that the existing forms of life are the descendants by true generation of pre-existing forms" (Darwin, 1866, p. xiii), Darwin explores in Chapter VIII the role that sterility plays in the so-called species problem (Olby, 2009).

On p. 311:

On Darwin's relationship with breeders, hybrists, and naturalists, reference to Secord (1985) should be added.

> As we shall see, in Chapter VIII, "On Hybridism," and along *Origin*, Darwin refers to authors against and for the theory of speciation through hybridization as well as of both traditions or schools – that of animal and plant breeders, or horticulturalists, and that of species hybridizers, or hybridists – and to other naturalists and botanists (Secord, 1985).

On p. 314:

Regarding Darwin's contemporaries who thought of each species as the product of a special act of creation and that all species were absolutely reproductively isolated from one another, reference to Reznick (2009) should be included, but differing from him in that not all contemporaries thought alike.

> Recall that, on the one hand, although many, but not all, of "Darwin's contemporaries thought that each species was the product of an act of special creation and that all species were absolutely reproductively isolated from one another" (Reznick, 2009, p. 190) – including Natural Theologists and some hybridists in the broad sense – this was not so for those who held a theory of speciation through hybridization, that is, hybridism in the narrow sense.

On p. 314:

> On the other hand, either the assumed sterility of hybrids (Kölreuter) or the gradual decreasing fertility of hybrid crosses (Gärtner) in contrast to the fertility of variety crosses (whose progeny is called "mongrels" by Darwin) were considered the Creator's (for Natural Theologists) (Olby, 2009) or Nature's (for Kölreuter and Gärtner) way of preserving the fixity of species.

On p. 314:

> Darwin "needed to explain why species, which, according to his theory, had descended from each other, usually could not interbreed" (Smocovitis, 2009, p. 230), given that, on the one hand, sterility or difference in fertility was not a property to be explained by natural selection, since it is not "a specially acquired or endowed quality, but is incidental on other acquired differences" (Darwin, 1859, 245), as already mentioned in Section 1. And on the other hand, the standard explanation of why species did not interbreed, namely, "that barriers to hybridization existed" (Smocovitis, 2009, p. 230) to "prevent the confusion of all organic forms" (Darwin, 1859, p. 245), "was [...] not a viable explanation" (Smocovitis, 2009, p. 231) for him.

On p. 314:

After establishing the fundamental distinction between the sterility of species crosses and hybrid crosses (Darwin 1859, 245–246), Darwin presented a brief overview "of the varied and complex patterns observed" (Smocovitis, 2009, p. 230) in the phenomena first of sterility and fertility of species crosses and then of hybrid crosses (Darwin 1859, 246–253) (Reznick, 2009).

On p. 314:

Darwin relied extensively on the insights and experimental studies of horticulturalists, plant and animal breeders, and hybridists (Smocovitis, 2009).

On p. 314:

Darwin compared his point of view with that of them, and used his interpretation of their experimental results, together with the results obtained by "the third most experienced hybridizer [. . .] Hon. and Rev. W[illiam]. Herbert" (Darwin, 1859, pp. 249–250), to oppose their views (Reznick, 2009).

On p. 315:

Darwin's conclusions of his overview are, besides that sterility is affected by "close interbreeding" (1859, p. 248) and is "capable of being removed by domestication" (1859, p. 254), "that some degree of sterility, both in first crosses and in hybrids, is an extremely general result; but that it cannot, under our present state of knowledge, be considered as absolutely universal" (1859, p. 254; this last citation also appears in Reznick, 2009, p. 196).

On p. 315:

These conclusions, which would show that "different species can sometimes interbreed and [. . .] produce" fertile hybrids, would go against those who defend the doctrine of special creation and would constitute an argument "that species cannot be defined as discrete units separated by absolute reproductive barriers" (Reznick, 2009, p. 196) and there were no clear lines between varieties, incipient species, and species.

On p. 315:

However, contrary to Gärtner's proposal, Darwin makes the following points (see Reznick, 2009, pp. 196–198).

On p. 315:

> Darwin indicates that there are more Gärtner's "rules," but the ones he has already discussed are sufficient to prove his own view (Reznick, 2009, p. 198).

On p. 316:

> There, Darwin also emphasizes that he "did not consider [. . .] hybridization itself to be the source of the resulting variation" (Olby, 2009, p. 38), either in general or in the vigour and fertility of the offspring of hybrids. Rather, the changes in conditions of life play in it a fundamental role.

On pp. 316–317:

The last paragraphs of the text, concerning Darwin's critique of hybridization as a theory of evolution, transcribes and paraphrases three paragraphs of Lenval A. Callender's well-known article, already referred to on p. 310, "Gregor Mendel: An opponent of descent with modification" (Callender, 1988, p. 49), as follows:

> Thus, as argued by Callender (1988), though Darwin accepted the existence of hybrid forms of plants completely fertile and relatively stable, he thought that absent other sources of variation, hybridization itself "could not account for the evolution of species" (Callender, 1988, p. 49). The main reason – which Darwin develops in other parts of the Origin as well as in other texts – is that hybridization "presupposes already existing differences", thus giving rise, once again, to the question of the origins of such differences (Callender, 1988, p. 49). By trying to explain, then, evolutionary change based on crossings without variation, Callender refers to Darwin's words: "we thus only push the difficulty further back in time, for what made the parents or their progenitors different?" (Darwin, 1868, v. II, p. 252, apud Callender, 1988, p. 49). In the case of the evolution of pigeons, for example, Darwin wrote, in a passage also cited by Callender and expanded here:
>
> ---Great as. . . . enormous crop? (Darwin, 1859, p. 23)
>
> That is, "*the fundamental objection to the doctrine of 'evolution' solely by means of hybridization* [. . .]: one is inevitably confronted either with an infinite regress or [with] some version of the doctrine of Special Creation" (Callender, 1988, p. 49; emphasis added). And either possibility is, according to Darwin, untenable.

On p. 317:

At the end of the text and as a conclusion, the following paragraph should be added:

> To summarize the conclusions of this chapter and explain its subtitle, while Darwin acknowledges the importance of "hybridism in the broad sense",

i.e., hybridism as the tradition that studies hybrids ("[t]he offspring of the union of two distinct species", Darwin, 1872, p. 435), he clearly rejects "hybridism in the narrow sense", i.e., hybridism as a theory of the origin of new species by hybridization, and defends his own theory, as he reaffirms elsewhere ("the whole subject of hybridism is one of the greatest obstacles to the general acceptance and progress of the great principle of evolution" Darwin, 1876, p. 27).

Author Index

M. Elice Brzezinski Prestes (ed.), *Understanding Evolution in Darwin's "Origin"*, History, Philosophy and Theory of the Life Sciences 34,
https://doi.org/10.1007/978-3-031-40165-7

V

W

Subject Index

M. Elice Brzezinski Prestes (ed.), *Understanding Evolution in Darwin's "Origin"*, History, Philosophy and Theory of the Life Sciences 34,
https://doi.org/10.1007/978-3-031-40165-7

The manufacturer's authorised representative in the EU is Springer Nature Customer Service Centre GmbH, Europaplatz 3, 69115 Heidelberg, Germany. If you have any concerns regarding our products, please contact ProductSafety@springernature.com

Printed and bound by CPI Group (UK) Ltd, Croydon, CR0 4YY
01/10/2026
02225902-0014